Biotechnological Advances in Horticultural Crops

The field of horticultural biotechnology has witnessed remarkable growth in recent years, offering profound insights and innovations that are reshaping agriculture's future. *Biotechnological Advances in Horticultural Crops* serves as a comprehensive guide to the latest innovations and applications. This meticulously curated volume bridges the gap between fundamental knowledge and cutting-edge advancements. From plant tissue culture to genomic approaches, this book spans a diverse array of topics, catering to a broad audience of students, researchers, and academics.

Key Features:

- **In-depth exploration** of plant tissue culture and its applications in horticultural crop improvement.
- Insightful coverage of **transgenic approaches** in fruit crops, including genetic modifications for enhancing fruit quality, yield, and resistance to pests and diseases.
- Detailed examination of the application of **omics sciences** in tropical fruits, opening up new perspectives for crop improvement and sustainable production.
- Extensive discussion on the potential of ***in vitro* production techniques** for underutilized and unconventional horticultural crops.
- Exploration of **sustainable agriculture** practices, including the use of beneficial microbes and the scientific validation of natural and organic farming practices.

This book is more than just a collection of topics; it is a gateway to the transformative potential of biotechnology in horticulture. It empowers readers (students, researchers, and academicians) to comprehend and harness these advancements, promoting innovation and sustainability in horticultural crop improvement.

Biotechnological Advances in Horticultural Crops

Edited by
Ajay Kumar Thakur, Pankaj Kumar, and
Dinesh Kumar Srivastava

CRC Press
Taylor & Francis Group
Boca Raton London New York

CRC Press is an imprint of the
Taylor & Francis Group, an **informa** business

First edition published 2025
by CRC Press
2385 NW Executive Center Drive, Suite 320, Boca Raton FL 33431

and by CRC Press
4 Park Square, Milton Park, Abingdon, Oxon, OX14 4RN

CRC Press is an imprint of Taylor & Francis Group, LLC

ISBN: 9781032381343 (hbk)
ISBN: 9781032381459 (pbk)
ISBN: 9781003343707 (ebk)

DOI: 10.1201/9781003343707

Typeset in Times New Roman
by Newgen Publishing UK

Dedication

This book is dedicated to Dr. Atul Mehta, a renowned plant biologist, and mentor to many students and grand students. Aptly nicknamed 'Guru' by his wide family of students, Atul Mehta still lives in their hearts.

Dr. Mehta was born to Jaikirti, a social worker and Ramakant, an accountant in Gaekwad state in Gujarat. Atul was very fond of sports and music since he was young, largely due to family influence. In his academic career too, he excelled in earning a teaching position at the MS University of Baroda after earning his master's degree. While teaching, he was awarded a scholarship for further study abroad and chose to work with the renowned botanist Prof. H E Street at the University of Wales in UK where a new technique of plant tissue culture was emerging. Thus, his background study on basics of botany ironically shifted to embarking upon a deeper insight into study on biological mechanisms. He did his doctoral research on growth patterns in crown gall and normal cells in in vitro cultures, which led to highly acclaimed and well-cited classic publications. Upon his return to his hometown in India in 1967, he established a tissue culture laboratory, the first one in the country, at Botany Department, MS University of Baroda. The initial basic research in plant physiology using organ and tissue culture techniques in Mehta's lab was not generated in a class 10 facility with air-conditioning and laminar workstations, but a simple inoculation chamber in the corner of the main physiology laboratory. This clearly indicated his thirst for excelling even under not so favorable and comforting circumstances available to him.

A large number of postgraduates, doctoral and postdoctoral students learnt their first lesson in plant tissue culture in this nest. Dr. Dinesh Kumar Srivastava, who is one of the Editors of this book, was mentored by Prof. Mehta on basics of plant tissue culture at MS University of Baroda, Vadodara, Gujarat.

To keep pace with advances in applied field of plant tissue culture, he later took a sabbatical leave for postdoctoral work with Prof. Staba at University of Minnesota to work on secondary product metabolism in cultured cells. The school of learning he established in the Botany Department inspired not only his own students and those of his colleagues who all worked as one team, but also was a seat of training courses for students from other universities in India and abroad. With the departmental research support awarded by UGC, distinguished scientists from acclaimed universities frequently engaged in interactions, leading to the flourishing of doctoral and postdoctoral research. He organized workshops, such as the UNESCO-ICRO sponsored workshop, which attracted researchers from several countries. Leading scientists from around the world were invited as experts in these workshops. Undoubtedly, this inspired young Indian scientists to pursue careers in plant biotechnology.

The vibrant environment of the centre buzzed with new ideas and enthusiasm, quickly establishing itself as a centre for plant tissue culture expertise. Funding from the Council of Scientific and Industrial Research and the University Grants Commission of the Government of India allowed the team to undertake a diverse variety of research projects involving in vitro techniques of plant tissue culture.

His visionary endeavours led him to receive the honour of being a recipient of the Jaikrishna Indrajit Award by the Hari Om Ashram in 1978 and Prof. Vishwambhar Puri Medal by the Indian Botanical Society in 1991. The scientific community had the privilege to appreciate him and select him to deliver Memorial lectures, notably among which were Prof. Philip R.White Memorial Lecture at Jadavpur University, Prof. H.E. Street Memorial Lecture at Nagpur University, and the Prof. A. R. Chavan Memorial Lecture at the MS University of Baroda. He earned the position and remained for many years as a National Correspondent of the International Association of Plant Tissue Culture (IAPTC) for South Asian region.

Besides being an eminent scientist, Prof. Mehta was also an efficient administrator. As a member of the academic senate of the university, he contributed significantly to the academic ambience and in fine-tuning the university administrative procedures and protocols. He also served on the national committees like DST, DBT, UGC, ICAR, and CSIR as an expert member. He also visited several universities and delivered lectures as a national professor. He was a core team member of the All India Coordinated Project on Medicinal and Aromatic Plants of ICAR. He was also an Honorary Advisor to a private biotech company engaged in commercial research and production in Vadodara. As Head of the Department, he brought many changes in the ambience and the discipline to make the education and research environment more conducive for students and staff, and all his students and colleagues now feel indebted to him for these values he inculcated, though he may have sounded harsh in those days. After his retirement from the university, he worked as a CSIR-Emeritus

Scientist at the Department of Plant Breeding and Genetics, Gujarat Agricultural University, Anand. He led a research team to conduct tissue culture work under the Greening of Gujarat and Rajasthan project funded by DBT, New Delhi on indigenous forest species such as guggal (Commiphora), Acacia, and soap nut tree (Sapindus spp.). Seeds were also sown here by his diversifying interest in employing fundamental techniques of plant tissue culture to commercial micro-propagation of economically important horticulture crops such as banana, while he was Professor Emeritus here.

He mentored 25 PhD students, authored more than 150 research papers in national and international journals of repute, and edited a book with P. N. Bhatt titled Handbook of Plant Tissue and Cell Culture. He worked on a wide array of topics such as cell differentiation, organ differentiation, the role of polyamines in morphogenesis, and plantlet regeneration in tree species like soap nut and recalcitrant species like guggal. He led his team to work on metabolism of many secondary plant products such as diosgenin, guggulsteroids, insect molting hormones, Lathyrus neurotoxin, and several phenolic compounds in cultured cells. His curiosity for unravelling the physiological and biochemical changes in haploid and diploid tissues of tobacco is exemplary. With his collaboration, stable somaclones of tobacco were selected for resistance to Fusarium oxysporum. He remained active until his physical health only permitted him to support his family and provide constant encouragement to his students.

Apart from his career as a scientist and administrator, Mehta was a fond lover of classical music and used to sing songs/ghazals. It used to be a unanimous demand on him to get on stage during the recreational evenings during symposia and scientific gatherings. Known for his cricketing talent, he represented the MS University cricket team and also the drama group. He was an excellent host along with his wife to all who visited his home. Mehta always attributed his success to his late wife Yashaswini Mehta. He is survived by a son and a daughter, wearing similar feathers.

The plant tissue culture/plant biotechnology fraternity misses his dynamic physical presence, but his inspiration, humility, and sense of humour still remain with us.

Ajay Kumar Thakur
Pankaj Kumar
Dinesh Kumar Srivastava

Contents

Preface

The field of horticultural biotechnology has evolved enormously in recent years, offering profound insights and innovations that continue to shape the future of agriculture. *Biotechnological Advances in Horticultural Crops* is a meticulously curated edited volume that delves into the multifaceted sphere of biotechnological applications in horticulture. This comprehensive collection encompasses a diverse array of topics, spanning plant tissue culture to nanotechnology, and from genomics to microbial consortiums. Designed to cater to a wide audience, including undergraduate and postgraduate students, as well as researchers, this book bridges the gap between fundamental knowledge and cutting-edge advancements in the field. The book commences with an in-depth exploration of plant tissue culture and its applications in horticultural crop improvement. It subsequently delves into the realm of transgenic approaches in fruit crops, offering insights into genetic modifications that hold promise for enhancing fruit quality, yield, and resistance to pests and diseases. The application of omics sciences in tropical fruits opens up new perspectives for crop improvement and sustainable production. Genomic approaches, including association mapping, provide valuable tools for deciphering the genetic basis of important traits in fruit crops, paving the way for precision breeding. Beyond fruit crops, this volume extends its reach into the realm of ornamental horticulture, elucidating how biotechnological approaches can be harnessed to improve floral attributes. Furthermore, it explores the potential of *in vitro* production techniques for underutilized and unconventional horticultural crops, fostering diversity in agricultural production. The latter part of this book delves into sustainable agriculture, exploring the use of beneficial microbes. It offers a scientific validation of natural and organic farming practices, demonstrating the versatility of biotechnology in the horticultural arena. As editors, we believe that this book will serve as a valuable resource, empowering students, researchers, and academics to comprehend and harness the transformative potential of biotechnology in horticulture. The broader advantages of this comprehensive compilation are bound to resonate across diverse academic and practical settings, promoting innovation and sustainability in the realm of horticultural crop improvement.

Shimla, India — Ajay Kumar Thakur

Solan, India — Pankaj Kumar

Solan, India — Dinesh Kumar Srivastava

Editor Biographies

Ajay Kumar Thakur has recently joined as Principal Scietist (Biotechnology) at ICAR-Central Potato Research Institute, Shimla, HP upon his transfer from ICAR-Directorate of Rapeseed-Mustard Research, Bharatpur, Rajasthan. He has obtained his graduation, PG and doctoral degrees from Dr. Y.S. Parmar University of Horticulture & Forestry, Solan, H.P. He was induced into Agricultural Research Services (ARS) in 2008. He has published 72 research/review papers in various journals of International and National repute, authored one book *Agricultural Biotechnology at a Glance*, edited 3 books and contributed 8 book chapters and 23 popular articles. He has developed high-efficiency plant regeneration and genetic transformation protocols in a number of crops including *Populus ciliata*, *P. deltoides*, *Punica granatum*, *Capsicum annum,* and *Cucumis sativus*. Dr. Thakur is part of the *Brassica juncea* improvement programme using biotechnological interventions from the last 14 years. He has developed a core set of SSR markers for *B. juncea* genomics and has worked on germplasm characterization and association mapping of various agronomically important traits in this oilseed crop. He has also been granted one patent from Indian Patent Office, New Delhi and has been part of the development of two high yielding Indian mustard varieties Giriraj & Brajraj, one gobhi sarson variety, AKMS-8141 and five disease resistant Indian mustard genetic stocks. Dr. Thakur has received many awards from various societies and scientific organizations for his scientific contributions. He is also an elected member of the Plant Tissue Culture Association of India (PTCAI) and selected member of the National Academy of Sciences-India (NASI), Allahabad.

Pankaj Kumar currently serves as an Assistant Professor in Biotechnology at the Department of Biotechnology, Dr. Yashwant Singh Parmar University of Horticulture and Forestry in Solan, Himachal Pradesh, India. His academic journey includes post-doctoral studies at the National Institute of Plant Genome Research in New Delhi and the Council of Scientific & Industrial Research – Institute of Himalayan Bioresource Technology in Palampur, Himachal Pradesh. Dr. Kumar boasts 12 years of expertise in horticulture biotechnology, focusing on plant secondary metabolite enhancement through cell and tissue engineering approaches. His research encompasses the identification of molecular cues associated with enhanced metabolites using a comparative transcriptomics approach. Additionally, he possesses experience in hydro-aero cultivation of medicinal plants for industrial significance.

During doctoral studies he (research team) has generated technology at Dr. Yashwant Singh Parmar University of Horticulture & Forestry, Nauni, Solan, for developing insect pest-resistant transgenic plants in economically important vegetable crops of cauliflower cv. Pusa Snowball, cabbage cv. Pride of India and broccoli cv. Solan Green Head (with *cryIAa* gene) for insect pest resistance. His accolades include the DST INSPIRE JRF/SRF Fellowship, ICAR AICE-SRF (PGS), and ICAR ASRB National Eligibility Test (NET). Dr. Kumar has an extensive publication record with more than 60 research/review papers, 23 book chapters, and 4 books. He has also filed an Indian patent and registered four apple mutants with unique, early color and early maturity characteristics in the Indian National Germplasm Repository (INGR 20103, INGR 20104, INGR 20105 & INGR 20106). Recognized for his scientific contributions, Dr. Kumar has received numerous awards, including the Young Scientists Fellowship Award-2021 (ICMR, New Delhi), Scientist's Pool Scheme Award-2020 (CSIR, New Delhi), and the SERB-ACS-NPDF Best Poster Award-2022, etc.

Dinesh Kumar Srivastava has retired as Director Extension Education. Prior to this he worked as Professor and Head in the Department of Biotechnology, Dr. Yashwant Singh Parmar University of Horticulture and Forestry, Solan, Himachal Pradesh, India. He did his M.Sc. and Ph.D. from Gujarat University, Ahmedabad, Gujarat. He learned plant tissue culture techniques at MS University of Baroda, Gujarat. He did his post-doctoral studies in the field of Plant Molecular Biology at the Institute of Molecular Genetics, USSR Academy of Sciences, Moscow, USSR and at Washington University, St. Louis, USA. He has 32 years of research experience and 29 years of teaching experience in the field of Plant Biotechnology and Molecular Biology. He had guided 32 students of M.Sc. and 9 students of Ph.D. for their research work. He has published 115 research papers in the journals of national and international repute, 40 research articles and contributed 15 chapters in the edited books, and also authored a book on biotechnology. He has delivered 47 invited lectures and attended 55 national and international conferences. His main area of research includes Plant tissue culture, Genetic transformation and Molecular characterization of plants. He is life member of various National and International Academic bodies/societies. He got an Indian patent on "Development of reproducible protocol for direct organogenesis in transgenic male plant of Himalyan poplar." He has received many awards for his scientific contributions including "MS Swaminathan Award" by the Society of Plant Research, Meerut, India.

Contributors

Héctor Adán Ruiz-Ortega
Centro de Investigacion en Alimentacion y Desarrollo, A.C
Carretera Gustavo E. Astiazarán Rosas, Col. La Victoria, Hermosillo, Sonora, Mexico

José María Anaya-Dyck
Centro de Investigacion en Alimentacion y Desarrollo, A.C
Carretera Gustavo E. Astiazarán Rosas, Col. La Victoria, Hermosillo, Sonora, Mexico

Tashi Angmo
Division of Fruit Science
Sher-e-Kashmir University of Agricultural Sciences & Technology of Jammu
Chatha, Jammu and Kashmir, India

Miguel Ánge Hernández-Oñate
CONACYT – Centro de Investigacion en Alimentacion y Desarrollo, A.C.,
Carretera Gustavo E Astiazarán Rosas, Col. La Victoria, Hermosillo, Sonora, México

Bhavya Bhargava
Floriculture Laboratory, Agrotechnology Division
Council of Scientific and Industrial Research (CSIR)
Institute of Himalayan Bioresource Technology, Palampur, Himachal Pradesh, India

Anjali Chandel
Floriculture Laboratory, Agrotechnology Division
Council of Scientific and Industrial Research (CSIR)
Institute of Himalayan Bioresource Technology, Palampur, Himachal Pradesh, India

Damaristelma De Jesús Campos
Departamento de Investigaciones Cientificas y Tecnologicas, Universidad de Sonora Blvd.
Luis Encinas y Rosales s/n. Colonia Centro, Hermosillo, Sonora, Mexico

Edmundo Dominguez-Rosas
Centro de Investigacion en Alimentacion y Desarrollo, A.C
Carretera Gustavo E. Astiazarán Rosas, Col. La Victoria, Hermosillo, Sonora, Mexico

Monica Durairaj
Department of Microbiology
Science Campus, Alagappa University, Karaikudi, Tamilnadu

Heriberto García-Coronado
Centro de Investigacion en Alimentacion y Desarrollo, A.C
Carretera Gustavo E. Astiazarán Rosas, Col. La Victoria, Hermosillo, Sonora, Mexico

Shivanshu Garg
Govind Ballabh Pant University of Agriculture and Technology
Pantnagar, Uttarakhand, India

Rucku Gupta
Division of Fruit Science
Sher-e-Kashmir University of Agricultural Sciences & Technology of Jammu
Chatha, Jammu and Kashmir, India

Deekshit HN
Floriculture Laboratory, Agrotechnology Division
Council of Scientific and Industrial Research (CSIR)
Institute of Himalayan Bioresource Technology, Palampur, Himachal Pradesh, India

José Ángel Huerta-Ocampo
CONACYT – Centro de Investigacion en Alimentacion y Desarrollo, A.C
Carretera Gustavo E. Astiazarán Rosas, Col. La Victoria, Hermosillo, Sonora, México

Surabhi Jasrotia
School of Biotechnology
Sher-e-Kashmir University of Agricultural Sciences & Technology of Jammu
Chatha, Jammu and Kashmir, India

Nisha Khatik
Department of Botany
Maharshi Dayanand Saraswati University, Ajmer, Rajasthan, India

Safina Kosser
Division of Fruit Science
Sher-e-Kashmir University of Agricultural Sciences & Technology of Jammu
Chatha, Jammu and Kashmir, India

Pankaj Kumar
Department of Biotechnology
Dr. Yashwant Singh Parmar University of Horticulture & Forestry, Nauni, Solan, HP, India

Pooja Nain
Govind Ballabh Pant University of Agriculture and Technology
Pantnagar, Uttarakhand, India

Deepesh Kumar Neelam
Department of Microbiology
JECRC University, Jaipur, Rajasthan, India

Ángel Javier Ojeda-Contreras
Centro de Investigacion en Alimentacion y Desarrollo, A.C
Carretera Gustavo E. Astiazarán Rosas, Col. La Victoria, Hermosillo, Sonora, Mexico

Sivasakthivelan Panneerselvam
Department of Agricultural Microbiology
Annamalai University, Chidambaram, Tamilnadu, India

Mahinder Partap
Floriculture Laboratory, Agrotechnology Division
Council of Scientific and Industrial Research (CSIR)
Institute of Himalayan Bioresource Technology, Palampur, Himachal Pradesh, India

Hardik Pathak
Department of Plant Biotechnology
JECRC University, Jaipur, Rajasthan, India

Pooja
Department of Fruit Science
Dr. YSP UHF, Nauni, Solan, Himachal Pradesh, India

Priti
Floriculture Laboratory, Agrotechnology Division
Council of Scientific and Industrial Research (CSIR)
Institute of Himalayan Bioresource Technology, Palampur, Himachal Pradesh, India

Himanshu Punetha
Govind Ballabh Pant University of Agriculture and Technology
Pantnagar, Uttarakhand, India

Sheikh Qurat
Division of Fruit Science
Sher-e-Kashmir University of Agricultural Sciences & Technology of Jammu
Chatha, Jammu and Kashmir, India

Vishal S Rana
Department of Fruit Science
Dr. YSP UHF, Nauni, Solan, Himachal Pradesh, India

Madhurjit Singh Rathore
Department of Biotechnology
Dr. YSP UHF, Nauni, Solan, Himachal Pradesh, India

Shiv Rattan
Floriculture Laboratory, Agrotechnology Division
Council of Scientific and Industrial Research (CSIR)
Institute of Himalayan Bioresource Technology, Palampur, Himachal Pradesh, India

Vidhyavathi RM
Department of Bioinformatics
Alagappa University
Karaikudi, Tamilnadu, India

Joseph Sahayarayan
Department of Bioinformatics
Alagappa University, Karaikudi, Tamil Nadu, India

Paramjeet Sajwan
Department of Fruit Science
Dr. YSP UHF, Nauni, Solan, Himachal Pradesh, India

Ankila Salathia
School of Biotechnology
Sher-e-Kashmir University of Agricultural Sciences & Technology of Jammu
Chatha, Jammu and Kashmir, India

Amit Sharma
Plant Tissue Culture and Biotechnology Laboratory
M N College and Research Institute, Bikaner, Rajasthan, India

Deepak Sharma
School of Agricultural Sciences
Jaipur National University, Jaipur, Rajasthan, India

Diksha Sharma
Floriculture Laboratory, Agrotechnology Division
Council of Scientific and Industrial Research (CSIR)
Institute of Himalayan Bioresource Technology, Palampur, Himachal Pradesh, India

Shabnam Sircaik
Brown University, Providence
RI, USA

Jebi Sudan
Department of Plant Biotechnology
JECRC University, Jaipur, Rajasthan, India

Lokesh Thakur
Department of Biotechnology
Dr. YSP UHF, Nauni, Solan, Himachal Pradesh, India

Meenakshi Thakur
Floriculture Laboratory, Agrotechnology Division
Council of Scientific and Industrial Research (CSIR)

Institute of Himalayan Bioresource Technology, Palampur, Himachal Pradesh, India

Kavitha Thangavel
Department of Microbiology
Science Campus, Alagappa University, Karaikudi, Tamilnadu

Martín E. Tiznado-Hernández
Centro de Investigacion en Alimentacion y Desarrollo, A.C
Carretera Gustavo E. Astiazarán Rosas, Col. La Victoria, Hermosillo, Sonora, Mexico

Eduardo Antonio Trillo-Hernández
Food Technology Unit-Secretariat of Research and Postgraduate Studies
Autonomous University of Nayarit, Ciudad de la Cultura S/N, Colonia Centro,
Nayarit, Mexico

Sapna Tyagi
Plant Tissue Culture and Biotechnology Laboratory
M N College and Research Institute, Bikaner, Rajasthan, India

Vipasha Verma
Floriculture Laboratory, Agrotechnology Division
Council of Scientific and Industrial Research (CSIR)
Institute of Himalayan Bioresource Technology, Palampur, Himachal Pradesh, India

Ujala
Floriculture Laboratory, Agrotechnology Division
Council of Scientific and Industrial Research (CSIR)
Institute of Himalayan Bioresource Technology, Palampur, Himachal Pradesh, India

Sajad Majeed Zargar
Division of Plant Biotechnology
Sher-e-Kashmir University of Agricultural Sciences & Technology of Jammu
Chatha, Jammu and Kashmir, India

1 Plant Tissue Culture and Its Applications in Horticultural Crop Improvement

Nisha Khatik

1.1 INTRODUCTION

Horticulture is the branch of plant agriculture used for the production and improvement of horticultural crops, such as fruits, vegetables, and ornamental plants. Accessibility of genomic assets in horticultural crops can be used for the productive abuse of the current research in creating further developed assortments and furthermore characterizing future objective nutritional security to individuals with developing population demands. In the past few decades because of land annihilation and impractical usage of regular assets numerous species including plants went under danger and their reality is currently under the control of people. Arid lands of The world's arid regions are dealing with major issues due to a lack of food, water, fuel wood and cruel natural conditions. Fortunately most of the normal plants of these spaces are multipurpose and on the off chance that they are appropriately used they could resolve a portion of these issues somewhat. The native plants of the locale are very much adjusted to abandon conditions and devour less water for the food. Propagation and cultivation of crop plants of this area through conventional and non-conventional methods will help in easing desertification dangers through obsession of sand, assurance and improvement of the soil, enhancement of environment and microenvironment and different conditions reasonable for maintainable horticulture. Plant tissue culture is as of now assuming an essential part in the improvement of horticultural crop plants generally. Plant tissue culture has extraordinary capability of tackling numerous issues relating to agriculture, industry, climate and wellbeing, and has a direct impact on the long-term development of a country's arid zones.'

To cater to the expanding demands of harvest plants in markets, protection and commercial production of these plants is vital. Since, the conventional methods for proliferation are for the most part sluggish, labour intensive, require a huge number of propagules and are not very successful; the use of current strategies for plant tissue culture ought to be embraced to accelerate the engendering of plants.

DOI: 10.1201/9781003343707-1

1.2 TISSUE CULTURE IN HORTICULTURE

Plant tissue culture is a fundamental part of plant biotechnology. It is the strategy of keeping up with and developing any plant cells, tissues, or organs on appropriate medium in test tube under aseptic and controlled physical conditions like light, temperature, and humidity. Aside from large-scale elite development, it also enables the multiplication and recovery of novel plants from genetically engineered cells. Tissue culture is essentially characterized as *in vitro* development of plantlets from isolated cells, embryonic tissues, organs, calli, or protoplasts on suitable nutrient medium. A wide extent of synthetic media like Murashige and Skoog (MS) medium, woody plant medium (WPM), Schenk-Hilderbrandt (SH), and the Nitsch and Nitsch (NN) and Linsmaier-Skoog (LS) mediums are used in the plant tissue culture. Gottlieb Haberlandt was the first scientist to produce entire plants from plant tissues and is thus prevalently called the "Father of Tissue Culture." Plant tissue culture is the best method for enormous scope production of propagules, particularly if there should arise an occurrence of significant cultivation of horticultural crop plant species where explant material is accessible in an exceptionally small amount.

Plant tissue culture includes the development of plant tissues or sections for creating or cloning a great deal of new cells, tissues, or plants; to consider advancement; to give components of hereditary design; or to deliver important synthetic substances found in plant cells.

Horticultural crops are a significant part of the diverse agro-environmental conditions in the Indian horticultural economy. They are a significant source of carbohydrates, organic acids, proteins, nutrients, and minerals for human sustenance. Exactly when individuals use plants or plant parts, regardless of whether for food or aesthetic purposes, there is reliably a postharvest component that prompts to loss (Fallik 2004). The rapid expansion in total populace to 8.9 billion individuals by 2050 will prompt higher demands for agricultural products (Lichtfouse et al. 2009). High food demands and the lack of new farming area upgrades will require multiplying crop yields utilizing sustainable means. Agricultural sustainability is characterized as the capacity of a crop production framework to consistently deliver food without natural degradation (Doran 2002). Plant tissue culture has the potential to meet these needs. Micropropagation of practically all the fruit crops and vegetables is conceivable at this point. The ability to produce virus-free planting material utilizing apical meristem culture has been made conceivable in many horticultural crops. A large number of the dry land legume species has been effectively recovered from leaf, petiole, hypocotyls, cotyledons, root, ovary, anthers, protoplast, and so on. Haploid generation through anther/pollen culture is perceived as one of the more significant region in crop improvement since it is fast and economical (Sopory and Munshi 1996). The preservation of *in vitro* germplasm is crucial for providing strategies and alternative ways to deal with obligations in the management of genetic resources. In crops which are propagated vegetative and which produce recalcitrant seeds and perennial crops which are profoundly heterozygous seed storage isn't reasonable (Berjak and Pammenter 2013). In such crops, *in vitro* storage is very useful. These procedures have effectively been used for various horticultural crops and are currently used in some germplasm collection centres. *In vitro* germplasm additionally guarantees the trading of pest and disease-free material and helps in better isolate. Researchers can also make a significant commitment to worldwide sustainability of the agricultural lands by interpreting scientific information on soil work into functional strategies that enhance cultivator's information to assess the supportability of their administration rehearses.

1.3 PLANT TISSUE CULTURE TECHNIQUES

Plant tissue culture techniques have benefited the development and improvement of many horticultural crop species. These have been used as an apparatus for the development of hereditarily controlled, unrivalled clones and conservation of ex situ germplasm for endangered and vulnerable plants species.

1.3.1 Micropropagation

Micropropagation in plant tissue culture is used to develop superior quality horticultural plants under aseptic and controlled conditions. The essential advantages include fast, large-scale proliferation of new genotypes and the use of small quantities of unique germplasm. The first effective use of micropropagation was shown by Morel (1964) in an orchid and *Cymbidium*, using meristem culture. Fruit crops like *Aegle marmelos*, *Emblica officinalis*, *Artocarpus hetrophyllus*, *Syzygium cumini*, and *Tamarindus indica* have been propagated using this method (Mishra et al. 2005).

Nonetheless, over the range of a large portion of the further assessments on increase of plants via tissue culture, the true basis of micropropagation - for example morphogenesis, had also regularly been overlooked, also with the results varying on numerous multiple times are either not reproducible at all or similarly as they have been reported earlier.

1.3.1.1 Stages of Micropropagation

Micropropagation is a complicated process and is in fact the miniature version of conventional propagation, which is done under aseptic conditions (Figure 1.1A and B). The ease with which plants can be micropropagated varies from species to species. The process of micropropagation aims to produce large numbers of crop plants. The development of plantlets is usually divided into the following stages:

1.3.1.2 Advantages of Micropropagation

Micropropagation has several advantages over traditional plant propagation methods, including:

- The key advantage of micropropagation is the ability to produce millions of clones in one year. With regards to conventional techniques, it requires a long time to produce an equal number of plants.
- Micropropagation can be used to produce disease-free plants. Meristem tip culture is used to accomplish this objective. For instance, virus-free potatoes must be obtained by the micropropagation of plants.

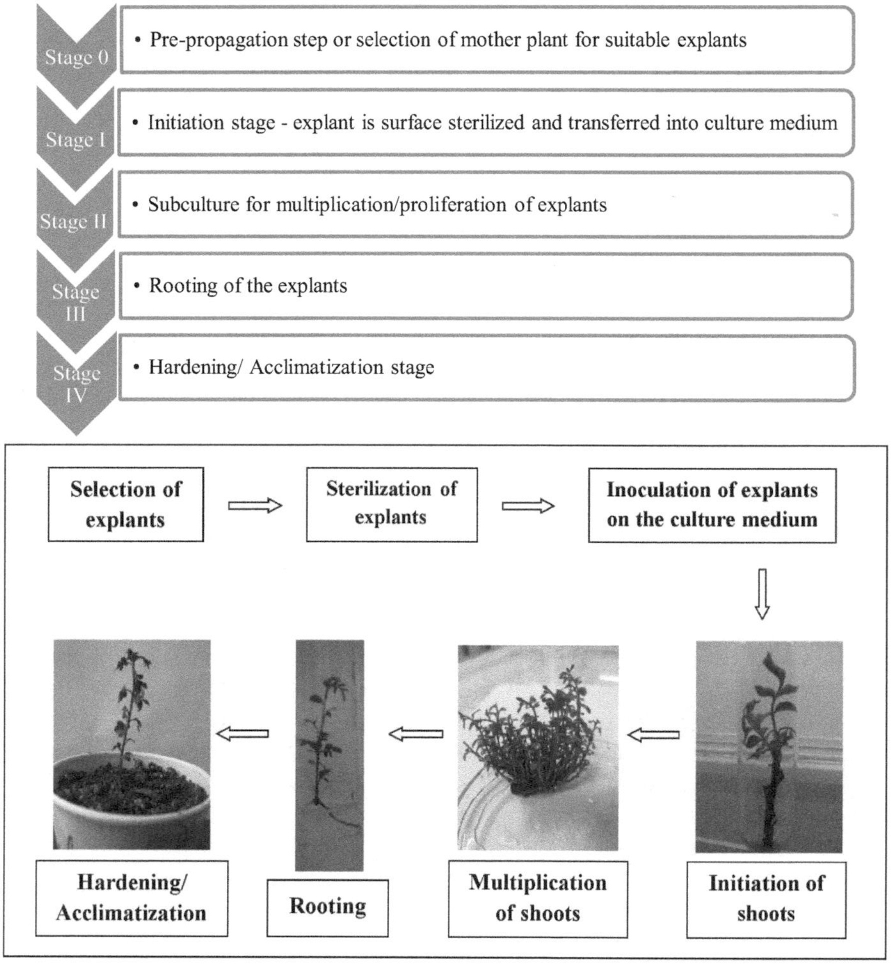

FIGURE 1.1 (A) Stages of Micropropagation; (B) Stages of Micropropagation.

- Micropropagation produces rooted plantlets prepared for development, saving time for the cultivator when seeds or cuttings are delayed to germinate or develop.
- The production of multiples of plants without any seeds or important pollinators to produce seeds.
- It is a suitable technique for recovering genetically modified cells or cells that have been fused with protoplasts.
- Seeds and vegetative organs of some types of plants have a limited storage life. It is useful in growing plants that yield inefficient amounts of seeds, or even when plants are sterile and therefore do not develop viable seeds, if the seeds can't be stored.

- Several other plants with very small seeds, such as most orchids, develop most reliably from seed in sterile culture.
- Micropropagation is used for germplasm storage and the protection of endangered species. Meristem cells are used for this reason since they are stable, make due under cryopreservation methods, and regenerate into pathogen-free plants whenever required.

1.3.2 Organogenesis

In vitro proliferation through meristem culture is an asset for producing a wide variety of disease-free plants in a limited period of time (Rout et al. 2006). At the point when organs such as shoots and roots are first initiated in plant tissue culture by stimulation caused by the genotype, physiological condition of the explants, age of the explants and the *in vitro* conditions like temperature, light and relative humidity and composition of medium, specifically concentration of plant growth regulators are produced in the culture by a method called organogenesis (Figure 1.2).

There are two forms of shoot organogenesis: Direct and indirect organogenesis. In indirect organogenesis, a plant's organ is developed from an intermediary callus, which first occurs on the explants, whereas direct shoot organogenesis results from the development of shoots directly from explants (Kane et al. 1994). The plants derived from calli may not be valid elites as a result of impressive morphological, physiological and genetic variations found inside the callus and there are moderately high occurrences of aneuploidy and polyploidy related with callus cells and plants got from it. One more disadvantage of the method of shoot multiplication including a callus stage is that it is not applicable for numerous significant crop species.

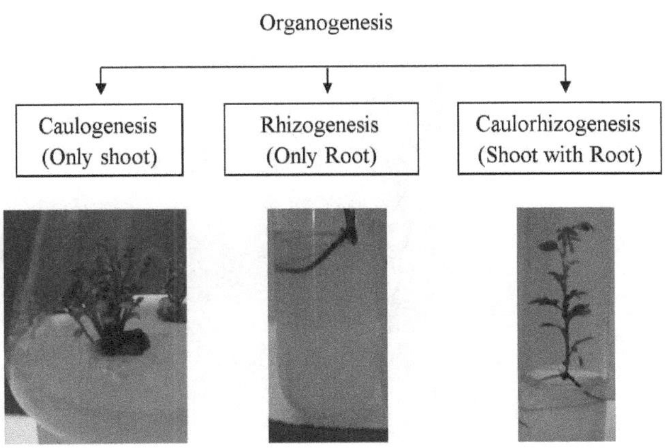

FIGURE 1.2 Development of organs through organogenesis.

1.3.2.1 Axillary Bud Proliferation

Plant tissue culture via axillary bud proliferation is the most powerful method for the mass production of many horticultural crop species. The benefit of this method is that it produces very little callus is and is also considered the most suitable to guarantee the genetic stability/fidelity of the regenerated plants obtained (Khatik and Joshi 2018, 2020). Axillary and terminal explants containing a solitary bud may also, contingent on the species and culture medium structure, produce a solitary shoot or multiple shoots (Figure 1.3 A). Once explants are set up and axillary bud proliferation is improved, the cultures could be subcultured for many generations, leading to increased shoot development. Crops propagated using this method include tomato, potato, sweet potato, cucumber, asparagus, broccoli, sprouts, cauliflower, brussels, cabbage, garlic, lettuce, etc. (George and Sherrington 1984). Axillary bud regeneration has been reported for many other crop species, like apple (McMeans et al. 1998), *Curcuma longa* (Panda et al. 2007), *Curcuma amada* (Mohanty et al. 2012), apple rootstock (Gupta et al. 2009), apple rootstock Merton 793 (Pathak and Dhawan 2012), *Allium ampeloprasum* L. (Gantait et al. 2010), barley (Bregitzer 2002), and rice (Medina et al. 2004).

1.3.2.2 Adventitious Shoot Formation

Adventitious shoots as well as relevant organs are structures that develop from tissues located in regions besides the axillary and terminal shoot tips. Adventitious shoots form via inter-nodal region, leaves, tubers, hypocotyls, bulbs, cotyledons, corms, root, or rhizomes (Figure 1.3 B). Callus that also serves as an intermediate stage among explants and production of plantlet can also give rise to adventitious shoots or organs. Adventitious shoot proliferation additionally has been used to multiply vegetable crops *in vitro*. Adventitious buds similarly can be developed on the stem cuttings and leaves of species that are normally not vegetative propagated, such as *Brassica* species and flax. Cabbage and lettuce are examples of vegetable harvests where

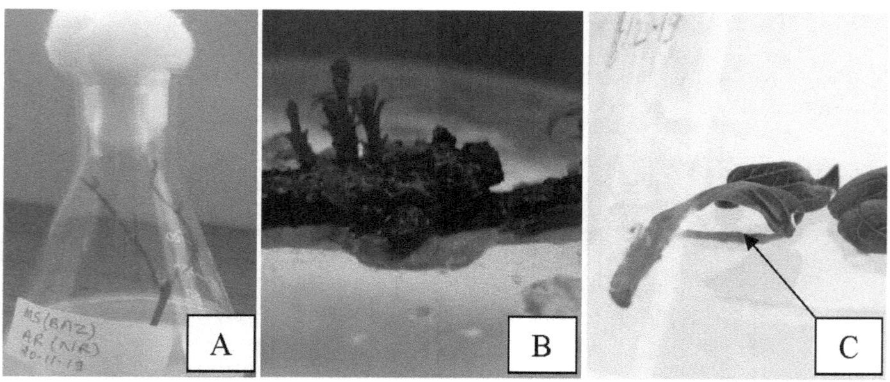

FIGURE 1.3 (A) Axillary bud proliferation from nodal region of *Asparagus*; (B) Adventitious shoot formation from intermodal region; (C) Adventitious root formation from leaf of *Vigna*.

adventitious plantlets have also grown straightforwardly as from primary explants. Callus-derived adventitious plant regeneration advancement has been reported in asparagus, cabbage, potato, tomato, sweet potato, broccoli, garlic, brussels sprouts, carrot, kale, and pepper.

The major drawback of adventitious plantlet development is that hereditary fluctuation tends to rise, especially once the plantlets are gotten from callus. This changeability by and large will in general broaden due to the expansion as the term of an age or the time allotment callus is in culture increments. The hereditary changeability typically found in some of these cultures includes variation in phenotypically expression, fluctuation in yield, and loss of organogenic potency, and therefore is usually a consequence of chromosome anomalies even as ploidy varies in number of chromosome (*Torres 1989*).

1.3.2.3 Adventitious Root Formation

Adventitious roots arise from any organ other than the radical or the root axis, usually leaves, nodes, internodes, etc., of the plant. Adventitious root formation *in vitro* has been studied in many species, including in micro-cuttings of grape vine (*Vitis vinifera* L.), apple (*Malus* spp.) and mung bean (*Vigna radiata* L.) (Figure 1.3 C).

1.3.3 Somatic Embryogenesis

Embryogenesis is the process of creating an embryo, whether from a zygote (zygotic embryogenesis, ZE) and from somatic cells (somatic embryogenesis, SE) (Loyola-Vargas and Ochoa-Alejo 2016). Somatic cell embryogenesis has the best chance of achieving quick micropropagation. Throughout this method, a single cell is stimulated to generate an embryo that also results in the formation of a complete plant. Somatic embryos are coordinated structures that grow from somatic cells but have the morphology of an emerging zygotic embryo. Somatic embryogenesis is a development process in which a somatic cell dedifferentiates into a totipotent embryonic stem cell that can bring about an embryo under suitable conditions (Guan et al. 2016). Steward et al. (1958) discovered somatic embryogenesis for the first time in the suspension cultures of *Daucus carota.* The induction of somatic embryo *in vitro* conditions can be developed from two pathways: 1) direct somatic embryogenesis, where somatic embryos form at the edge of an explant; 2) indirect somatic embryogenesis, where somatic embryos develop from a callus, which is a dedifferentiated mass of parenchymatous tissues (Quiroz-Figueroa et al. 2006).

Somatic embryogenesis has had a major impact on horticultural crops. Somatic embryogenesis throughout fruit crops is starting to emerge as an attractive and, at times, necessary supplement to traditional plant breeding. Due to poor germination of somatic embryos and a low frequency of recovering somatic seedlings, the use of somatic embryogenesis in fruit crop improvement has indeed been bounded.

A single embryogenic cell divides transversely and longitudinally, transferring through the globular, torpedo, and cotyledonary stages in dicots and the globular, scutellar, and coleoptilar stages in monocotyledons. Eventually, it develops a bipolar structure with a shoot and a root on different poles of the same axis, capable of reproducing whole plants (Arnold et al. 2002). Somatic embryos generate reserves and develop desiccation tolerance during the maturation stage, which is essential for appropriate germination and growth of seedlings (Ammirato 1974). The phase of maturation in somatic embryogenesis was recently identified as a result of analyses, also the percentage of germination and regeneration in well-developed somatic embryos is frequently very low (Bhojwani and Razdan 1996).

Mango, citrus spp., loquat (from nuclear tissue), date palm (from shoot tips), banana, and plantain (from leaf tissue) have all shown regeneration via somatic embryogenesis (Litz and Jaiswal 1991). Somatic embryogenesis has also piqued the interest of fruit breeders and pomologists for horticultural crops improvement as a framework for huge quantities in bioreactors via repeating induction (Litz and Gray 1992).

1.3.4 Androgenesis

The term androgenesis refers to plant regenerate from either anther or pollen grains (microspores) under *in vitro* conditions. Haploid plants developed from microspore culture is an important part of creating homozygous lines because of the huge amounts of microspores that are formed from a single plant. However, androgenic development was first described by Guha and Maheshwari (1964, 1966) in *in vitro* anther culture of *Datura innoxia*. The haploid plants were produced through the anther, pollen, unfertilized ovary, ovule, etc., under *in vitro* conditions (Basu et al. 2011; Tazeb 2017). Haploid plants play an important role for plant breeders because they can be effortlessly doubled due to the colchicines treatment when needed to make every locus homozygous in an exceptionally brief timeframe (Basu et al. 2011; Tazeb 2017). Haploids have been produced quickly and developed from plant tissue culture and afterward used for crop enhancements in rice, wheat, maize, potato, barley, asparagus, tobacco, sunflower, etc. (Bajaj 1990).

1.3.5 Protoplast Culture

Protoplasts are naked cells that have had their cell walls completely removed by mechanical or enzymatic means (Eeckhaut et al. 2013). On an essential level, every individual protoplast has been totipotent, meaning that it has the ability to regenerate a cell wall and then grow rapidly into different organs under essential compound and physical conditions (Khatri et al. 2010). The protoplast is a valuable biological framework used to study cell wall, cell division, and proliferation (Aoyagi 2011). All of this employs plant hereditary methods such as electroporation, somatic hybridization, DNA microinjection and micro-protoplast interceded chromosome transfer (Rezazadeh and Niedz 2015). Many natural product species like banana (Haicour et al. 2009), guava (Rezazadeh and Niedz 2015), grape (Yu et al. 2013), and mango (Rezazadeh et al. 2011) have been studied for protoplast isolation and regeneration.

Protoplast fusion is a novel technique for transferring desired genes form one species to another. The *fusion* products are cultured to produce somatic hybrid plants. Somatic hybrids are created by combining the cytoplasm and nuclei of two different plant species. Purified protoplasts acquired from any two unique sources (e.g., elite

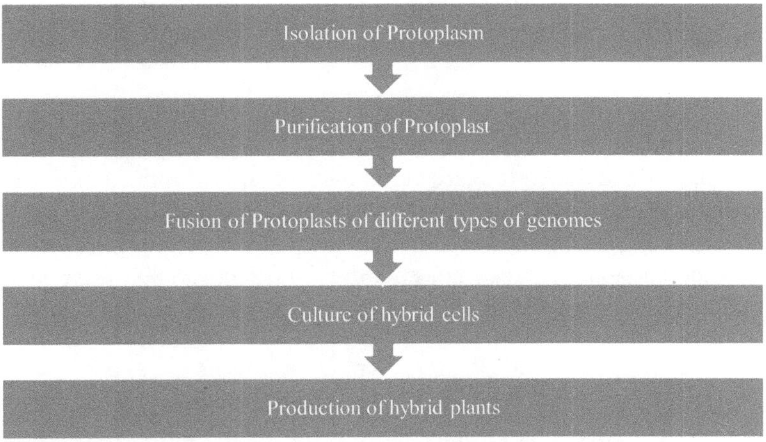

FIGURE 1.4 Production of hybrid plant through protoplast culture.

plants, species, genera) can be fused together to form somatic hybrids (Figure 1.4). Somatic hybridization is a significant part of plant breeding and crop improvement because it produces inter-generic and inter-specific hybrids (Brown and Thorpe 1995).

Somatic hybridization was first reported for the fusion of *Nicotiana glauca* and *N. Langsdorfi* by Carlson et al. (1972). Somatic hybrids were developed by fusing protoplasts from rice and ditch reed and treating them with electrofusion for salt resistance (Mostageer and Elshihy 2003). The procedure was used in the horticultural industry to develop new hybrids with higher natural product quality and high infection resistance. Citrus protoplasts have been fused with other related citrinae species, creating effective hybrid plants (Motomura et al. 1997).

1.4 CURRENT DEVELOPMENTS (NEW CULTURE SYSTEMS)

1.4.1 Bioreactors

A bioreactor is a manufactured device used for secondary metabolite production (Panda et al. 1989). Bioreactors for plant culture have always been the most visible, with different species being accommodated. Since Murashige (1974) showed the fundamental micropropagation plan, one of the most significant advancements, mostly in plant tissue culture sector, was the use of bioreactors. In contrast to conventional tissue culture strategies, bioreactor frameworks have a few benefits: they save time and work, are moderately easy to increase, permit improved development and multiplication (e.g., by constrained air circulation), and work on supplement accessibility because of the utilization of liquid medium. Some new procedures have been adjusted to create bioreactors appropriate for different plant species and their particular prerequisites (Aitken-Christie et al. 1995; Paek et al. 2001). The principal systems include:

1) Aeration-agitation bioreactor
2) Spin filter bioreactor

3) Gaseous phase bioreactor
4) Rotating drum bioreactor
5) Air-driven bioreactor (Lee 2004)

The above frameworks have been successfully used for large-scale production of more than 80 crops and therefore are currently used in the production of numerous different plant species (Takayama 1991; Paek et al. 2001). Initially used for *Begonia* micropropagation, bioreactors were later applied to a variety of horticultural crop species including apple, grape, pineapple, *Stevia rebaudiana, Chrysanthemum*, garlic, and *Phalaenopsis* (Paek et al. 2005).

1.4.2 *In vitro* Mycorrhization

Tissue culture-raised plants are adversely influenced by water stress, because of the low water absorption limit of their own root systems or the stomata's insufficient regulate loss of water (Bonga 1977; Flick et al. 1983; Nowak 1998). Hardening/acclimatization of tissue culture-raised plants relates to a period during which roots adjust to a substrate with a less accessible nutrient content, which leads to an autotrophic condition. At this step, the presence of mycorrhizae may increase the accessibility of restricting supplements like phosphorus (P) and nitrogen (N), assisting with assimilation. Mycorrhization in micropropagation, especially the utilisation of arbuscular mycorrhizal fungi (AMF), is presently more popular because of its exhibited positive effect on post-transfer execution of *in vitro* developed plants (Lovato et al. 1996; Rai 2001). Increased nutrient uptake, air circulation, water relations, balance of soil pH, and their potential can be used as bioregulators had also currently increased research interest in AMF, contributing to the development of powerful strategies of production of AMF, mycorrhization of in cultured cells plants, and evaluating for effective strains of AMF (Sylvia 1998; Lovato et al. 1996). *Piriformospora indica*, for example, promotes explant hardening; *Psuedomonas spp., Bacillus pumilus,* and *Alcaligenes faecalis* promote multiplication of shoot; and *Psuedomonas spp.* may also decrease hyperhydricity (Sahay and Varma 1999; Monier et al. 1998; Bela et al. 1998).

1.5 APPLICATIONS OF TISSUE CULTURE IN HORTICULTURAL CROP IMPROVEMENT

The use of plant tissue culture can improve quality of life by growing new varieties of horticultural crop plants that include are most commonly herbicide resistance (soybean), insect resistance (cotton, corn), and delayed fruit ripening (tomato), give more significant returns with less information sources, can be filled in a more broad extent of conditions, give better rotations to conserve natural resources, nutritious procured things that keep any more extended away and transport, and proceed with minimal expense food supplies to consumers.

1.5.1 Production of Elite and High-Quality Healthy Plants

Tissue culture technology provides a unique opportunity to produce elite and high-quality types of plants through vegetative propagation. An entire plant can be recovered from a small tissue in an appropriate culture medium under controlled environment. The development of high planting material of horticultural food crops and fruits developed from vegetative parts has also opened up new avenues for international trade and assisted cultivators and nursery owners. Nonetheless, there are endless possibilities to produce and sell high-quality planting material, including horticultural crops like potato, banana, date palm, sweet potato, sugarcane, pineapple, cassava, ornamentals crops, fruit, and woody plants.

1.5.2 Development of Homozygous Lines

Haploid plants have been produced from either anther culture or pollen culture under sterile conditions. These plants contain a haploid number of chromosomes. The plants can be made fertile through doubling their chromosome number, and the resulting plants will be homozygous diploid or isogenic diploid. Haploids and double haploids have significance in inherited characteristics and programmes of plant breeding (Hu and Guo 1999; Maluszynski et al. 2003). The ability to achieve mostly homozygozity in only one advance through doubling the chromosome number is a significant benefit of haploids in plant reproducing programmes. Haploids are most commonly used in F1 hybrids for recombination fixing to develop heterozygous hybrids, resulting in the rapid selection of phenotypes for quantitative and qualitative characteristics. In comparison to traditional methods, which involve time-consuming, labour-intensive, and recurrent backcrossing, the doubled haploid technique saves a lot of time in establishing new hybrid crop varieties over 2-4 years. Pollen-derived doubled haploids have been widely used in crop improvement programmes, assisting in the innovation of a few more developed cultivars (Maluszynski et al. 2003; Palmer et al. 2005). The haploid set of chromosomes in microspore- determined plants spontaneously doubles under culture conditions. However, the percentage of doubling changes among the crops, including various genotypes of a cultivar. So far, barley has shown the most noteworthy rate of spontaneous doubling (up to 87%), followed by rice (up to 72%), wheat (up to 50%), and maize (6.3% and quite unpredictable) (Hoekstra et al. 1993; Cho and Zapata 1990; de Buyser and Henry 1986; Ku et al. 1978). In general, colchicine is used for chromosome doubling at the entire plant level for barley, rice, and maize. Colchicine is sometimes used in mechanism for chromosome doubling in wheat (Jahne and Lorz 1995). Colchicine is used in the medium before the first microspore mitosis, which can contribute to a significant increase in gametophytic chromosome number.

1.5.3 Production of Pathogen-Free Crop Plants

Plant pests and pathogens cause extensive harm to the growth of many horticultural crop species. Plant tissue culture is important in phytopathology for producing pathogen-free plants. Pathogen-free plants develop from apical shoot meristem. It was proved that growing pathogen-free crops yields a greater yield than growing field-developed crops. Virus infections infect a large portion of the yield. The crop plants, like banana, cassava, sugarcane, potato, sweet potato, horticultural crops (e.g., citrus, stone fruits, pome), and ornamentals that are generally proliferated by vegetative methods, are particularly vulnerable to losses caused by viruses that are passed down to the next generation. Potato Virus Y (PVY) and Potato Leafroll Virus (PLRV) are the most damaging potato viruses, with PLRV alone reducing total yield by approximately 60% and marketable yield by up to 88% (Loebenstein et al. 1997; Brunt and Loebenstein 2001; Hamm and Hane 1999). The plant tissue culture has taken advantage of the property of totipotency to maintain the whole development potential of a single cell or group of cells, which, under suitable conditions, can multiply and give rise to pathogen-free plants (Brown and Thorpe 1995; Basavaraju 2011; Tazeb 2017). The procedure has supported the production of vegetative proliferated strawberry plants (Brown and Thorpe 1995).

1.5.4 Somaclonal Variation

Genetic variability plays an enormous part in the improvement of horticultural crop plants. Significant advances in horticultural production have been made with the emergence of new technologies like plant tissue culture and recombinant DNA innovation. Somaclonal variations seem to be hereditary changes that occur during in vitro stress. These might be epigenetic or non-genetic changes that are heritable or not. Somaclonal variations in crops cause a variety of hereditary changes that are used to modify plant characteristics like plant tallness, grain quality, and yield, as well as provide resistance to pests, diseases, and drought (Patnaik et al. 1999). Somaclonal variations have also been accounted for in numerous crop plants like apple, potato, banana, *Heliconia bihai*, and sugarcane (Rosati and Predieri 1990; Das et al. 2000; James et al. 2007; Rodrigues 2008; Sengar et al. 2009). Sugarcane somaclonal variations have vastly increased sugar yield and continue to improve resistance to eye-spot disease (Sengar et al. 2009). Additionally, it was also used in potatoes to improve tuber shape and consistency, as well as resistance to late blight disease (Das et al. 2000). Different new traits that were absent in the natural gene pool have been recovered, such as increased methionine and lysine content in cereals, glyphosate resistance in tobacco, atrazine resistance in maize, and fungal resistance in alfa alfa (Sengar et al. 2009). Screening of plants developed from selected calli resulted in the isolation of a tomato line resistant to *Pseudomonas solanacearum*-caused bacterial wilt. A line of maize with Southern leaf blight is caused by *Helminthosporium maydis*, which also produces a toxin that binds to mitochondria. Somaclonal variant selection has resulted in the delivery of a predetermined growth of emerging cultivars of important horticultural crop species (Table 1.1).

1.5.5 Production of Secondary Metabolites through *in vitro* Cultures

Plant secondary metabolites have been the most important phytochemicals of plant secondary metabolism and have adequate compound or fundamental intricacy with

TABLE 1.1
Somaclonal variants delivered as new cultivars in different horticultural crops

Crop	Somaclonal trait	Reference
Banana	*Fusarium* wilt	Hwang and Ko (2004)
Brown mustard	Yield	Katiyar and Chopra (1995)
Sweet potato	Tolerant to salinity	Anwar et al. (2010)
Sugarcane	High yielding and resistant to red rot and smut disease	Jalaja et al. (2006)
	High sugar and cane yield, high sucrose content, and moderate resistance to smut, grassy shoot, and red rot diseases	Tawar et al. (2016)

TABLE 1.2
Some secondary metabolites produced through plant tissue culture

Secondary metabolite	Plant species	Type of Culture	References
Anthocyanin	*Vitis vinifera*	Suspension	Qu et al. (2006)
Capsaicin and dihydrocapsaicin	*Capsicum chinense*	Shoot	Kabita et al. (2020)
Mulberroside A	*Morus alba*	Shoot	Komaikul et al. (2019)
L-Dopa	*Hybanthus enneaspermus*	Adventitious root	Sathish et al. (2020)
Oleanolic acid glycosides	*Calendula officinalis*	Hairy root	Alsoufi et al. (2019)

the goal that artificial synthesis is troublesome or not as of now conceivable. Plant cell culture has huge potential as an in vitro method for producing extremely complicated secondary metabolites. Secondary metabolites like alkaloids, phenols, terpenoids, glycosides, tanins, etc., are used in drugs, flavors, scents, cosmetics, food added substances, agrochemicals, insecticides, and raw materials in industries. Also, with effective foundation of cell lines able to produce high yields and quality secondary compounds in suspension cultures, a major development in cell culture procedure occurred (Zenk 1978). Different studies on secondary metabolite production using in vitro are listed in Table 1.2.

1.5.6 *Agrobacterium*-Mediated Genetic Transformation

Agrobacterium-mediated transformations are some of the most commonly implemented crop plant gene transfer strategies. *Agrobacterium* has two strains: *A. tumefaciens* and *A. rhizogene*. *A. tumefaciens* contamination causes crown or gall diseases, while *A. rhizogene* causes hairy root culture in plants (Bhatia 2015). *Agrobacterium tumefaciens*, a plant pathogen, is frequently used as a vector for

TABLE 1.3
***Agrobacterium*-mediated transformation in several horticultural crops species**

Crop	Remarks	Reference
Malus micromalus	*Agrobacterium*-mediated transformation system using cotyledons as explants	Dai et al. (2014)
Lycopersicon esulentum	*Agrobacterium*-mediated transformation of tomato plants expressing defensin gene	El-Siddig et al. (2011)
Citrus maximma	In planta transformation method of 3–4-week old seedlings using *Agrobacterium tumefaciens*	Zhang et al. (2017)
Prunus domestica	The phosphomannose-isomerase (PMI) gene and green fluorescent protein (GFP) gene for *Agrobacterium*-mediated transformation using leaf as explants	Sidorova et al. (2017)
Rubus fruticosus	Genetic transformation of a stilbene synthase gene in somatic embryos from whole flower cultures	Dai et al. (2015)

introducing desired foreign genes into crops and tends to result in transgenic plant regeneration (Table 1.3). *A. tumefaciens* is naturally able to infect dicotyledonous plant injury sites and causes crown gall disease. Because the bacterium appears to have a large plasmid which induces induction of tomours, this was given the name tumour inducing (Ti) plasmid. Tumors form at the infection site as a result of Ti-plasmid T-DNA gene expression in plant cells. Hairy root culture is notable for its increased scale-up development of biomass, metabolite release in culture medium, production of secondary metabolite, and sensitivity to external stimuli.

1.5.7 Production of Synthetic Seeds

An artificial or synthetic seed has been characterized as a somatic embryo encapsulated inside a coating and is viewed as practically equivalent to a zygotic seed (Redenbaugh 1993). The famous scientist Murashige introduced the concept of artificial seeds and called them "synthetic seeds." Synthetic seeds are artificially encapsulated plant propagation material. This material could be shoot buds, somatic embryos, cell aggregates, or whatever other tissue that we can use as a seed for propagation. Synthetic seed technology primarily includes encapsulating somatic embryos in a hydrogel like sodium alginate, carrageenan, sodium pectate, potassium alginate, or sodium alginate with gelatine. Encapsulation or synthetic seeds are used to give protection to the artificially produced propagules. Somatic embryos have been the most widely used micropropagule for artificial seed production even though their structures can produce the plumule and radical axis, which can advance into the shoot and root in a single step (Ara et al. 2000). Artificial seeds produced through somatic embryos can likewise give undeniable degrees of generation. Plant lines generated from somatic embryos have been prepared to save their renewing limit for an extended period of time, achieving uniform plant production; it evades the dedifferentiation callus stage, as well as consistent hereditary structure production (Leroy et al. 2000;

Deng 2009). Use of such synthetic seeds as an enhancement for further traditional micropropagation methods in vegetative propagated crop varieties could result in tissue culture, avoiding the laborious process of transferring plants from in vitro to soil/field. Artificial seed production using somatic embryos has been examined in different of horticultural crop species, including *Daucus carota*, *Brassica campestris*, *Medicago sativa*, *Vitis vinifers*, *Mangifera indica*, *Oryza sativa* (hybrid rice), *Citrus reticulata*, and *Saccharum officinarum* (Kitto and Janick 1982; Latif et al. 2007; Kitto and Janick, 1985; Ara et al. 1999; Kumar et al. 2005; Antonietta et al. 1999; Nieves et al. 2003).

1.5.8 Germplasm Conservation

Plant tissue culture severely impacts germplasm conservation. First of all, it gives occasionally options for conserving entire whole organisms. Second, it can assist with the exchange of germplasm. Third, the strategies of molecular biology can be applied to the issues of managing and utilizing germplasm. The fourth impact results from the expanded interest in germplasm and conservation services by biotechnologists. Conservation of germplasm is primarily concerned with ensuring the legal conservation and storage of germplasm storage from economically important crops (Gosal et al. 2010). International Plant Genetic Resources Institute (IPGRI) is an international scientific organization and the Indian Council of Agriculture research Centre (ICAR) in New Delhi (1976) additionally to advance exercises of germplasm like collection, conservation, documentation, and utilization. It additionally organizes courses on crop genetic resource of collection, conservation, evaluation, and use (Quazi et al. 2021). By reducing the water content of agricultural crop germplasm, seeds can be stored for a longer period of time. For many horticulture crops, the National Bureau of Plant Genetic Resources has designated National Active Germplasm sites. For example, NRS Banana in Tiruchirapalli, NRC for grapes in Pune, the Central Institute of Arid Horticulture for Arid fruits in Bikaner, the Central Institute of Temperate Horticulture for Temperate horticulture crops in Srinagar, the Indian Institute of Horticultural Research for Tropical fruits in Bangalore, and so on (Tripathi et al. 2018). Cryopreservation is a method of preserving cells, tissues, organelles, or any other biological material by storing it at an extremely low temperature of about -196°C. All metabolic actions in the cell cease at this low temperature, allowing the material to be stored for an extended period of time with no changes or alterations. Different horticultural crop plants have been recovered completely from tissues that had been stored at extremely low temperatures for years (Ding et al. 2008). Cryopreservation has also been used to eradicate viruses from affected plants through cryotherapy (Bettoni et al. 2016).

1.6 CONCLUSIONS AND FUTURE PROSPECTS

Food security is a looming problem in the twenty-first century as the world's populations grows and competition for arable land and water increases. In addition to these problems with hunger, many simply lack the ability to purchase enough food to meet their nutritional needs. In order to meet the needs of a growing population, a sustainable food supply must be created. This can be accomplished by creating fortified staple foods or crops. The beneficial effects to modern agricultural advances, changing climate, growing population, and pollution threaten the effectiveness of current capital assets for high-yield agricultural production. To feed the world's rapidly growing population, mass propagation of horticulture crops is required. Tissue culture is an effective technique for improving crop quality and nutrition. The utilization of plant tissue culture techniques for commercial propagation of horticultural crop species and for the production of bioactive components from them has become a profitable industry around the world. Tissue culture has been used to increase genetic variability in horticultural crop plants, improve overall plant health, and expand the number of desirable germplasms accessible to plant breeders. At present, thin cell layer technology has opened additional opportunities for ornamental and floricultural crop improvement.

REFERENCES

Aitken-Christie, J., Kozai, T., Takayama, S. 1995. Automation in plant tissue culture- general introduction and overview. In: Aitken-Christie, J., Kozai, T. and Smith, M. A. L. (ed.) *Automation and environmental control in Plant Tissue Culture.* Springer, Dordrecht, pp. 1–8.

Alsoufi, A. S. M., Paczkowski, C., Szakiel, A. and Dlugosz, M. 2019. Effect of jasmonic acid and chitosan on triterpenoid production in *Calendula officinalis* hairy root cultures. *Phytochem. Lett.* 31: 5–11.

Ammirato, P. V. 1974. The effects of abscisic acid on the development of somatic embryos from cells of caraway (*Carum carvi* L.). *Bot. Gaz.* 135: 328–337.

Antonietta, G. M., Emanuele, P. and Alvaro, S. 1999. Effects of encapsulation on *Citrus reticulata* Blanco somatic embryo conversion. *Plant. Cell Tissue Organ. Cult.* 55: 235–237.

Anwar, A., Akira Kikuchi, A. and Watanabe, K. N. 2010. Assessment of somaclonal variation for salinity tolerance in sweet potato regenerated plants. *Afr. J. Biotechnol.* 9: 7256–7265.

Aoyagi, H. 2011. Application of plant protoplasts for the production of useful metabolites. *Biochem. Eng. J.* 56: 1–8.

Ara, H., Jaiswal, U. and Jaiswal, V. S. 1999. Germination and plantlet regeneration from encapsulated somatic embryos of mango (*Mangifera indica* L.). *Plant Cell Rep.* 19: 166–170.

Ara, H., Jaiswal, U. and Jaiswal, V. S. 2000. Synthetic seed: Prospects and limitation. *Curr. Sci.* 78: 1438–1444.

Arnold, S. V., Sabala, I., Bozhkov, P. and Dyachok, J. 2002. Developmental pathways of somatic embryogenesis. *Plant Cell Tissue Organ. Cult.* 69: 233–249.

Bajaj, Y. P. S. 1990. Haploids in crop improvement. *Biotechnol. Agric. Forest.*12. doi:10.1007/978-3-642-61499-6.

Basavaraju, R. 2011. Plant tissue culture-agriculture and health of man. *Ind. J. Sci. Technol.* 4(3): 333–335.

Basu, S. K., Datta, M., Sharma, M. and Kumar, A. 2011. Haploid plant production technology in wheat and some selected higher plants. *Aust. J. Crop. Sci.* 5(9): 1087–1093.

Bela, J., Ueno, K. and Shetty, K. 1998. Control of hyperhydricity in anise (*Pimpinella anisum*) tissue culture by *Pseudomonas* spp. *J. Herbs Spices Med. Plants* 6: 57–67.

Berjak, P. and Pammenter, N. W. 2013. Implications of the lack of desiccation tolerance in recalcitrant seeds. *Front. Plant Sci.* 4: 478.

Bettoni, J. C., Costa, M. D., Gardin, J. P. P., Kretzschmar, A. A. and Pathirana, R. 2016. Cryotherapy: a new technique to obtain grapevine plants free of viruses. *Rev. Bras. Frutic.* 38: 833.

Bhatia, S. 2015. Modern applications of plant biotechnology in pharmaceutical sciences. *Appl. Plant Biotechnol.*, 157–207. doi:10.1016/B978-0-12-802221-4.00005-4.

Bhojwani, S. S. and Razdan, M. K. 1996. Somatic embryogenesis. In: Bhojwani, S. S. and Razdan, M. K. (2nd ed.) *Plant Tissue Culture: Theory and Practice*, Elsevier Science, Amsterdam BV, the Netherlands, pp. 125–166.

Bonga, J. M. 1977. Applications of tissue culture in Forestry. In: Reinert, J. and Bajaj, Y. P. S. (ed.) *Plant Cell Tissue and Organ Culture*, Springer-Verlag, New York, pp. 93–107.

Bregitzer, P., Zhang, S., Chob, M. J. and Lemaux, P. G. 2002. Reduced somaclonal variation in barley is associated with culturing highly differentiated, meristematic tissues. *Crop Sci.* 42(4): 1303–1308.

Brown, D. W. and Thorpe, T.A. 1995. Crop improvement through tissue culture. *World J. Microbiol. Biotechnol.* 11: 409–415.

Brunt, A. A. and Loebenstein, G. 2001. Potato leafroll virus (PLRV; genus Polerovirus; family Luteoviridae). In: Loebenstein, G., Berger, P. H., Brunt, A. A., and Lawson, R. H. (ed.), *Virus and Virus-like Diseases of Potatoes and Production of Seed Potatoes*, Kluwer Academic Publisher, Dordrecht, The Netherlands, pp. 65–134.

Carlson, P. S., Smith, H. H., and Dearing, R. D. 1972. Parasexual interspecific plant hybridization. *Proc. Nat. Acad. Sci. (Wash.)* 69: 2292–2294.

Cho, M. S. and Zapata, F. J. 1990. Plant regeneration from isolated microspore of indica rice. *Plant Cell Physiol.* 31: 881–885.

Dai, H. Y., Li, W. R., Mao, W. J., Zhang, L., Han, G. F., Zhao, K., Liu, Y. and Zhang, Z. 2014. Development of an efficient regeneration and *Agrobacterium*-mediated transformation system in crab apple *(Malus micromalus*) using cotyledons as explants. *In Vitro Cell. Dev. Biol. Plant* 50: 1–8.

Dai, L. M., Zhou, Q., Li, R. M., Du, Y. J., He, J., Wang, D., Cheng, S., Zhang, J. and Wang, Y. 2015. Establishment of a picloram-induced somatic embryogenesis system in *Vitis vinifera* cv. chardonnay and genetic transformation of a stilbene synthase gene from wild-growing *Vitis* species. *Plant Cell Tissue Organ Cult.* 121: 397–412.

Das, A., Gosal, S. S., Sidhu, J. S. and Dhaliwal, H.S. 2000. Induction of mutations for heat tolerance in potato by using *in vitro* culture and radiation. *Euphytica* 114: 205–209.

de Buyser, J. and Henry, Y. 1986. Wheat production of haploids, performance of doubled haploids, and yield trials. In: Bajaj, Y. P. S. (ed.), *Biotechnology in Agriculture and Forestry, Transgenic Crops I,* vol 2, Springer-Verlag, Berlin, pp. 73–88.

Deng, W. 2009. A novel method for induction of plant regeneration via somatic embryogenesis. *Plant Sci.* 177: 43–48.

Doran, J. W. 2002. Soil health and global sustainability: translating science into practice. *Agric. Ecosyst. Environ.* 88: 119–127.

Eeckhaut, T., Lakshmanan, P. S., Deryckere, D., Van Bockstaele, E. and Van Huylenbroeck, J. 2013. Progress in plant protoplast research. Planta 238: 991–1003.

El-Siddig, M. A., El-Hussein, A. A. and Saker, M. M. 2011. *Agrobacterium*-mediated transformation of tomato plants expressing defensin gene. *Int. J. Agric. Res.* 6(4):323–334.

Fallik, E. 2004. Pre-storage hot water treatments immersion, rinsing and brushing. *Postharvest Biol. Technol.* 32: 125–134.

Flick, C. E., Evans, D. A. and Sharp, W. R. 1983. Organogenesis. In: Evans, D. A., Sharp, W. R., Ammirato, P. V., Yamada, Y. (ed.), *Handbook of Plant Cell Culture.* vol I – *Techniques for Propagation and Breeding*, McMillan Publishing Co., New York, pp. 13–81.

Gantait, S., Mandal, N. and Das P. K. 2010. Field evaluation of micropropagated vs. conventionally propagated elephant garlic. *J. Agric. Technol.* 7(1): 97–103.

George, E. F. and Sherrington, P. D. 1984. *Plant Propagation by Tissue Culture.* pp. 703. Exgetics Ltd., Eversley, England.

Gosal, S. S., Wani, S. H. and Kang, M. S. 2010. Biotechnology and crop improvement. *J. Crop. Improv.* 24 (2): 153–217.

Guan, Y., Li, S. G., Fan, X. F. and Su, Z. H. 2016. Application of somatic embryogenesis in woody plants. *Front. Plant Sci.* 7: 938. doi:10.3389/fpls.2016.00938

Guha, S. and Maheshwari, S. C. 1964. *In vitro* production of embryos from anthers of *Datura. Nature* 204: 497.

Guha, S. and Maheshwari, S. C. 1966. Cell division and differentiation of embryos in the pollen grains of *Datura. In vitro. Ibid.* 212: 97–98.

Gupta, R., Modgil, M. and Chakrabarti, S. K. 2009. Assessment of genetic fidelity of micropropagated apple rootstock plants, EMLA 111, using RAPD markers. *Indian J. Exp. Biol.* 47(11): 925–928.

Haicour, R., Assani, A., Trang, V. B. and Guedira, A. 2009. Protoplast isolation and culture for banana regeneration via somatic embryogenesis. *Fruits* 64: 261–269.

Hamm, P. B. and Hane, D. C. 1999. Effects of seed borne potato virus Y infection in two potato cultivars expressing mild disease symptoms. *Plant Dis.* 83(12): 43–45.

Hoekstra, S., van Zijderveld, M. H., Heidekamp, F. and van der Mark, F. 1993. Microspore culture of *Hordeum vulgare* L.: the influence of density and osmolality. *Plant Cell Rep.* 12: 661–665.

Hu, H. and Guo, X. 1999. *In vitro* induced haploids in plant genetics and breeding. In: Soh, W. Y. and Bhojwani, S. S. (ed.), *Morphogenesis in plant tissue cultures.* Kluwer Academic Publishers, Dordrecht, pp. 329–361.

Hwang, S. C. and Ko, W. H. 2004. Cavendish banana cultivars resistant to *Fusarium* wilt acquired through somaclonal variation in Taiwan. *Plant Dis.* 88(6): 580–588.

Jahne, A. and Lorz, H. 1995. Cereal microspore culture. *Plant Sci.* 109: 1–12.

Jalaja, N. C., Sreenivasan, T. V., Pawar, S. M., Bhoi, P. G. and Garker, R. M. 2006. Co 94012-a new sugarcane variety through somaclonal variation. *Sugar Tech.* 8(2–3): 132–136.

James, A. C., Peraza-Echeverria, S., Peraza-Echeverria, L. and Herrera-Valenci, V. A. 2007. Variation in micropropagated plants. *Acta Hort.* 748: 55–63.

Kabita, K. C., Sanatombi, K. and Sharma, S. K. 2020. Efficient enhancement of capsaicinoids biosynthesis in cell suspension cultures of *Capsicum chinense* Jacq. cv. 'Umorok' by elicitors and differential gene expression analysis of elicited cultures. *Plant Cell Tissue Organ Cult.* 141: 145–154.

Kane, M. E., Philman, N. J. and Jenks, M. A. 1994. A laboratory exercise to demonstrate direct and indirect shoot organogenesis using internodes of *Myriophyllum aquaticum. Hort Tech.* 4: 317–320.

Katiyar, R. K. and Chopra, V. L. 1995. A somaclone of *Brassica juncea* is processed into a variety and is released for commercial cultivation in India. *Cruciferae Newslett.* 17: 92–93.

Khatik, N. and Joshi, R. 2018. *In vitro* propagation of *Murraya koenigii* by axillary bud proliferation using mature explants. *Indian J. Biotechnol.* 17(2): 379–382.

Khatik, N. and Joshi, R. 2020. Assessment of genetic fidelity in microclones of curry leaf plants [*Murraya koenigii* (L.) Spreng.] using ISSR markers. *Indian J. Exp. Biol.* 58(4): 286–291.

Khatri, A., Dahot, M. U., Khan, I. A. and Nizamani, G. S. 2010. An efficient method of protoplast isolation in banana (*Musa* spp.). *Pak. J. Bot.* 42(2): 1267–1271.

Kitto, S. L. and Janick, J. 1982. Polyox as an artificial seed coat for a sexual embryos. *Hort Sci.* 17: 488.

Kitto, S. L. and Janick, J. 1985. Production of synthetic seeds by encapsulating asexual plant embryos of carrot. *J. Am. Soc. Hort. Sci.* 110: 277–282.

Komaikul, J., Kitisripanya, T., Inyai, C., Likhitwitayawuid, K., Sritularak, B., Tanaka, H. and Putalun, W. 2019. Phytostilbenoid production in white mulberry (*Morus alba* L.) cell culture using bioreactors and simple deglycosylation by endogenous enzymatic hydrolysis. *Vitr. Cell Dev. Biol. Plant.* 55: 199–208.

Ku, M. K., Cheng, W. C., Kuo, L. C., Kuan, Y. L., An, H. P. and Huang, C. H. 1978. Induction factors and morphocytological characteristics of pollen-derived plants in maize (*Zea mays*), Proceedings of Symposium on Plant Tissue Culture, Science Press, Peking, pp. 35–41.

Kumar, M. B. A., Vakeswaran, V. and Krishnasamy, V. 2005. Enhancement of synthetic seed conversion to seedlings in hybrid rice. *Plant Cell Tissue Organ Cult.* 81: 97–100.

Latif, Z., Nasir, I. A. and Riazuddin, S. 2007. Indigenous production of synthetic seeds in *Daucus carota*. *Pak. J. Bot.* 39: 849–855.

Lee, J. M. 2004. Plant cell culture and its applications. In: Goodman, R. M. (ed.), *Encyclopaedia of Plant and Crop Sciences*, Marcel Dekker, USA, pp. 931–933.

Leroy, X. J., Leon, K., Charles, G. and Branchard, M. 2000. Cauliflower somatic embryogenesis and analysis of regenerant stability by ISSRs. *Plant Cell Rep.* 19: 1102–1107.

Lichtfouse, E., Navarrete, M., Debaeke, P., Souchere, V., Alberola, C. and Menassieu, J. 2009. Agronomy for sutainable agriculture. A review. *Agron. Sustain. Dev.* 29: 1–6.

Litz, R. E. and Gray, D. J. 1992. Organogenesis and somatic embryogenesis. In: Hammerschlag, F. A. and Litz, R. E. (ed.), *Biotechnology of Perennial Fruit Crops*, CAB International, Wallingford, UK, pp. 3–34.

Litz, R. E. and Jaiswal, V. S. 1991. Micropropagation of tropical and subtropical fruits. In: Debergh, P. C. and Zimmerman, R. H. (ed.), *Micropropagation: Technology and Application*, Kluwer Academic Publishers, Dordrecht, pp. 247–263.

Loebenstein, G., Akad, F., Filatov, V., Sadvakasova, G., Manadilova, A., Bakelman, H., Teverovsky, E., Lachmann, O. and David, A. 1997. Improved detection of potato leafroll luteovirus in leaves and tubers with a digoxigenin-labeled cRNA probe. *Plant Dis.* 81: 489–491.

Lovato, P. E., Gianinazzi-Pearson, V., Trouvelot, A. and Gianinazzi, S. 1996. The state of mycorrhizas and micropropagation. *Adv. Hort. Sci.* 10: 46–52.

Loyola-Vargas, V. M. and Ochoa-Alejo, N. 2016. Somatic embryogenesis: an overview. In: Loyola-Vargas, V. M. and Ochoa-Alejo, N. (ed.), *Somatic Embryogenesis. Fundamental Aspects and Applications*, Springer Cham, Switzerland, pp. 1–10. doi:10.1007/978-3-319-33705-0_1

Maluszynski, M., Kasha, K. J., Forster, B. P. and Szarejko, L. 2003. *Doubled Haploid Production in Crop Plants*. Kluwer Academic Publishers, Dordrecht.

McMeans, O., Skirvin, R. M. Otterbacher, A. and Mitiku, G. 1998. Assessment of tissue culture-derived 'Gala' and 'Royal Gala' apples (*Malus x domestica* Borkh.) for somaclonal variation. *Euphytica* 103(2): 251–257.

Medina, R., M., Faloci, M., Marassi, A. and Mroginski, L. A. 2004. Genetic stability in rice micropropagation. *Biocell* 28(1): 13–20.

Mishra, M., Chandra, R., Tiwari, R. K., Pati, R. and Pathak, R. K. 2005. Micropropagation of certain underused fruit crops. *Small Fruits Review*, 4(4): 7–18. doi:10.1300/J301v04n04_03

Mohanty, S., Joshi, R. K., Subudi, E., Sahoo, S. and Nayak, S. 2012. Genetic stability assessment of micropropagated mango Ginger (*Curcuma amada* Roxb.) through RAPD and ISSR markers. *Res. J. Med. Plant* 6(7): 529–536.

Monier, C., Bossis, E., Chabanet, C. and Samson, R. 1998. Different bacteria can enhance the micropropagation response of *Cotoneaster lacteus* (Rosaceae). *J. Appl. Microbiol.* 5: 1047–1055.

Morel, G. A. 1964. A new means of clonal propagation of orchids. *Am. Orchid Soc. Bull.* 31: 473–477.

Mostageer, A. and Elshihy, O. M. 2003. Establishment of salt tolerant somatic hybrid through protoplast fusion between rice and ditch reed. *Arab. J. Biotech.* 6(1): 1–12.

Motomura, T., Hidaka, T., Akihama, T. and Omura, M. 1997. Protoplast fusion for production of hybrid plants between citrus and its related genera. *J. Japan. Soc. Hort. Sci.* 65: 685–692.

Murashige, T. 1974. Plant propagation through tissue cultures. *Ann. Rev. Plant Physiol.* 25: 135–166.

Nieves, N., Zambrano, Y., Tapia, R., Cid, M., Pina, D. and Castillo, R. 2003. Field performance of artificial seed derived sugarcane plants. *Plant Cell Tissue Organ Cult.* 75: 279–282.

Nowak, J. 1998. Benefits of *in vitro* "biotization" of plant tissue cultures with microbial inoculants. *In vitro Cell. Dev. Biol. Plant* 34: 122–130.

Paek, K. Y., Chakrabarty, D. and Hahn, E. J. 2005. Application of bioreactor systems for large scale production of horticultural and medicinal plants. *Plant Cell Tiss. Organ Cult.* 81: 287–300. doi:10.1007/s11240-004-6648-z

Paek, K. Y., Hahn, E. J. and Son, S. H. 2001. Application of bioreactors for large scale micropropagation systems of plants. *In vitro Cell. Dev. Biol. Plant* 37: 149–157.

Palmer, C., Keller, W. A. and Kasha, K. J. 2005. *Haploids in Crop Improvement II*. Springer, Berlin.

Panda, A. K., Mishra, S., Bisaria, V. S. and Bhojwani, S. S. 1989. Plant cell reactors: a perspective. *Enzyme Microb. Technol.* 11: 386–397.

Panda, M. K., Mohanty, S., Subudi, E., Acharya, L. and S. Nayak, 2007. Assessment of genetic stability of micropropagated plants of *Curcuma longa* L. by cytophotometry and RAPD analyses. *Int. J. Integr. Biol.* 1(3): 189–195.

Pathak, H. and Dhawan, V. 2012. ISSR assay for ascertaining genetic fidelity of micropropagated plants of apple rootstock Merton 793. *In vitro Cell. Dev. Biol.* 48(1): 137–143.

Patnaik, J., Sahoo, S. and Debata, B. K. 1999. Somaclonal variation in cell suspension culturederived regenerants of *Cymbopogon martinii* (Roxb.) Wats var. motia. *Plant Breed.* 118: 351–354.

Qu, J. G., Yu, X. J., Zhang, W. and Jin, M. F. 2006. Significantly improved anthocyanins biosynthesis in suspension cultures of *Vitis vinifera* by process intensification. *Sheng Wu Gong Cheng Xae Bae* 22: 299–305.

Quazi, S., Golani, T. and Capuzzo, A. M. 2021. Germplasm Conservation, Endangered Plants. Kumar, S. IntechOpen doi:10.5772/intechopen.96184

Quiroz-Figueroa, F. R., Rojas-Herrera, R., Galaz-Avalos, R. M. and Loyola-Vargas, V. M. 2006. Embryo production through somatic embryogenesis can be used to study cell differentiation in plants. *Plant Cell Tiss. Org. Cult.* 86: 285–301. doi:10.1007/s11240-006-9139-6

Rai, M. K. 2001. Current advances in mycorrhization in micropropagation. *In vitro Cell. Dev. Biol. Plant* 37: 158–167.

Redenbaugh, K. 1993. *Synseeds: Applications of Synthetic Seeds to Crop Improvement*, CRC Press, Boca Raton, FL, pp. 38–46.

Rezazadeh, R. and Niedz, R. P. 2015. Protoplast isolation and plant regeneration of guava (*Psidium guajava* L.) using experiments in mixture-amount design. *Plant Cell Tiss. Organ Cult.* 122: 585–604.

Rezazadeh, R., Williams, R. and Harrison, D. 2011. Factors affecting mango (*Mangifer indica* L.) protoplast isolation and culture. *Sci Hortic.* 130: 214–221.

Rodrigues, P. H. V. 2008. Somaclonal variation in micropropagated *Heliconia bihai* cv. Lobster claw I plantlets (Heliconiaceae). *Sci. Agric.* 65: 681–684.

Rosati, P. and Predieri, S. 1990. *In vitro* selection of apple rootstock somaclones with *Phytophthora cactorum* culture filtrate. *Acta Hort.* 280: 409–413.

Rout, G. R., Mohapatra, A. and Jain, S. M. 2006. Tissue culture of ornamental pot plant: a critical review on present scenario and future prospects. *Biotechnol. Adv.* 24: 531–560.

Sahay, N. S. and Varma, A. 1999. *Piriformospora indica*: a new biological hardening tool for micropropagated plants. *FEMS Microbiol. Lett.* 181: 297–302.

Sathish, S., Vasudevan, V., Karthik, S., Elayaraja, D., Pavan, G., Ajithan, C. and Manickavasagam, M. 2020. Elicitors induced L-Dopa accumulation in adventitious root cultures of *Hybanthus enneaspermus* (L.) F. Muell. *Vegetos* 33: 304–312.

Sengar, A. S., Thind, K. S., Kumar, B., Mittal, P. and Gosal, S. S. 2009. *In vitro* selection at cellular level for red rot resistance in sugarcane (*Saccharum* sp.). *Plant Growth Regul.* 58: 2001–2009.

Sidorova, T., Mikhailov, R., Pushin, A., Miroshnichenko, D. and Dolgov, S. 2017. A non-antibiotic selection strategy uses the phosphomannose-isomerase (PMI) gene and green fluorescent protein (GFP) gene for *Agrobacterium*-mediated transformation of *Prunus domestica* L. leaf explants. *Plant Cell Tiss. Organ Cult.* 128: 197–209.

Sopory, S. K. and Munshi M. 1996. Anther culture. In: Jain, S. M., Sopory, S. K. and Veilleux, R. E. (ed.), *In vitro Haploid Production in Higher Plants. Current Plant Science and Biotechnology in Agriculture*, vol 1, Kluwer Academic Publishers, Dordrecht, 145–176.

Steward, F., Mapes, M. O. and Mears, K. 1958. Growth and organized development of cultured cells. II. Organization in cultures grown from freely suspended cells. *Am. J. Bot.* 45: 705–708.

Sylvia, D. 1998. The promise (and obstacles) of AMF inoculation. In: Gianinnazzi, S. and Schuepp, H. (ed.), *Arbuscular Mycorrhizas in Sustainable Soil-Plant Systems*. Report of 1997 activities, Cost Action 821, Iceland, p. 152.

Takayama, S. 1991. Mass production of plants through shake-and bioreactor-culture techniques. In: Bajaj, Y. P. S. (ed.), *Biotechnology in Agriculture and Forestry High-Tech and Micropropagation 1*, vol 17, Springer-Verlag, Berlin, pp. 495–515.

Tawar, P. N., Sawant, R. A., Sushir, K. V., Devarumath, R. M., Hapase, R. S., and Meti, N. 2016. VSI 434: new sugarcane variety obtained through somaclonal variation. *Agric. Res.* 5: 127–136.

Tazeb, A. 2017. Plant tissue culture technique as a novel tool in plant breeding: a review. *American-Eurasian J. Agric. Environ. Sci.* 17(2): 111–118.

Torres, K. C. 1989. *Tissue Culture Techniques for Horticultural Crops*. AVI Book, Van Nostrand Reinhold, New York.

Tripathi, P. C., Yogeesha, H. S., Kanupriya, and Rajashankara. 2018. Management of genetic resources of perennial horticultural crops: a review. *Current Horticult.* 6(1): 3–14.

Yu, C., Yang, X., Wang, Z., Qi, J. and Wu, Y. 2013. Research on preparation method of protoplasts from callus of grape. *J Fruits Sci.* 3.

Zenk, M. H. 1978. The impact of cell culture on industry. In: Thorpe, T. A. (ed.), *Frontiers of Plant Tissue Culture*. University of Calgary, Calgary, Canada, pp. 1–13.

Zhang, Y. Y., Zhang, D. M., Zhong, Y., Chang, X. J., Hu, M. L. and Cheng, C. Z. 2017. A simple and efficient in planta transformation method for pommelo (*Citrus maxima*) using *Agrobacterium tumefaciens*. *Sci. Hortic.* 214: 174–179.

2 Advances in Transgenic Approaches in Fruit and Vegetable Crops

Surabhi Jasrotia, Ankila Salathia, Tashi Angmo, Sheikh Qurat, Safina Kosser, Hardik Pathak, Deepesh Kumar Neelam, Jebi Sudan, and Rucku Gupta

2.1 INTRODUCTION

In comparison to fruit and nut species, cereal grains, root and tuber crops, and legumes often lack essential nutrients, leading to food insecurity and malnutrition exacerbated by environmental stressors and population growth. With the global population expected to reach 8.5 billion by 2025, there's a pressing need to increase fruit production to meet demand and mitigate chronic diseases (Drewnowski & Fulgoni, 2008). Biotechnology offers avenues for improving crop quality and quantity, addressing challenges such as insect resistance and nutritional enhancement (Dias & Ortiz, 2014). However, conventional breeding faces limitations due to genetic constraints and extended timelines (Ahmar et al. 2020). Transgenic research has emerged as a promising solution, although it is not without legal and safety concerns (Dias & Ortiz, 2014. Despite obstacles, countries like China, Canada, the United States, Brazil, and Argentina are legalizing transgenic crop cultivation (Mathur et al. 2017). Fruit crop cultivation is increasing due to their nutritional value and health benefits, with various strategies employed for improvement (Prasanna et al. 2015). Challenges include time-consuming breeding methods, difficulty in transferring traits, and the complexity of traits like abiotic stress tolerance (Krishna et al. 2019). Recombinant DNA technology, often known as genetic engineering, is an alternative to traditional breeding that overcomes all of these restrictions (GE). It has minimal production costs, great effectiveness, ease of interspecies gene transfers, and has no substantial effect on the cultivar's genetic background. It has the extra benefit of being able to be employed in vegetatively propagated crops. Transgenic technology has the potential to greatly enhance food, nutrition, and agricultural sustainability. A transgenic plant is one whose genome has been changed by the inclusion of one or more transgenes. It is based on the identification and characterization of the gene responsible for the desired characteristic. The gene to be introduced is chosen for its potential to battle the target disease as well as the sort of resistance mechanism to be employed. These genes can then be used to modify and change any current crop variety. Moreover, it

DOI: 10.1201/9781003343707-2

bypasses the species barrier, allowing genes cloned from one species to be transferred into other agriculturally important species. When compared to conventional plant breeding, this method reduces the time necessary to generate new kinds by nearly half. By addressing the majority of the limitations of traditional breeding and agricultural procedures, GE has provided new strategies to combat plant viruses. Genetically modified plants may be used to research and characterize how gene networks for abiotic stress tolerance interact and operate, in addition to developing new cultivars (Kissoudis et al. 2015).

2.2 LIST OF FRUIT CROPS IN WHICH TRANSGENICS HAVE BEEN DEVELOPED

Several fruit crops with genetically modified features have been granted regulatory permission for commercialization in various countries around the world and are aimed for production as either human food or animal feed (Gomez et al. 2021). These are tomato (*Solanum lycopersicum*), papaya (*Carica papaya*), pepper (*Capsicum annuum*), plum (*Prunus domestica*), eggplant (*Solanum melongena* L.), apple (*Malus domestica Borkh.*), melon (*Cucumis melo* L.), pineapple (*Ananas comosus*), and many more.

2.2.1 Eggplant

Eggplant (*Solanum melongena*), also known as brinjal or aubergine, belongs to the Solanaceae family (nightshade family) and because of its edible fruit, it is grown in most parts of the world. The major factors for its limited productivity despite the increased production in recent years are a broad range of plant diseases, insects, pests, and abiotic stress. Some common diseases that occur in eggplant are verticillium wilt, bacterial wilt, Alternaria rot, phytophthora blight, fruit rot, damping off, leaf spot, mosaic, etc. (Rotino et al. 1997). Abiotic stresses like drought, flooding, or extreme temperature conditions can reduce the fruit quality and yield. Thus, the eggplant breeding programmes aim to develop improved high-yielding varieties with high fruit quality, increased shelf life, and resistance to major diseases, insects, and abiotic stresses. Intensive breeding programmes for the improvement of cultivated eggplant have been undertaken by various research institutes in the last two to three decades resulting in open-pollinated as well as F1 hybrids with differentiated phenotypes (Alam and Salimullah 2021). The key objectives of breeding are high yield, tolerance to biotic and abiotic stress, and high quality. Due to sexual incompatibilities, crossing of eggplant with its wild relatives had limited success rates. Moreover, conventional techniques for breeding crop plants with abiotic stress tolerance and improved pest resistance have a narrow range of success so far because of their time-consuming and laborious nature. On the other hand, new generation approaches such as genetic engineering offer new and improved traits to eggplant as it breaks the barrier of gene(s) transfer from heterologous or unrelated species and expands the gene pool beyond what has been available to conventional breeding approaches.

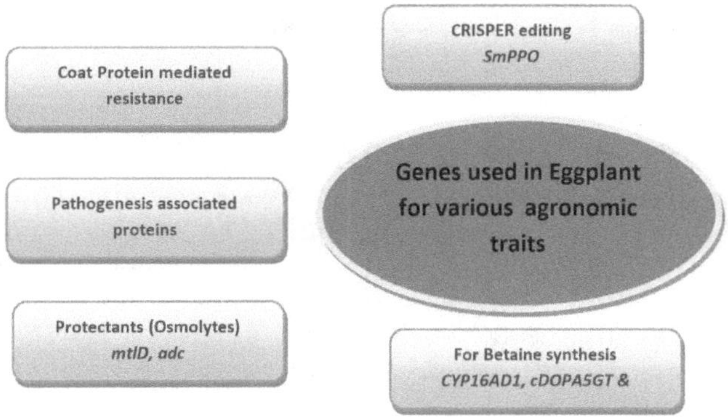

FIGURE 2.1 Genes incorporated in eggplant through transgenic technology to confer new traits.

Many new traits have been offered by genetic engineering in eggplant that do not occur naturally as well as those difficult to incorporate by classical breeding methods that led to varieties that are resistant to biotic and abiotic stresses, have seedless parthenocarpic fruits, and improved fruit quality. In eggplant, genetic transformation is mainly carried out using *Agrobacterium tumefaciens*, though a biolistic approach has also been used in a couple of reports. Guri and Sinks (1988) reported the first successful genetic transformation in eggplant through Agrobacterium. Since then many genes have been incorporated into eggplant through transgenic technology to confer agronomically important traits; some of them are shown in Figure 2.1.

2.2.1.1 Genetic Engineering in Eggplant for Biotic Stress

A significant loss of eggplant yield is caused by insects, nematodes, and plant diseases every year. Some of the major insects are fruit and shoot borers (FSB), stem borers, epilachna beetles, red spider mites, etc. High amounts of chemicals such as pesticides and insecticides have been used conventionally to limit damage caused by insects. Thus, over the last decades, eggplant has been genetically modified to develop insect resistance. For example, EFSB resistance transgenic eggplant lines are developed that express insecticidal crystal protein gene (Bt. gene obtained from *Bacillus thuringiensis*) (Kumar et al. 1988; Pal et al. 1998; Rai et al. 2013). CaMV35S is the only promoter used to express Bt. genes and different versions of the *cry* gene have been transferred to several eggplant varieties. Some of the examples are given in Table 2.1 and Table 2.2.

In addition, a number of parasitic nematode species such as *M. javanica* and *Meloidogyne* also infect eggplant and cause substantial yield losses as eggplant crop becomes more susceptible to other diseases like bacterial and fusarium wilt due to nematode infection (Firoz et al. 2010; Koenning et al. 1999). A couple of reports have been registered in which eggplant has been engineered to inhibit the growth of nematodes inside the plant. In one of them, eggplant is genetically modified to express a rice cystatin gene (OC-IΔD86) that inhibited nematode growth by 78% (Papolu et al. 2016). In the other study, nematode growth in eggplant was restricted

TABLE 2.1
Introduction of insect-resistant traits in eggplant

S. No.	Gene	Eggplant Variety	Resistance Against	Reference
1.	***cry1Ab***	Pusa Purple Long	FSB (*Leucinodes orbonalis*)	Kumar et al. (1988)
2.	***cry1Ac***	cv Kashi Taru	FSB (*Leucinodes orbonalis*)	Pal et al. (2009)
3.	***cry1F***	Manjarigota, Ruchira, Poona selection, and Krishna kathi	FSB (*Leucinodes orbonalis*)	Jadhav et al. (2015); Shruti et al. (2015)

TABLE 2.2
Transgenes introduced in Eggplant for developing insect resistance variety

Gene	Insect Targeted	Key Action	Reference
cry1Aa3	EFSB (*Leucinodes orbonalis*) Lepidoptera	ESFB larval mortality	Rai et al. (2013)
cry1Ab	EFSB (*Leucinodes orbonalis*) Lepidoptera	Significant insecticidal activity	Kumar et al. (1988)
cry1Fa1	EFSB (*Leucinodes orbonalis*) Lepidoptera	Significant larval mortality and reduced growth of any surviving larva	Pal et al. (1998)
cry3A	Colorado potato beetle (*Leptinotarsa decemlineata*)	Most larvae dead within 72 h of feeding	Arpaia et al. (1997)

by host-induced gene silencing (HIGS). Two genes, *msp-18* and *msp-20,* have been targeted to silence those specific to dorsal and subventral oesophageal of *M. incognita*, respectively, in transgenic eggplant (Shivakumara et al. 2017). In eggplants, some of the PR (pathogenesis-related) genes have been incorporated or overexpressed to enhance the disease resistance in plants as shown in Table 2.3. PR proteins govern the plant defense response through various mechanisms including synthesis of secondary metabolites, production of reactive oxygen species (ROS) & antimicrobial peptides, rigidification of cell walls, and sometimes programmed cell death.

2.2.1.2 Genetic Engineering in Eggplant for Abiotic Resistance

Eggplant production is quite vulnerable to climatic conditions such as drought, low or high temperature, and salinity. Some of the key regulator genes that could potentially be involved in the process of tolerance against abiotic stress have been identified by the researchers. In a few reports, eggplants developed through genetic engineering exhibitedtolerance against abiotic stresses such as heat, cold, osmotic stress, and salinity as listed below in Table 2.4. For this, molecular mechanisms like synthesis of osmolytes, modulation of ion transport, and transcriptional regulation have been exploited.

TABLE 2.3
Transgenes introduced in eggplant for developing disease resistance variety

Genes	Trait targeted	Key Result	References
SmMYB44	Bacterial wilt tolerance	• Overexpression of *SmMYB44* induced spermidine synthase gene expression and spermidine content, increasing the resistance to bacterial (*Ralstonia solanacearum*) wilt.	Qiu et al. 2019
Chi	Fungal wilts tolerance	• Showed higher chitinase activity. • Expressed higher level of fungal resistance to the wilts caused by *Verticillium dahlia* and *Fusarium oxysporum*.	Singh et al. 2015
Glu	Fungal wilts tolerance	• Expressed higher glucanase activity. • Showed lower wilting in transgenic plants challenged by *V. dahlia* and *F. oxysporum* spores	Singh et al. 2014
mtlD	Fungal wilt resistance	• Increased resistance against fungal wilts caused by *F. oxysporum*, *V. dahliae* and *R. solaniin vitro* and *ex vitro*	Prabhavathi and Rajam 2007
CMV-CP	Virus resistance (Cucumber mosaic virus)	• Transgenic plants (65–70%) were resistant to Cucumber mosaic virus	Pratap et al. 2011

TABLE 2.4
Abiotic stress-tolerant transgenic eggplants developed through transgenic approach

Gene	Trait targeted	Key Result	References
AtCBF3 & AtCOR15A	Chilling tolerance	• After exposure to cold (2 ± 1 ◦C), slower withering of the leaves of the homozygous transgenic plants was observed. • Proline content as well as catalase and peroxidase activities were considerably enhanced.	Wan et al. 2014
DREB1A	Water stress tolerance	• The transformed plants were in normal condition after fifteen-day of water stress whereas control plants were completely dried.	Sagare and Mohantys 2012
mtlD	Multiple abiotic stresses tolerance	• Transgenic seeds shown higher rate of germination under 200 mM salt-amended MS medium. • Better growth under chilling (6–8 ◦C for 10 days), and salt (200 mM NaCl in hydroponics)	Prabhavathi et al. 2002
Adc	Multiple abiotic stresses tolerance	• In the transformed plants, seed germination, shoot length and fresh weight were significantly higher than non transformed plants after being exposed to heat (45 ◦C), cold (4–6 ◦C), salt (150 mM), osmotic (7.5% PEG) or cadmium (1 mM)	Prabhavathi and Rajam 2007

2.2.1.3 Genetic Engineering with Traits Associated with the Fruit Quality of Eggplant

Undesirable environmental conditions such as excessive humidity, sub- or supra-optimal temperature, drought, etc., adversely affect the quality as well as productivity of fruits in eggplant crops (Karapanos et al. 2008). Under uncertain environmental conditions, parthenocarpy may play a vital role in sustaining productivity either by regulating plant growth hormones by gene manipulation or by inserting genes, as shown in Table 2.5. CRISPR-Cas is a new technology that is successfully being used to engineer human, animal, and plant genomes. With this technique precise gene modifications are possible including gene knock-in, deletion or mutation (Yin et al. 2017). In eggplant, CRISPR-Cas technology has been used to knock out three PPO genes (SmelPPO4, SmelPPO5, and SmelPPO6) to improve the fruit quality. These genes are responsible for catalyzing the oxidation of polyphenols that lead to the browning of the eggplant fruit flesh after cutting. Thus, this modification results in reduced browning, which is a desirable trait for both fresh as well as industrial consumption (Maioli et al. 2020). Some examples of transgenic eggplant with better quality and productivity are listed in Table 2.5.

2.2.2 Tomato

Cultivatedtomatoes (*Solanum lycopersicum*) belong to the Solanaceae family andinclude more than 3,000 species, 12of which are wild relativesoftomatoes. These species have a wide range of adaptations to a variety of habitats, plant morphology, fruit size, and color.The latter varies from green to white, yellow, pink, red,and brown, mainly depending on the metabolite content of the fruit. The consumption of tomatoes has further increased over the years due to its antioxidant properties of lycopene that could protect against cancer and cardiovascular diseases. Tomato fruits are particularly rich in nutrients such as lycopene and alpha-carotene, vitamin C, flavonoids, and

TABLE 2.5
Improved fruit quality in eggplant through transgenic approach

Gene	Trait Targeted	Key Results	References
***SmARF8* (RNAi)**	Parthenocarpic fruit	• Seedless parthenocarpic fruit was obtained by silencing *SmARF8* gene in eggplant	Rotino et al. (1997)
Cas9-SmPPO	Fruit flesh browning	• Editing of three PPO genes (*SmelPPO4, SmelPPO5*, and *SmelPPO6*) using CRISPR/Cas9 results in reduced fruit flesh browning	Maioli et al. (2020)
crtB	β-carotene accumulation	• In transformed plants the accumulation of β-carotene in fruits was 30-fold higher than that of the untransformed plant	Mishiba et al. (2020)

hydroxycinnamic acid andtheirinclusionmayhave health benefits Antonio Di Matteo et al. 2011). However, its productivity is affected by various abiotic stresses. Tomatoes grown intraditionalwaysalso soften during the ripening process and are lesstolerantof many biotic and abiotic stresses. In order to provide plants the ability to tolerate a wider range of climatic conditions and to boost productivity, genetic engineering offers a solution. As a result, many of these strategies concentrate on traits that lowerproduction costs or assure consistent production of great yieldsof fruits ofexcellent quality.

2.2.2.1 Approaches Used in Developing Transgenic Tomato

In general, plant alteration techniques have three basic goals: improving desirable features, reducing the expression of undesirable traits, and developing new traits. Many genes have been defined for their function using recombinant DNA technology and can be used to generate novel transgenic tomatoes with increased quality and resilience to biotic and abiotic stressors. Transformation in tomatoes is accomplished by co-culturing with *Agrobacterium tumifaciens*, a particularly successful technique for plant transformation due to its repeatability, and has been employed in early gene transfer research to introduce foreign genes into plants (Kumar et al. 2018). Conversely, the biolistic technique has arisen as a basic and potentially transformative tool. It allows for the effective transformation of a wide range of plant tissues. In this approach, DNA-coated metal particles are swiftly blasted into plant tissues, and the resultant transformants are typically stable. An efficient plant regeneration system for the creation of stable transgenic plants from explant tissue is the most crucial condition for a successful genetic engineering programme. The technique for introducing a foreign or mutated host gene into the plant must also be optimized.

2.2.2.2 Applications of Transgenic Tomatoes

2.2.2.2.1 Fruit Quality Enhancement

The Food and Drug Administration authorized and released the first genetically modified fruit (Flavr Savr tomato) in 1994 (Baranski et al. 2019). The Flavr Savr tomato (CGN-89564) was the first genetically altered fruit that was commercially cultivated and approved for human consumption. In the transgenic tomato, a gene that induces pectin solubilization was downregulated, resulting in delayed fruit ripening and a longer shelf life. Polygalacturonase (Langley et al. 1994), pectin methylesterase (Tieman & Handa 1994), expansin (Brummell et al. 1999), and β- galactosidase (Smith et al. 2002) genes, in particular, have shown efficiency in modulating fruit firmness and softness in transgenic tomato plants. Some examples of transgenic tomatoes with improved fruit quality are listed in Table 2.6.

2.2.2.2.2 Resistance to Abiotic Stresses

Transgenic techniques have been proposed to boost abiotic stress tolerance. Numerous research have been undertaken to better understand how plants respond to various stimuli such as heavy metals, drought, salt, heat, cold, oxidative stress, and hypoxia caused by waterlogging. Several genes expressing stress-relieving enzymes have been inserted into regulatory and signaling networks in order to develop plants with

TABLE 2.6
Fruit and nutritional quality traits enhancement in tomato through transgenic approach

Trait	Inserted gene	Traits developed	References
Firmness	PG	Reduced softening	Langley et al. (1994)
	PME	Reduced shelf life	Tieman and Handa (1994)
	EXP1A	Reduced firmness	Brummell et al. (1999)
	β- galactosidase	Increased firmness	Smith et al. (2002)
Size	fw 2.2	Increased size	Frary et al. (2000)
Flavour	Thaumatin	Enhanced flavor	Bartoszewski et al. (2003)
Parthenocarpy	*Arf8* *IAA9*	Induced parthenocarpy	Goetz et al. (2007)
Carotenoid content	*PSY-1*	Increased Carotenoids	Römer et al. (2000)
	CrtB, CrtI, CrtY	Increased Carotenoids	Wurbs et al. (2007)
	CYC-B, *LCY-B*	Increased lycopene & carotene	Rosati et al. (2000) D'Ambrosio et al. (2004)
	CHS, CHI, F3H, FLS	increased flavonols	Colliver et al. (2002)
	CUL4	Increased carotenoid and flavonoid	Wang et al. (2008)
	Spermidine Synthase	increased lycopene	Neily et al. (2011)
	STS, CHS, CHR, FNS, CHI	accumulation of resveratrol, deoxychalcones, & flavones	Schijlen et al. (2006)
	MYB12	accumulation of flavonols	Adato et al. (2009)
Flavonoid content	*Del, Ros1*	high levels of anthocyanins	Butelli et al. (2008)
	GME	increased/ decreased fruit ascorbic acid	Zhang et al. (2011)
Ascorbic acid content	*GCHI* and/or *ADCS*	Increased fruit folate	de la Garza et al. (2004);

enhanced stress tolerance against salt, oxidative stress, high and low temperatures, and drought (Karkute et al. 2019). Scientists have targeted one or a few genetic alterations to modify important metabolites such as glycine betaine and proline (Table 2.7).

2.2.2.2.3 Resistance to Biotic Stresses

Plant resistance mechanisms have been studied at the molecular level, and genes that might boost resistance to infections have been identified. Moreover, genes involved

TABLE 2.7
Transgenes introduced in tomatoes for developing abiotic stress resistance variety

Stress Response	Gene	Mechanism of action	Reference
Drought tolerance	*LeNCED1*	Increases in abscisic acid (ABA) accumulation. Thus stomatal closure and increased water-use efficiency	Tung et al. (2008)
	PtADC	Improve dehydration and drought tolerance	Wang et al. (2011)
	ZAT12	Enhanced expression of downstream stress-related genes confers drought tolerance	Rai et al. (2013)
	AtGAMT1	*GAMT1* overexpression inhibited the expansion of leaf-epidermal cells	Nir et al. (2014)
	SlNAC4	Modulation of ABA-independent signaling networks	Zhu et al. (2014)
	GalUR	Detoxifies superoxide anion radical and hydroxyl radical, and also plays a crucial role in scavenging ROS	Lim et al. (2016)
Heat stress tolerance	*Hsp101*	Acquired thermo-tolerance	Nieto-Sotelo (2002)
	LeUCP	Increase redox level and antioxidant capacity	Chen et al. (2013)
	LeCDJ1	Overexpression facilitated heat tolerance	Kong et al. (2014)
	SlCDJ2	Protects Rubisco activity under heat stress	Wang et al. (2015)
Salt stress tolerance	*TaNHX2*	Na+/ H+ antiporters are involved in intracellular ion (Na+), pH regulation	Yarra et al. (2012)
	Tas14	Overexpression of dehydrin *tas14* gene improves the osmotic stress imposed by drought and salinity	Muñoz- Mayor et al. (2012)
	GlyI, GlyII	Decrease oxidative stress	Álvarez-Viveros et al. (2013)
	HAL5	Maintenance of Na+ and K+ transporter levels such as *SlHKT1* and *2*, and *SlHAK5*	García-Abellan et al. (2014)
	MdSOS2L1	*MdSOS2L1* protein kinase physically interacts with *MdCBL1*, *MdCBL4* and *MdCBL10* proteins to increase tolerance	Hu et al. (2016)
	Coda	Glycine betaine enhanced NaCl induced expression of genes encoding the K+ transporter, Na+/H+ antiporter, and H+-ATPase	Wei et al. (2017)

TABLE 2.8
Successful genetic engineering for increased resistance to biotic stresses in tomato

Inserted genes	Developed resistance against disease	Reference
Bs2,CHI	Bacterial spot	Horvath et al. (2012)
hCAP18/LL-37	Bacterial spot and Bacterial soft rot	Jung (2013)
LF	Bacterial wilt	Lee et al. (2002)
AOX	tomato spotted wilt virus	Ma et al. (2011)
Cry1Ac	Fruit Borer	Rashid and Bal (2011)
Cry 2Ab	Lepidopterous insects	Saker et al. (2011)
ech42	Fungal pathogens	Shah et al. (2010)
amiR-AV1-3	Tomato leaf curl virus	Vu et al. (2013)

in plant susceptibility to infections might be prospective targets for RNA interference (RNAi) methods. To inhibit viral *AV2/AV1* (coat proteins) transcripts, tomato plants overproduced amiRs (artificial micro RNAs). Because this study revealed that some of the transformants developed tolerance or resistance to tomato leaf curl New delhi virus, it is considered that the amiR approach may be successfully employed to defend agricultural plants against viruses (Vu et al. 2013). Another approach for improving plant defense against infections such as fungi and bacteria is to create genetically altered plants that produce antimicrobial peptides. Jung (2013) developed transgenic tomato lines that produce a human antimicrobial peptide (hCAP18/ll-37)-cathelicidin and exhibited high expression of PR proteins (Pathogenesis-related), lipid transfer proteins, and antifungal proteins, as well as strong resistance to bacterial infections. Many attempts have been undertaken to develop transgenic tomato plants with fungal pathogen resistance (Table 2.8). Shah et al. (2010) created transgenic tomato lines that expressed the *Trichoderma virens* endochitinase (*ech42*) gene. As compared to control plants, these transgenic lines demonstrated improved resistance to fungal infections.

2.2.3 Squash

Cucurbita pepo, commonly known as zucchini, yellow crookneck squash, or acorn squash, is susceptible to viral infections transmitted by aphids, leading to significant economic losses (Schultheis and Walters, 1998). The primary viruses affecting cucurbits include watermelon mosaic virus 2 (WMV2), papaya ringspot virus (PRSV, formerly WMV1), cucumber mosaic virus (CMV), and zucchini yellow mosaic virus (ZYMV), with unpredictable spread and symptoms often appearing too late for effective pesticide control (Price, 1940; USDA, 1994; Tricoli et al., 1995). Traditional breeding and transgenic technologies offer potential solutions, with early successes in developing resistant cultivars such as Tigress and CZW-3 (Kyle, 1993; USDA, 1994; Fuchs and Gonsalves, 1995; Kreuze and Valkonen, 2017). These transgenic lines have significantly benefited growers, representing a substantial portion of squash production in the United States (Johnson and Strome, 2006). Concerns regarding transgenic squash safety, particularly with incorporated viral and bacterial genes, have been raised, although regulatory bodies like the FDA have deemed them safe for consumption (Krimsky, 2000). While the possibility of transferring antibiotic resistance markers from transgenic crops to bacteria exists, it is considered minimal and manageable (Nap et al., 1992; Fuchs et al., 1993; Kunkel et al., 1999). Therefore, ongoing research is essential to explore alternative marker options for future genetically engineered crops.

2.2.4 Papaya

Papaya is a popular fruit produced in tropical and subtropical plains. It is a member of the genus Carica and has just one species, *Carica papaya* L. It is a herbaceous plant with a single stem that has a crown of palmate-shaped leaves. It is widely planted in home gardens since it is reasonably easy to cultivate from seed. The first mature fruits may be harvested nine months after the seeds are sown, and fruit is produced continually all year. The papaya fruit is high in vitamins and minerals while being low in salt, fat, and calories (Chukwuka et al. 2013. While the ripe fruit is mainly eaten raw, green papaya is used as a salad or vegetable. Traditional breeding limits in papaya improvement can be addressed by biotechnology approaches such as embryo rescue and genome editing. Certain features can be added into elite varieties by genetic engineering of papaya without adversely affecting the current phenotypes. A functional cell culture system for plant regeneration and gene insertion technologies are required for effective gene transfer. PRSV is the most prevalent and dangerous virus that affects papaya. Genetically modified papaya has been developed to improve features that are difficult to include through normal breeding, such as PRSV resistance, insect resistance, and shelf-life extension (Fitch et al. 1993; Gonsalves 1998). PRSV is spread by aphids, which have a limited host range of cucurbits and papaya. Also, pesticides are ineffective in controlling the aphid vectors that transmit the virus (Conover 1964), resulting in the abandonment of various orchards due to PRSV infestation (Fitch et al. 1992).

2.2.4.1 Approaches to Develop Transgenic Papaya

In 1992, Cornell University, the University of Hawaii, and Asgrow collaborated to create the first PRSV-resistant papaya (Fitch et al. 1992). The procedure for papaya transformation by particle bombardment utilizing zygotic embryos was developed at the University of Hawaii (Fitch et al. 1992, Fitch et al. 1990), whereas Huanong No. 1 papaya was created using an *Agrobacterium*-mediated transformation. (Chen et al. 2001). In Hawaii, two genetically modified cultivars called "SunUp" (red fleshed) and "Rainbow" (yellow-fleshed) that are resistant to the PRSV virus have been successfully developed and are now utilized for human consumption. These transgenic papaya cultivars were released commercially in Hawaii in 1998, effectively saving Hawaii's papaya industry from the devastation being caused by PRSV. In these two transgenic virus-resistant cultivars, a pathogen-derived resistance approach was used, in which the PRSV coat protein (CP) is expressed via an epigenetic gene silencing mechanism (Tennant et al. 2001). "Rainbow"papaya with yellow flesh was created by crossing "SunUp"papaya with a non-GMO cultivar named "Kapoho"(Gonsalves et al. 2003). The coat protein gene of the mild *PRSV HA 51* isolate is expressed in the papaya cultivars "SunUp" and "Rainbow"(Fitch et al. 1992). The replicase protein domain (NIb) from the PRSV Ys isolate, which is the most prevalent strain in China, is present in transgenic Huanong No.1 papaya, similar to the "SunUp"cultivar. This transgenic papaya is resistant to the four most common PRSV strains found in South China, which are Vb, Lc, Ys, and Sm (Ye & Li 2010; Ivanov et al. 2014). Also, it yields larger fruits with thicker flesh than the original variety (Ye & Li 2010).

2.3 POTENTIAL RISKS ASSOCIATED WITH TRANSGENIC CROPS

Despite the clear advantages of GE for viral resistance, there are possible hazards if precautions are not taken prior to applying this technology. Horizontal genetic erosion, gene transfer, transgenic recombination with infectious viruses, and transgene-related injury to humans and animals are all potential concerns. As stated further below, it has been scientifically established that GE crops, especially those resistant to viruses, provide no unacceptable, new, or additional risks to humans, animals, or the environment as compared to traditionally cultivated crops. Furthermore, these worries are, tragically, based on the idea that transgenes are unnatural and therefore dangerous. Concerns have been raised about the possibility of transgenes and selectable markers spreading to other crops or wild species through cross-pollination. The selectable marker gene can also be passed to bacteria, especially in the root system, where gene transfer with plants is common (Sengelov et al. 2001). The majority of virus resistance engineering uses viral genes or genome sequences that naturally appear in infected plants but are not integrated into the genome. As a result, such transgenes are extremely unlikely to generate an environmental imbalance, at least not any more than viruses that damage cultivated crops do. Furthermore, in order to assure safety, the complex regulatory approval procedure for GM crops may result in an unneeded obstruction to the introduction of new GM crops.

2.4 IMPROVEMENT OF FRUIT CROPS THROUGH CRISPR-CAS TECHNOLOGY

CRISPR (clustered regularly interspaced short palindromic repeat) technology may be a preferable solution for improving several qualities in crops quickly and correctly, such as abiotic stress tolerance, quality, disease resistance, yield, and nutritional factors. Because of its ease of usage and high mutation effectiveness, this method has been utilized to create new germplasm resources through gene-directed mutations. The US FDA authorized the CRISPR-edited mushroom in 2016 without applying the onerous rules that typical GM crops do, which has transformed the rate of agricultural improvement using this method (Kim and Kim 2016). CRISPR-Cas9 technology was used to edit the genomes of numerous economically significant fruit varieties (Table 2.9).

TABLE 2.9
Genetic improvement of economically significant fruit crops through CRISPR-Cas9 technology

Crop species	Target genes	Mutation	Transformation method	Traits modified	Reference
Tomato	*ClDMR6*	Loss of function	*Agrobacterium-mediated* transformation	Resistance against downy mildew	Paula de Toledo Thomazella et al. (2016)
Tomato	*ClMl01*	Loss of function	*Agrobacterium-mediated* transformation	Resistance against downy mildew	Nekrasov et al. (2017)
Tomato	*Solyc08g075770*	Loss of function	*Agrobacterium-mediated* transformation	Fusarium wilt susceptibility	Cahya et al. (2018)
Tomato	*CCD8*	Loss of function	*Agrobacterium-mediated* transformation	Resistance against *Phelipanche aegytiaca*	Bari et al. (2021)
Tomato	*SGR1,LCY-E, Blc, LCY-B1, LCY-B2*	Loss of function	*Agrobacterium-mediated* transformation	Lycopene content	Li et al. (2018)
Watermelon	*ClPSK1*	Loss of function	*Agrobacterium-mediated* transformation	Resistance to Fusarium oxysporum f.sp. niveum	Zhang et al. (2020)
Watermelon	*ClALS*	Site directed mutagenesis	*Agrobacterium-mediated* transformation	Herbicide resistance	Tian et al. (2018)
Watermelon	*ALS*	Loss of function	*Agrobacterium-mediated* transformation	Herbicide resistance	Tian, S. (2018)
Banana	*ORF region of virus*	Loss of function	*Agrobacterium-mediated* transformation	Resistance against banana streak virus	Tripathi et al. (2019)
Grape	*MLO7*	Loss of function	*Agrobacterium-mediated* transformation	Resistance against powdery mildew	Malnoy, M. et al (2016)
Grape	*WRKY52*	Loss of function	*Agrobacterium-mediated* transformation	Resistance against gray mold disease	Wang et al. (2018)
Apple	*DIPM1,2,4*	Loss of function	*Agrobacterium-mediated* transformation	Resistance against fire blight disease	Malnoy, et al. (2016)
Papaya	*alEPIC8*	Loss of function	*Agrobacterium-mediated* transformation	Resistance against *Phytophthora palmivora*	Gumtow et al. (2018)
Kiwifruit	*CEN4, CEN*	Loss of function	*Agrobacterium-mediated* transformation	A compact plant with rapid terminal flower and fruit development	Varkonyi-Gasic et al. (2019)

2.5 CONCLUSION

Transgenic technology offers enormous promise to assist farmers in generating crops that will be needed to fulfill future food demands as the world population expands, as well as crops that will be adaptable to future conditions as the climate changes. Some countries have exempted several of these genome-edited cultivars from GMO laws. Although the genetic study in many fruit crops has made tremendous progress, there is still more work to be done to make it a favored approach to fruit enhancement. Consumer acceptability and environmental safety are two important challenges that must be addressed. The recent advent of genome editing tools such as CRISPR/Cas9 and other nucleases has allowed for precise manipulation of target genes and is expected to provide much-needed fuel to fruit improvement projects. These methods allow for precise modifications at pre-selected genomic regions while leaving no genetic footprints or off-targets.

REFERENCES

Adato, A., Mandel, T., Mintz-Oron, S., Venger, I., Levy, D., Yativ, M., Dominguez, E., Wang, Z., De Vos, R. C., Jetter, R., Schreiber, L., Heredia, A., Rogachev, I., and Aharoni, A. 2009. Fruit-surface flavonoid accumulation in tomato is controlled by a SlMYB12-regulated transcriptional network. *PLoS Genetics* 5(12):e1000777.

Ahmar, S., Gill, R.A., Jung, K.H., Faheem, A., Qasim, M.U., Mubeen, M. and Zhou, W. 2020. Conventional and molecular techniques from simple breeding to speed breeding in crop plants: recent advances and future outlook. *International journal of molecular sciences 21*(7):2590.

Alam, I. andSalimullah, M. 2021. Genetic engineering of eggplant (*Solanum melongena* L.): progress, controversy and potential. *Horticulturae* 7: 78.

Álvarez-Viveros, M. F., Inostroza-Blancheteau, C., Timmermann, T., González, M., and Arce-Johnson, P. 2013. Overexpression of GlyI and GlyII genes in transgenic tomato (Solanum lycopersicum Mill.) plants confers salt tolerance by decreasing oxidative stress. *Mol. Biol. Rep.*.4:3281–3290.

Antonio,D. M., Maria,M. R., and Adriana,S.,Luigi, F., and Amalia, B. 2011. Genetic transformation in tomato: Novel tools to improve fruit quality and pharmaceutical production, Genetic Transformation, Prof. MarÃa Alvarez (Ed.),ISBN: 978-953-307-364-4, InTech.

Anwar, F.,Sharmila, P., andSaradhi, P.P. 2010.No more recalcitrant: Chickpea regeneration and genetic transformation. *Afr. J. Biotechnol.* 9: 782–797.

Arpaia, S.,Mennella, G.,Onofaro, V.,Perri, E.,Sunseri, F., andRotino, G.L. 1997. Production of transgenic eggplant (*Solanum melongena* L.) resistant to Colorado potato beetle *(Leptinotarsa decemlineata* Say). *Theor. Appl. Genet.*95:329–334.

Baranski, R., Klimek-Chodacka, M.,andLukasiewicz, A. 2019. Approved genetically modified (GM) horticultural plants: a 25-year perspective. *Folia Horticurae.* 31:3–49.

Bari, V. K.,Nassar, J.A., andAly, R. 2021. CRISPR/Cas9 mediated mutagenesis of MORE AXILLARY GROWTH 1 in tomato confers resistance to root parasitic weed Phelipanche aegyptiaca. *Sci. Rep.*11:3905.

Bartoszewski, G., Niedziela, A., Szwacka, M., and Niemirowicz-Szczytt, K. 2003. Modification of tomato taste in transgenic plants carrying a thaumatin gene from *Thaumatococcus daniellii* Benth. *Plant Breeding* 122(4):347–351.

Brummell, D. A., Harpster, M. H., Civello, P. M., Palys, J. M., Bennett, A. B., and Dunsmuir P. 1999. Modification of expansin protein abundance in tomato fruit alters softening and cell wall polymer metabolism during ripening. *The Plant Cell.* 11:2203–2216.

Butelli, E., Titta, L., Giorgio, M., Mock, H. P., Matros, A., Peterek, S., Schijlen, E. G., Hall, R. D., Bovy, A. G., Luo, J., and Martin, C. 2008. Enrichment of tomato fruit with healthpromoting anthocyanins by expression of select transcription factors. *Nat. Biotechnol.* 26:1301–1308.

Cahya, P., Barbetti, M.J., and Barker, S.J. 2018. A novel tomato fusarium wilt tolerance gene. *Front. Microbiol.* 9: 1226.

Chen, G., Ye, C. M., Huang, J. C., Yu, M.,and Li, B. J. 2001. Cloning of the papaya ringspot virus (PRSV) replicase gene and generation of PRSV-resistant papayas through the introduction of the PRSV replicase gene. *Plant Cell Rep.* 20:272–277.

Chen, S., Liu, A., Zhang, S. Li, C., Chang, R., Liu, D., Ahammed, G.J. and Lin, X. 2013. Overexpression of mitochondrial uncoupling protein conferred resistance to heat stress and *Botrytis cinerea* infection in tomato. *Plant Physiol.Biochem.* 73:245–253.

Chukwuka, K.S., Iwuagwu, M. and Uka, U.N. 2013. Evaluation of nutritional components of Carica papaya L. at different stages of ripening. *IOSR Journal of Pharmacy and Biological Sciences* 6(4):13–16.

Colliver, S., Bovy, A., Collins, G., Muir, S., Robinson, S., de Vos, C. H. R., and Verhoeyen, M. E. 2002. Improving the nutritional content of tomatoes through reprogramming their flavonoid biosynthetic pathway. *Phytochem. Rev.* 1:113–123.

Conover, R. A. 1964. Distortion ringspot, a severe virus disease of papaya in Florida. *Proc. Fl. State Hortic. Soc.* 77:440–444.

D'Ambrosio, C., Giorio, G., Marino, I., Merendino, A., Petrozza, A., Salfi, L., Stigliani, A. L., Cellini, F. 2004. Virtually complete conversion of lycopene into beta-carotene in fruits of tomato plants transformed with the tomato lycopene β-cyclase (tlcy-b) cDNA. *Plant Sci* .166:207–214.

Drewnowski, A., Fulgoni III, V. 2008. Nutrient profiling of foods: creating a nutrient-rich food index. *Nutrition reviews*. 66(1):23–39.

Dias, J.S. & Ortiz, R. 2014. Advances in transgenic vegetable and fruit breeding. *Agricultural Sci.* 5(14):1448.

Fitch, M., Manshardt, R., Gonsalves, D., Slightom, J.,andSanford, J. 1992. Virus resistant papaya plants derived from tissues bombarded with the coat protein gene of papaya ringspot virus. *Nat. Biotechnol.* 10:1466–1472.

Fitch, M. M. M.,Manshardt, R.M., Gonsalves, D., Slightom, J. L.,andSanford, J. C. 1990. Stable transformation of papaya via microprojectile bombardment. *Plant Cell Rep.* 9:189–194.

Frary, A., Nesbitt, T. C., Grandillo, S., van der Knaap, E., Cong, B., Liu, J., Meller, J., Elber, R., Alpert, K. B., and Tanksley, S. D. 2000. fw2.2: A quantitative trait locus key to the evolution of tomato fruit size. *Science* 289(5476):85–88.

Fuchs, M. and Gonsalves, D. 1995. Resistance of transgenic squash Pavo ZW-20expressing the coat protein genes of zucchini yellow mosaic virus and watermelon mosaic virus 2 to mixed infections by both potyviruses. *Bio/Biotechnol.* 13:1466–1473.

Fuchs, R. L., Ream, J. E., Hammond, B. G., Naylor, M. W., Leimgruber, R. M., and Berberich, S. A.1993. Safety assessment of the neomycin phosphotransferase II (NPTII) protein. *BioTechnology (N Y).*11(13):1543–1547.

de la Garza, R. I. D., Gregory, J. F., Hanson, A. D. 2007. Folate biofortification of tomato fruit. *Proc. Natl. Acad. Sci.USA* 104(10): 4218–4222.

García-Abellan, J.O., Egea, I., Pineda, B., Sanchez-Bel, P., Belver, A., Garcia-Sogo, B., Flores, F.B., Atares, A., Moreno, V. and Bolarin, M.C. 2014. Heterologous expression of the yeast HAL5 gene in tomato enhances salt tolerance by reducing shoot Na+ accumulation in the long term. *Physiologia plantarum* 152(4):700–713.

Goetz, M., Hooper, L. C., Johnson, S. D., Rodrigues, J. C., Vivian-Smith, A., and Koltunow, A. M. 2007. Expression of aberrant forms of AUXIN RESPONSE FACTOR8 stimulates parthenocarpy in Arabidopsis and tomato. *Plant Physiol.* 145:351–366.

Gonsalves, D. 1998. Control of papaya ringspot virus in papaya: A case study. *Annu. Rev. Phytopathol.* 36:415–437.

Gonsalves, D., Basin, P., Ferreira, S., Sciences, E. P., and Resources, H. 2003. Transgenic papaya: A case for managing risks of papaya ringspot virus in Hawaii plant health progress. *Plant Manag. Netw.* 1:1–5.

Gumtow, R., Wu, D., Uchida, J., and Tian, M. 2018. A phytophthora palmivora extracellular cystatin-like protease inhibitor targets papain to contribute to virulence on papaya. *Mol. Plant Microbe.* 31:363–373.

Guri, A. and Sinks, K.C. 1988. Agrobacterium transformation of eggplant. *J. Plant Physiol.*133:52–55.

Horvath, D. M., Stall, R. E., Jones, J. B., Pauly, M. H., Vallad, G. V., Dahlbeck, D., Staskawicz, B. J., and Scott, J. W. 2012. Transgenic resistance confers effective field level control of bacterial spot disease in tomato. *PLoSONE* 7(8):e42036.

Hu, D-G., Ma, Q-J., Sun, C-H., Sun, M.H., You, C.X. and Hao, Y.J. 2016. Overexpression of MdSOS2L1, a CIPK protein kinase, increases the antioxidant metabolites to enhance salt tolerance in apple and tomato. *Physiol. Plant.* 156:201–214. https://doi.org/10.1111/ppl.12354

Ivanov, K. I., Eskelin, K., Lõhmus, A., and Mäkinen, K. 2014. Molecular and cellular mechanisms underlying potyvirus infection. *J. Gen. Virol.* 95:1415–1429.

Jadhav, M., Jadhav, A., Pawar, B., Kale, A., and Kutes, N. 2015. Agrobacterium-mediated genetic transformation of Brinjal with cry1F gene for resistance against shoot and fruit borer. *J. Crop Improv.*29:518–527.

Johnson, S. and Strome, S. 2006. *Quantification of the Impacts on US Agriculture of Biotechnology-Derived Crops Planted in 2006*. Washington, DC:National Center for Food and Agricultural Policy.

Jung, Y. J. 2013. Enhanced resistance to bacterial pathogen in transgenic tomato plants expressing cathelicidin antimicrobial peptide. *Biotechnol. Bioprocess Eng.* 18:615–624.

Karapanos, I.C., Mahmood, S., and Thanopouloss, C. 2008. Fruit set in solanaceous vegetable crops as affected by floral and environmental factors. *Eur. J. Plant Sci. Biotechnol.*2:88–105.

Kim, J. and Kim, J. 2016. Bypassing GMO regulations with CRISPR gene editing. *Nat. Biotechnol.* 34: 1014–1015.

Kissoudis, C., Chowdhury, R., van Heusden, S., van de Wiel, C., Finkers, R., Visser, R. G. F., Bai, Y., and van der Linden, G. 2015. Combined biotic and abiotic stress resistance in tomato. *Euphytica.* 202:317–332.

Koenning, S.R., Overstreet, C., Noling, J.W., Donald, P.A., Becker, J.O., and Fortnum, B.A. 1999. Survey of crop losses in response to phytoparasitic nematodes in the United States for 1994. *J. Nematol.*31:587–618.

Kong,F., Deng, Y., Wang, G. et al. 2014. LeCDJ1, a chloroplast DnaJ protein, facilitates heat tolerance in transgenic tomatoes. *J. Integr. Plant Biol.* 56:63–74. https://doi.org/10.1111/jipb.12119

Kreuze, J. F. and Valkonen, J. P. 2017. Utilization of engineered resistance to viruses in crops of the developing world, with emphasis on sub-Saharan Africa. *Curr. Opin. Virol.* 26:90. doi:10.1016/j.coviro.2017.07.022

Krimsky, S. 2000. Risk assessment and regulation of bioengineered food products. *International Journal of Biotechnology* 2(1-3):231–238.

Krishna, R., Karkute, S. G., Ansari, W. A., Jaiswal, D. K., Verma, J. P., and Singh, M. 2019. Transgenic tomatoes for abiotic stress tolerance: Status and way ahead. *3 Biotech.* 9(4):143.

Kumar, K., Aggarwal, C., Sapna, I. and Yadava, P. 2018.Microbial genes in crop improvement. In: *Crop Improvement Through Microbial Biotechnology*. Elsevier, pp.39–56.

Kumar, P. A., Mandaokar, A., Sreenivasu, K., Chakrabarti, S. K., Bisaria, S., Sharma, S., R., Kaur, S., and Sharma, R.P. 1998. Insect-resistant transgenic brinjal plants. *Mol. Breed.* 4:33–37.

Kunkel, T., Niu, Q. W., Chan, Y. S., and Chua, N. H. 1999. Inducible isopentenyl transferase as a high-efficiency marker for plant transformation. *Nat. Biotechnol.* 17:916–919.

Kyle, M. 1993. *Resistance to Viral Diseases of Vegetables: Genetic and Breeding*. Portland, OR: Universidad Autónoma Chapingo, p. 278.

Langley, K. R., Martin, A., Stenning, R., Murray, A. J., Hobson, G. E., Schuch, W. W., and Bird, C. R.1994. Mechanical and optical assessment of the ripening of tomato fruit with reduced polygalacturonase activity. *J. Sci. Food Agric.* 66(4):547–554.

Lee, T. J., Coyne, D. P., Clemente, T. E., and Mitra, A. 2002. Partial resistance to bacterial wilt in transgenic tomato plants expressing antibacterial lactoferrin gene. *J. Am. Soc. Hortic. Sci.* 2:150–164.

Li, X., Wang, Y., Chen, S., Tian, H., Fu, D., Zhu, B., Luo, Y., and Zhu, H. 2018. Lycopene is enriched in tomato fruit by CRISPR/Cas9-mediated multiplex genome editing. *Front. Plant Sci.* 9: 559.

Lim, M. Y., Jeong, B. R., Jung, M., and Harn, C. H. 2016. Transgenic tomato plants expressing strawberry d-galacturonic acid reductase gene display enhanced tolerance to abiotic stresses. *Plant Biotechnol. Rep.*10:105–116.

Ma, H., Song, C., Borth, W., Sether, D., Melzer, M., and Hu, J. 2011. Modified expression of alternative oxidase in transgenic tomato and petunia affects the level of tomato spotted wilt virus resistance. *BMC Biotechnol.* 11:96. Maioli, A.,Gianoglio, S.,Moglia, A.,Acquadro, A.,Valentino, D.,Milani, A.M.,Prohens, J.,Orzaez, D.,Granell, A.,Lanteri, S. 2020. Simultaneous CRISPR/Cas9 editing of three PPO genes reduces fruit flesh browning in *Solanum melongena* L. *Front. Plant Sci.* 11 : 1883. [PubMed]

Malnoy, M. Viola, R., Jung, M.H., Koo, O.J., Kim, S., Kim, J.S., Velasco, R. and Nagamangala K.C. 2016. DNA-free genetically edited grapevine and apple protoplast using CRISPR/Cas9 ribonucleoproteins. *Front Plant Sci.* 7: 1904.

Mathur, V., Javid, L., Kulshrestha, S., Mandal, A. and Reddy, A.A. 2017. World cultivation of genetically modified crops: opportunities and risks. *Sustainable agriculture reviews* 8:45–87.

Mishiba, K.-I., Nishida, K., Inoue, N., Fujiwara, T., Teranishi, S., Iwata, Y., Takeda, S., and Koizumi, N. 2020. Genetic engineering of eggplant accumulating β-carotene in fruit. *Plant Cell Rep.* 39:1029–1039. [PubMed]

Muñoz-Mayor, A., Pineda, B., Garcia-Abellan, J. O., Anton, T., Garcia-Sogo, B., Sanchez-Be,l P., Flores, F. B., Atares, A., Angosto, T., Pintor- Toro, J. A., Moreno, V., and Bolarin, M. C. 2012. Overexpression of dehydrin tas14 gene improves the osmotic stress imposed by drought and salinity in tomato. *J. Plant Physiol.* 169:459–468. Nap, J-P, Bijvoet, J., and Stiekema, W. J.1992. Biosafety of kanamycin-resistant transgenic plants. *Transgenic Res.* 1:239.

Neily, M. H., Matsukura, C., Maucourt, M., Bernillon, S., Deborde, C., Moing, A., Yin, Y.-G., Saito, T., Mori, K., Asamizu, E., Rolin, D., Moriguchi, T., and Ezura, H. 2011.Enhanced polyamine accumulation alters carotenoid metabolism at the transcriptional level in tomato fruit over-expressing spermidine synthase. *J. Plant Physiol.* 168(3):242–252.

Nekrasov, V., Wang, C., Win, J., Lanz, C., Weigel, D., and Kamoun, S. 2017. Rapid generation of a transgene-free powdery mildew resistant tomato by genome deletion. *Sci. Rep.* 7:1–6.

Nieto-Sotelo, J. 2002. Maize HSP101 plays important roles in both induced and basal thermotolerance and primary root growth. *Plant Cell Online* 14:1621–1633. Nir, I., Moshelion, M., and Weiss, D. 2014. The Arabidopsis gibberellins methyl transferase 1 suppresses gibberellin activity, reduces whole-plant transpiration and promotes drought tolerance in transgenic tomato.*Plant Cell Environ.* 37:113–123.

Pal, J.K., Singh, M., Rai, M., Satpathy, S., Singh, D.V., and Kumars, S. 2009. Development and bioassay of CryiAc-transgenic eggplant (*Solarium melongena* L.) resistant to shoot and fruit borer. *J. Hortic. Sci. Biotechnol.* 84: 434–438.

Papolu, P.K., Dutta, T.K., Tyagi, N., Urwin, P.E., Lilley, C.J., and Rao, U. 2016. Expression of a cystatin transgene in eggplant provides resistance to root-knot nematode, Meloidogyne incognita. *Front. Plant Sci.* 7: 1122. [PubMed]

Paula de Toledo Thomazella, D., Brail, Q., Dahlbeck, D., and Staskawicz, B. CRISPR-Cas9 mediated mutagenesis of a DMR6 ortholog in tomato confers broad-spectrum disease resistance. *BioRxiv* 15:064824.

Prabhavathi, V. and Rajam, M. 2007. Mannitol-accumulating transgenic eggplants exhibit enhanced resistance to fungal wilts. *Plant Sci.*173:50–54.

Prabhavathi, V., Yadav, J., Kumar, P., and Rajam, M. 2002. Abiotic stress tolerance in transgenic eggplant (*Solanum melongena* L.) by introduction of bacterial mannitol phosphodehydrogenase gene. *Mol. Breed.*9:137–147.

Prabhavathi, V. R. and Rajams, M. V. 2007.Polyamine accumulation in transgenic eggplant enhances tolerance to multiple abiotic stresses and fungal resistance. *Plant Biotechnol.* 24:273–282.

Prasanna, H. C., Sinha, D. P., Rai, G. K, Krishna, R., Kashyap, S. P., Singh, N. K., Singh, M., and Malathi, V. G. 2015. Pyramiding T y-2 and T y-3 genes for resistance to monopartite and bipartite tomato leaf curl viruses of India. *Plant Pathology* 64(2):256–264.

Pratap, D., Kumar, S., Raj, S.K., and Sharmas, A.K. 2011. Agrobacterium-mediated transformation of eggplant (*Solanum melongena* L.) using cotyledon explants and coat protein gene of cucumber mosaic virus. *Indian J. Biotechnol.* 10:19–24.

Price, W. C. 1940. Comparative host ranges of six plant viruses. *Am. J. Bot.* 27:530–541.

Qiu, Z.,Yan, S., Xia, B., Jiang, J., Yu, B., Lei, J., Chen, C., Chen, L., Yang, Y., Wang, Y., and Tian, S. 2019. The eggplant transcription factor MYB44 enhances resistance to bacterial wilt by activating the expression of spermidine synthase. *J. Exp. Bot.* 70:5343–5354.

Rai, N.P., Rai, G.K., Kumar, S., Kumari, N., and Singh, M. 2013. Shoot and fruit borer resistant transgenic eggplant (*Solanum melongena* L.) expressing cry1Aa3 gene: Development and bioassay. *Crop Protect.*53:37–45.

Rashid, R., and Bal, S. S.2011. Agrobacterium-mediated genetic transformation of tomato (*Solanum lycopesricum* L.) with Cry1Ac gene for resistance against fruit borer. *J. Trop. Agric.* 49(1–2):110–113.

Römer, S., Fraser, P. D., Kiano, J. W., Shipton, C. A., Mills, P. B., Drake, R., Schuch, W., and Bramley, P. M. 2000. Elevation of the provitamin: A content of transgenic tomato plants. *Nat. Biotechnol.*18:666–669.

Rosati, C., Aquilani, R., Dharmapuri, S., Pallara, P., Marusic, C., Tavazza, R., Bouvier, F., Camara, B., and Giuliano, G. 2000. Metabolic engineering of beta-carotene and lycopene content in tomato fruit. *Plant J.* 24:413–420.

Rotino, G.L., Perri, E., Zottini, M., Sommer, H., and Spena, A. 1997. Genetic engineering of parthenocarpic plants. *Nat. Biotech.*15:1398–1401. [PubMed]

Sagare, D.B. and Mohantys, I. 2012. Development of moisture stress tolerant brinjal cv. Utkal Anushree (*Solanum melongena* L.) using Agrobacterium mediated gene transformation. *J. Agric. Sci.* 4:141.

Saker, M. M., Salama, H. S., Salama, M., El-Banna, A., and AbdelGhany, N. M. 2011. Production of transgenic tomato plants expressing Cry 2Ab gene for the control of some lepidopterous insects endemic in Egypt. *J. Genet. Eng. Biotechnol.* 9:149–155. Schijlen, E., de Vos,C. H. R, Jonker, H., van den Broeck, H., Molthoff, J., van Tunen, A., Martens, S., and Bovy, A. 2006. Pathway engineering for healthy phytochemicals leading to the production of novel flavonoids in tomato fruit. *Plant Biotechnol. J.* 4:433–444.

Schultheis, J. R. and Walters, S. A.1998. Yield and virus resistance of summer squash cultivars and breeding lines in North Carolina. *Hortic. Technol.* 8:31–39.

Sengeløv,G., Kristensen, K.J., Sørensen, A.H., Kroer, N., and Sørensen S.J. 2001. Effect of genomic location on horizontal transfer of a recombinant gene cassette between pseudomonas strains in the rhizosphere and spermosphere of barley seedlings. *Curr. Microbiol.* 42:160–167. doi:10.1007/s002840010197

Shah, M. R., Mukherjee, P. K., and Eapen, S. 2010. Expression of a fungal endochitinase gene in transgenic tomato and tobacco results in enhanced tolerance to fungal pathogens. *Physiol. Mol. Biol. Plants* 1:39–51.

Shivakumara, T.N., Chaudhary, S., Kamaraju, D., Dutta, T.K., Papolu, P.K., Banakar, P., Sreevathsa, R., Singh, B., Manjaiah, K.M., and Rao, U. 2017. Host-induced silencing of two pharyngeal gland genes conferred transcriptional alteration of cell wall-modifying enzymes of meloidogyne incognita vis-á-vis perturbed nematode infectivity in eggplant. *Front. Plant Sci.*8:8. [PubMed]

Shruti, B.K., Roy, R., Kumar, P.A., and Roy, S.P. 2015. Gene integration of cry1f in, brinjal for resistance against fruit and shoot borer. *J. Adv. Zool* 36:83–89.

Singh, D., Ambroise, A., Haicour, R., Sihachakr, D., and Rajam, M.V. 2014. Increased resistance to fungal wilts in transgenic eggplant expressing alfalfa glucanase gene. *Physiol. Mol. Biol. Plants* 20: 143–150.

Singh, D., Haicour, R., Sihachakr, D., and Rajams, M.V. 2015. Expression of rice chitinase gene in transgenic eggplant confers resistance to fungal wilts. *Indian J. Biotechnol.*14:233–240.

Smith, D. L., Abbott, J. A., and Gross, K. C. 2002. Down-regulation of tomato betagalactosidase 4 results in decreased fruit softening. *Plant Physiol.* 129:1755–1762.

Tennant, P., Fermin, G., Fitch, M.M., Manshardt, R.M., Slightom, J.L., and Gonsalves, D. 2001.Papaya ringspot virus resistance of transgenic Rainbow and SunUp is affected by gene dosage, plant development, and coat protein homology. *Eur. Eur. J. Plant Pathol.* 107:645–653.

Tian, S., Jiang, L., Cui, X., Zhang, J., Guo, S., Li, M., Zhang, H., Ren, Y., Gong, G., and Zong, M. 2018. Engineering herbicide-resistant watermelon variety through CRISPR/Cas9-mediated base-editing. *Plant Cell Rep.* 37:1353–1356.

Tieman, D. M. and Handa, A. K. 1994. Reduction in pectin methyltransferase activity modifies tissue integrity and cation levels in ripening tomato (*Lycopersicon esculentum Mill.*) fruits. *PlantPhysiol.* 106:429–436.

Tricoli, D. M., Carney, K. J., Russell, P. F., McMaster, J. R., Groff, D. W., Hadden, K. C., Himmel, P. T., Hubbard, J. P., Boeshore, M. L., and Quemada, H. D. 1995. Field evaluation of transgenic squash containing single or multiple virus coat protein gene constructs for resistance to cucumber mosaic virus,watermelon mosaic virus 2, and zucchini yellow mosaic virus. *Nat. Biotechnol.* 13:1458–1465.

Tripathi, J. N., Ntui, V.O., Ron, M., Muiruri, S.K. Britt, A and Tripathi, L. 2019. CRISPR/Cas9 editing of endogenous banana streak virus in the B genome of Musa spp. overcomes a major challenge in banana breeding. *Commun. Biol.* 2:46.

Tung, S. A., Smeeton, R., White, C. A., Black, C.A., Taylor, I.B. Hilton, H.W. and Thompson, A.J. 2008. Over-expression of LeNCED1 in tomato (*Solanum lycopersicum* L.) with the rbcS3C promoter allows recovery of lines that accumulate very high levels of abscisic acid and exhibit severe phenotypes. *Plant Cell Environ.* 31:968–981.

USDA (US Department of Agriculture). 1994. Response to Upjohn Company/Asgrow Seed Company Petition 92-204-01 for Determination of Nonregulated Status for ZW-20 Squash. Washington, DC: USDA, Animal and Plant Health Inspection Service.

Varkonyi-Gasic, E., Wang, T., Voogd, C., Jeon, S., Drummond, R.S., Gleave, A.P. and Allan, A.C. 2019 Mutagenesis of kiwifruit CENTRORADIALIS-like genes transforms a climbing woody perennial with long juvenility and axillary flowering into a compact plant with rapid terminal flowering. *Plant Biotechnol. J.* https://doi.org/10.1111/pbi.13021.

Vu, T., Choudhury, N. R., and Mukherjee, S. K. 2013. Transgenic tomato plants expressing artificial microRNAs for silencing the pre-coat and coat proteins of a begomovirus, Tomato leaf curl New Delhi virus, show tolerance to virus infection. *Virus Res.*172:35–45.

Wan, F., Pan, Y., Li, J., Chen, X., Pan, Y., Wang, Y., Tian, S., and Zhang, X. 2014. Heterologous expression of Arabidopsis C-repeat binding factor 3 (AtCBF3) and cold-regulated 15A (AtCOR15A) enhanced chilling tolerance in transgenic eggplant (*Solanum melongena* L.). *Plant Cell Rep.* 33:1951–1961.

Wang, B. Q., Zhang, Q. F., Liu, J. H., and Li, G. H. 2011. Overexpression of PtADC confers enhanced dehydration and drought tolerance in transgenic tobacco and tomato: Effect on ROS elimination. *Biochem BiophysResCommun.*413:10–16.

Wang, G., Kong, F., Zhang, S. Meng, X., Wang, Y. and Meng, Q. 2015. A tomato chloroplast-targeted DnaJ protein protects Rubisco activity under heat stress. *J. Exp. Bot.* 66:3027–3040.

Wang, S., Liu, J., Feng, Y., Niu, X., Giovannoni, J., and Liu, Y. 2008. Altered plastid levels and potential for improved fruit nutrient content by downregulation of the tomato DDB1-interacting protein CUL4. *Plant J.* 55(1):89–103.

Wurbs, D., Ruf, S., and Bock, R. 2007. Contained metabolic engineering in tomatoes by expression of carotenoid biosynthesis genes from the plastid genome. *Plant J.* 49:276–288.

Wei, D., Zhang, W., Wang, C. Meng, Q., Li, G., Chen, T.H. and Yang, X. 2017. Genetic engineering of the biosynthesis of glycinebetaine leads to alleviate salt-induced potassium efflux and enhances salt tolerance in tomato plants. *Plant Sci.*257:74–83. Yarra, R., He, S. J., Abbagani, S. Ma, B., Bulle, M. and Zhang, W.K.. 2012. Overexpression of a wheat Na+/H+ antiporter gene (*TaNHX2*) enhances tolerance to salt stress in transgenic tomato plants (*Solanum lycopersicum* L.). *Plant Cell Tissue Organ Culture* 111:49–57.

Ye, C. and Li, H. 2010. 20 years of transgenic research in china for resistance to papaya ringspot virus. *Transgenic Plant J.* 4: 58–63.

Yin, K., Gao, C., and Qiu, J.-L. 2017. Progress and prospects in plant genome editing. *Nat. Plants* 3:17107.

Zhang, C., Liu, J., Zhang, Y., Cai, X., Gong, P., Zhang, J., Wang, T., Li, H., and Ye, Z. 2011. Overexpression of SlGMEs leads to ascorbate accumulation with enhanced oxidative stress, cold, and salt tolerance in tomato. *Plant Cell Rep.*30:389–398.

Zhang, M., Liu, Q., Yang, X., Xu, J., Liu, G., Yao, X., Ren, R., Xu, J., and Lou, L. 2020. CRISPR/Cas9-mediated mutagenesis of Clpsk1 in watermelon to confer resistance to Fusarium oxysporum f.sp. niveum. *Plant Cell Rep.* 39:589–595.

Zhu, M., Chen, G., Zhang, J. Zhang, Y., Xie, Q., Zhao, Z., Pan, Y. and Hu, Z. 2014. The abiotic stress-responsive NAC-type transcription factor SlNAC4 regulates salt and drought tolerance and stress-related genes in tomato (Solanum lycopersicum). *Plant Cell Rep.* 33:1851–1863.

3 Omics Sciences Advances in Tropical Fruits and Perspectives for Crop Improvement

Heriberto García-Coronado, José Ángel Huerta-Ocampo, Damaristelma De Jesús Campos, Héctor Adán Ruiz-Ortega, Ángel Javier Ojeda-Contreras, José María Anaya-Dyck, Edmundo Dominguez-Rosas, Eduardo Antonio Trillo-Hernández, Martín E. Tiznado-Hernández, and Miguel Ánge Hernández-Oñate

3.1 INTRODUCTION

Since humans started plant cultivation more than 10,000 years ago, it has been domesticated with different species, establishing several crops in different regions of the world. The cultivation of plants has represented an essential activity for humanity, mainly as a source of food and metabolites with pharmacological activity (Gepts 2008). Further, tropical fruits are an important nutritional source for the human diet, mainly because they contain many bioactive compounds, such as phenolic compounds, carotenoids, flavonoids, vitamins, and fiber that benefit human health (Sayago-Ayerdi et al. 2021). Tropical fruits include a wide number of species that grow in the world area enclosed by the tropic of Cancer and the tropic of Capricorn which includes regions located in Africa, Asia, Africa, Central America, North America, South America, Caribbean, and Oceania (Vázquez-Manjarrez et al. 2020; Gepts 2008). The wide diversity of tropical fruits is well known. In America, Asia, and Africa; more than 1000, 500 and 1200 species of fruits have been described, respectively; however, only a small percentage of these are marketed locally and a smaller percentage are exported (Paull and Duarte 2011). The best-known tropical fruits because of their greatest presence in the market are the banana, pineapple, mango, papaya, and avocado (Altendorf 2017). The cultivation of these fruits represents an important source of revenue and employment that contributes positively to the economy of regions where they are grown. According to the Food and Agriculture Organization (FAO), in 2019, world production for these fruits was about 221,729,019 tons, which

DOI: 10.1201/9781003343707-3

represented a gross production value of around 86 million dollars (FAO 2019). In addition, it is estimated that tropical fruits cultivation is one of those that will have the fastest growth (OECD/FAO 2020; Altendorf 2017).

Despite the importance of tropical fruit crops and the significant biotechnological advances, considerable post-harvest losses of these fruits have been reported that are around 20–40% (Rajapaksha et al. 2021). This may be due to different circumstances, including the physiological conditions of each crop, poor postharvest handling techniques, utilization of obsolete equipment and the susceptibility to different biotic and abiotic factors (Mathiazhagan et al. 2021). Postharvest technologies have been established to extend the fruits shelf life in optimal conditions (Indiarto et al. 2020). Further, although these technologies have represented great advances in the objective of maintaining the quality of the fruits for a longer time, there is still the need to establish programs for crop improvement. Therefore, efforts should be directed to the improvement of crops to obtain resistance varieties to biotic and abiotic factors, as well as to improve the quality and postharvest life of the fruits. In this sense, the long-life cycles from seed to fruit production complicate the classic strategy of crop improvement through the genetic improvement program by Mendelian segregation (Yamamoto 2016). However, these limitations have been minimized by advances in massive DNA sequencing technologies, bioinformatics, and data computer storage (Mathiazhagan et al. 2021).

In the last 15 years, omics sciences have played an important role in generating data that have helped to improve knowledge about the biology of tropical fruits. An important number of nuclear, mitochondrial and chloroplast genomes, transcriptomes, proteomes, and metabolomes have been analyzed, which has led to the identification of key genes, regulatory networks and metabolic pathways that play a role in the agronomic characteristics of interest such as the development, maturation, size, flavor and color of the fruit, as well as resistance to biotic and abiotic stresses (Mathiazhagan et al. 2021). This knowledge represents an excellent tool and will help researchers and breeders to address the challenge of fruit crop improvement in a more efficient and faster way (Scossa et al. 2021). The scientific information generated by the utilization of different omics approaches carried out and the advances in the improvement of tropical fruits will be described in this chapter.

3.2 THE CURRENT STATE OF TROPICAL FRUIT CROPS

The world tropical fruit trade has expanded to unprecedented heights in recent years, tropical fruits constitute a comparatively new group in the global commodity trade (Altendorf 2017). As consumer health consciousness becomes more widespread, international awareness campaigns that highlight the benefits of nutrient-rich tropical fruits induce consumer willingness to pay high prices to include them in their diet (Anaya-Dyck et al. 2021). According to the FAO, in 2018, the total production of citrus was 133,805.5 thousand tons, mangoes 52,084.1 thousand tons, pineapple 28,252.4 thousand tons, papaya 13,581.6 thousand tons, and avocado 6,323.0 thousand tons (Table 3.1). Moreover, it is estimated that tropical fruit cultivation is one of those that will have the fastest growth (Altendorf 2017).

TABLE 3.1
Total production of fruits by region/production area in 2018 (thousands of tons)

Fruit	Africa	Asia	Caribbean and Central América	Europe	Northern América	South América	Oceanía	World production
Citrus	14,813.5	65,335.0	9,992.0	11,087.1	14,744.8	26,502.4	0.0	133,805.5
Oranges	9,436.0	25,222.9	6,028.3	6,465.4	3,560.0	19,741.0	388.0	70,841.5
Tangerines	3,160.4	25,680.1	547.9	2,901.1	729.0	2,793.1	145.2	35,964.9
Lemons and Limes	1,429.4	7,804.4	2,743.6	1,624.1	806.0	3,695.5	40.2	18,143.2
Grape fruits	787.6	6,627.6	601.9	96.3	469.0	272.9	6.9	8,863.3
Mangoes	8,209.1	37,813.7	3,344.9	-	1.2	2,661.4	53.8	52,084.1
Pineapples	5,735.4	12,168.1	5,387.5	-	148.6	4,687.0	0.9	28,252.4
Avocados	725.8	679.0	3,232.2	102.5	106.1	1,391.6	85.3	6,323.0
Papaya	1,428.2	8,198.1	2,322.0	-	10.3	1,609.6	13.5	13,581.6

Source: FAO (2019). Major tropical fruits- statistical compendium 2018. Rome.

Although the production of tropical fruits shows an increasing trend, postharvest losses remain a challenge to produce tropical fruits worldwide (Cruz et al. 2021). Postharvest losses of horticultural crops average 45 percent in both developing and developed countries (Rajapaksha et al. 2021). Losses of this magnitude, represent a significant loss of food and considerable economic damage for traders, and the general population. There are several causes leading to postharvest fruit losses besides fruit senescence (Cruz et al. 2021). Biotechnology has been used to solve problems in all aspects of agricultural production, including plant breeding to increase and stabilize yields and improve resistance to pests, diseases, and adverse abiotic conditions such as drought and cold (Rasmussen 2020; Singh et al. 2020). In recent decades, biotechnology through omics sciences has gained greater relevance in studies that seek to improve crops because they can accelerate genetic improvement processes (Scossa et al. 2021). Hence, it is necessary to identify the genes, regulatory networks, and biological processes involved in the response of fruits to different stress types during postharvest.

3.3 OMICS FOR GENE DISCOVERY OF AGRONOMIC IMPORTANCE

Plants are constantly exposed to various biotic and abiotic stressors in the environment where they live (Baweja and Kumar 2020). Although plants manage to adapt to their harsh environments, these conditions represent a challenge to agriculture due to significant reductions in crop yields (Calanca 2017). In response to stress conditions, plants can activate complex response mechanisms that include changes in gene expression levels, protein, and metabolite abundances, which promotes their adaptation to these harsh conditions (Baillo et al. 2019; Sun et al. 2019b). Therefore, the identification of key genes involved in the plant response to stress factors is of vital importance for the improvement of crops focused on greater resistance to various stress types.

The rapid advance of DNA sequencing technologies and increasingly powerful bioinformatics tools have allowed the sequencing of genomes and transcriptomes of many crops which has led to the discovery of a large number of genes involved in essential biological processes such as plant growth, fruit development, photosynthesis and tolerance to abiotic and biotic stress (Scossa et al. 2021; Shalini et al. 2018). Proteomics and metabolomics have contributed to the identification of proteins and metabolites associated with interesting traits in crops. Proteomics involving the identification and characterization of protein sets encoded in the genome has shown to be effective in improving the understanding of the biological processes and interaction of crops with their environment (Shalini et al. 2018). Meanwhile, metabolomics has made it possible to evaluate the metabolic characteristics of crops, identifying primary and secondary metabolites associated with the essential biological processes of crops, as well as those associated with crop resistance mechanisms (Scossa et al. 2021).

DNA sequencing technology has also allowed the identification of a great number of molecular markers, mainly SNPs, which has increased the available information on the genetic diversity of crops and the identification of those associated with important traits in crops, which is essential for breeding programs (Scossa et al. 2021; Le Nguyen et al. 2019). Multiple databases have been created to store and organize the plant genomics data. Table 3.2 shows the main databases containing the plant omics data.

TABLE 3.2
List of main databases containing plant omics data

Database	URL	Description
GenBank	https://www.ncbi.nlm.nih.gov/genbank/	GenBank is one of many databases available in the National Center for Biotechnology Information (NCBI) website. This database contains information of DNA sequences from more than 260,000 different species (animals, bacteria, plants, fungi) in addition to bioinformatic tools such as BLAST to analyze and compare this data.
The Arabidopsis Information Resource (TAIR)	https://www.arabidopsis.org	Contains molecular and genetic information of *Arabidopsis thaliana* genes. Also provides information about the expression of certain genes in different tissues and the phenotypic characteristics of various mutants of this plant.
The Sol Genomics Network (SGN)	https://solgenomics.net	This database contains genomic and genetic information of Solanaceae plants, which includes economically important crops such as tomato, tobacco, potato, and pepper. SGN also offers phenotypic data and metabolic or enzymatic reactions information as well.
iTAK	http://itak.feilab.net/	iTAK is a database dedicated to store information and classification of transcription factors, transcriptional regulators, and protein kinases from 196 different plan species.
PlanTFDB	http://planttfdb.gao-lab.org	Contains genetic and molecular information and classification of 320,370 sequences coding for transcription factors from 165 plant species
miRbase	https://www.mirbase.org	This database contains miRNA sequences as well as their respective annotations. Is reported to contain around 39,000 hairpin sequences, which produced 48,000 miRNA sequences from 271 different organisms.

(*continued*)

TABLE 3.2 (Continued)
List of main databases containing plant omics data

Database	URL	Description
Pfam	http://pfam.xfam.org	Contains protein sequences, whose domains are generated through hidden Markov models and are classified in families depending on the function that their domain possesses.
phytozome	https://phytozome-next.jgi.doe.gov	Phytozome is a resource focused on comparative genomics that contains information of a great number of plants with sequenced genomes
Plant rDNA Database	https://www.plantrdnadatabase.com	This database contains relevant information of ribosomal DNA from 1,609 different plant species.

3.4 TROPICAL FRUITS GENOMICS ADVANCES

Genomics, encompassing sequencing, assembly, and analysis of entire genomes, offers multifaceted insights into living organisms (Moore and Ming, 2008a). However, traditional breeding for tropical fruits faces challenges due to limited germplasm availability (Schnell and Priyadarshan, 2012). Advancements in biotechnology and bioinformatics enhance non-model plant genome analyses (Thottathil et al., 2016), promising insights for tropical fruit improvement. Genome analysis, augmented by marker-assisted selection (MAS), accelerates breeding by identifying desirable progeny early. Quantitative trait loci (QTL) and single nucleotide polymorphisms (SNP) aid in identifying chromosomal regions and allelic associations governing agronomically significant traits (Schnell and Priyadarshan, 2012). Post-assembly genome characterization involves assessing size, sequencing depth, GC content, gene content, polymorphisms, repetitive sequences, and polyploidy events. Integrated, these methodologies offer a comprehensive approach to harnessing genomic data for tropical fruit improvement, thereby fostering food security and development in lesser developed nations (Mathiazhagan et al., 2021). A list of tropical fruit tree genomes published is shown in Table 3.3. It is made of 15 horticultural commodities including papaya (*Carica papaya*) (Ming et al. 2008; Moore and Ming 2008b), date palm (*Phoenix dactylifera*) (Al-Dous et al. 2011; Al-Mssallem et al. 2013), cacao (*Theobroma cacao*) (Argout et al. 2011; Motamayor et al. 2013; Argout et al. 2017), banana (*Musa acuminata*) (Wang et al. 2019b), pineapple (*Ananas comosus*) (Ming et al. 2015), coconut (*Cocos nucifera*) (Xiao et al. 2017), durian (*Durio zibethinus*) (Teh et al. 2017), avocado (*Persea americana*) (Rendón-Anaya et al. 2019), mango (*Mangifera indica*) (Wang et al. 2020a), soursop (*Annona muricata*) (Strijk et al.

2021), guava (*Psidium guajava*) (Feng et al. 2021), passion fruit (*Passiflora edulis*) (Xia et al. 2021), rambutan (*Nephelium lappaceum*) (Zhang et al. 2021) and lychee (*Litchi chinensis*) (Hu et al. 2022). Information about the sequencing platforms, assembling software, metrics, and statistics of the genomes is included in Table 3.3.

Whole-Genome Duplication (WGD) events and gene duplication events contribute to generating genetic variation, which is an important source of protein diversity in the plant kingdom that contributes to the development of novel functions for adapting to biotic and abiotic stress (Schnell and Priyadarshan 2012). To investigate subgenome evolution and functional divergence of bananas, the genome of *Musa acuminata* and *M. balbisiana* were assembled (Wang et al. 2019b). Genomic analysis suggested that three WGD events have occurred in *Musa* spp., followed by a chromosome rearrangement and gene loss (D'hont et al. 2012). Altogether, these analyses provide insights into the implication of functional divergence of banana subgenomes, and its relationship with the fruit ripening process (Wang et al. 2019b). The availability of the pineapple (*Ananas comosus*) genome, has allowed the development of molecular markers, which are useful to improve pineapple molecular-breeding. These findings include 296,896 segregating SNPs, 10,151 alternative splicing events (Ming et al. 2015), and 278,245 SSRs located in genes involved in fruit development (Fang et al. 2016).

3.5 TROPICAL FRUITS TRANSCRIPTOMICS ADVANCES

Fruit quality, a multifaceted trait influenced by genetic and environmental factors, underscores the importance of developing resistant varieties (Ding, 2017). Recent advancements in molecular techniques, particularly transcriptomics, enable the identification and quantification of transcripts, including non-coding RNAs (ncRNAs), crucial for breeding efforts (Haas et al., 2010; Hrdlickova et al., 2017). Transcriptomic analysis offers two main strategies for transcript reconstruction: alignment with reference genomes and de novo assembly. By studying gene expression in tissues under various growth stages or stress conditions, insights into metabolic pathways and potential markers for fruit quality improvement can be gleaned (Vergara-Pulgar et al., 2019). Ge et al. (2019) identified isoforms of carotenoid biosynthesis genes in avocado fruit, establishing a correlation between gene expression and carotenoid levels in mesocarp, enhancing understanding of carotenoid biosynthesis pathways. Transcriptomic analysis of date palm fruit development revealed heightened metabolic activity during early stages, emphasizing nutrient reservoir utilization, while later stages saw increased carbohydrate, lipid, and amino acid metabolism as the fruit matures (Yin et al., 2012).

The comprehensive analysis of the transcriptome of two banana genotypes allowed us to establish that the *MAPKKK-MAPKK* genes (mitogen-activated protein kinases), identified in the genome are involved in tissue development, fruit development and maturation, as well as in the response to abiotic stress (Wang et al. 2017). Through a comparative transcriptomic analysis between guava leaf and fruit, genes involved in the pathways to fruit ripening were identified. Also, signaling cascades, ABA, ethylene biosynthesis and accumulation of secondary metabolites were

TABLE 3.3
Genome projects of tropical fruits

	Cp	*Tc*	*Ma*	*Pd*	*Ac*	*Dl*
Sequencing technology	WGS and BAC-end	Roche/454, Illumina, Sanger and PacBio	Sanger, Roche/454 and Illumina	Roche/454, SOLiD, and BAC-end	Illumina, Roche/454, PacBio and BACs	Illumina HiSeq 2000
Total raw data (Gb)	-	26	~37	~255	-	316.84
Assembler software	Arachne	Newbler and SOAP	Newbler	SOAPdenovo and BAMBUS	ALLPATHS-LG	SOAPdenovo and SSPACE
Chromosomes (1n)*	9	10	11	18	25	15
Genome size	372 Mb	326.9 Mb	523 Mb	605.4 Mb	382 Mb	471.88 Mb
Scaffold number	17,764	554	7,513	82,354	3,133	17,367
N50 scaffolds	1 Mb	6.5 Mb	1.3 Mb	0.33 Mb	11.8 Mb	0.56 Mb
Sequencing Depth	3X	0.2-44X	20.5-50X	17.3X	~2-150X	~20X
GC %	35.3	34.38**	35.70	36.4	38.3	33.7
Number of genes	24,746	28,798**	36,542	41,660	27,024	31,007
Average gene lenght (bp)	2,373	3,346**	~3,600	3,971	4,893	3,266
Repetitive sequences %	58.3	35.4	41.85	38.41	51.84	52.87
Transposable elements %	51.9	25.7	33.3	~22.3	~44	36.54***
Reference	Ming et al. 2008	Argout et al. 2011; Argout et al. 2017	D'Hont et al. 2012	Al-Mssallem et al. 2013	Ming et al. 2015	Lin et al. 2017

Notes: **Cp:** Papaya (*Carica papaya*). **Tc:** Cacao (*Theobroma cacao*). **Ma:** Banana (*Musa acuminata*). **Pd:** Date palm (*Phoenix dactylifera*). **Ac:** Pineapple (*Ananas comosus*). **Dl:** Longan (*Dimocarpus longan*). **Cn:** Coconut (*Cocos nucifera*). **Dz:** Durian (*Durio zibethinus*). **Mb:** Banana (*Musa balbisiana*) **Pa:** Avocado (*Persea americana*) **Mi:** Mango (*Mangifera indica*). **Am:** Soursop (*Annona muricata*).

WGS: Whole-Genome Shotgun. **BAC:** Bacterial Artificial Chromosome. **PacBio:** Pacific Biosciences. *Haploid chromosome number. **Data corresponding to an earlier version of the genome assembly. *** Data corresponding to retrotransposons.

Cn	*Dz*	*Mb*	*Pa*	*Mi*	*Am*
Illumina HiSeq 2000	PacBio and Illumina	PacBio and Illumina	Illumina, Roche/454, PacBio and BACs	PacBio and Illumina	PacBio, Illumina, 10× Genomics, and Bionano
419.67	~307	145.33	75.3	86.5	444.32
SOAPdenovo2	CANU and FALCON	wtdbg, Arrow, and SSPACE	Newbler and FALCON	CANU, ALLPATHS-LG and Pilon	SOAPdenovo2 and Pilon
16	28	11	12	20	7
2.42 Gb	715 Mb	430 Mb	~920 Mb	392.9 Mb	639.6 Mb
-	677	-	42,722	252	949
~0.42 Mb	22.7 Mb	5.05 Mb	0.32Mb	17.6 Mb	3.43 Mb
173.32X	~153X	113-166X	3.3-39X	~240X	676X
-	32.5	37.90	~31.5	-	35.4–37.6
28,039	45,335	35,148	24,616	41,251	23,375
-	1,700	~5,000	-	3,503	5,208
74.48	54.8	55.75	~50	40.5	54.87
72.75	29.4***	-	-	39.9***	41.28
Xiao et al. 2017	Teh et al. 2017	Wang et al. 2019b	Rendón-Anaya et al. 2019	Wang et al. 2020	Strijk et al. 2021

detected too. This study also allowed the detection of transcription factor families belonging to *MYB, MADS, HB, WRKY, ARF, bHLH, AP/EREBP, bZIP, NAC, AUX/IAA, B3*, Jumonji and Polycomb involved in fruit development (Mittal et al. 2020, Mathiazhagan et al. 2021). Li et al (2019b) identified ERF and bHLH family transcription factors related to banana peel softening, genes related to auxin signaling, and members of the xyloglucan endotransglucosylase/hydrolase (XTH) family were also observed (Mathiazhagan et al. 2021).

The introduction of diseased exotic plants, climate change, and pesticides use, among other factors, induce the development of pests and resistant pathogens that can affect the yield and quality of fruits. Therefore, it is essential to understand the molecular basis underlying the successful response of resistant genotypes. In a transcriptomic analysis of the germplasm defense response of two citrus species, one tolerant and the other susceptible to *Phytophthora nicotianae*, it was observed that the necrotrophic phase of infection is a transition point that includes a stronger modulation of genes related to pathogen response that in the biotrophic phase, in addition, in the tolerant species the ability to perceive the pathogen was reflected even after its change to necrotrophic phase, which determines its ability to deal with the infection (Ajengui et al. 2018). In a banana cultivar resistant to *Fusarium oxysporum*, genes associated with the synthesis of defense-related metabolites, such as NB-LRR proteins, calmodulin-binding protein, and phenylpropanoid biosynthesis genes, were found to be induced significantly (Sun et al. 2019a).

Plants that in natural conditions are found in severe environments (abiotic stress) are important genetic resources to study the resistance genes that allow them to survive under these conditions. For instance, in the date palm, naturally subjected to abiotic stress (long periods of UV radiation, saline soils, low humidity), it was observed that under saline stress, the genes of the salt hypersensitivity pathway regulate this response; furthermore, the splicing factors were found to be involved in RNA splicing regulation under this kind of stress (Xu et al. 2021). In ultraviolet radiation stress, a greater modulation of genes related to UV radiation absorption and synthesis of protective agents was observed (Maher et al. 2021). Transcriptomic analysis allows us to analyze a specific group of transcripts, including those that do not code for any protein called non-coding RNAs (ncRNA) which was involved in the regulation of fundamental biological processes such as development, maturation and fruit softening (Wang et al. 2018a, Wang et al. 2020b, Wang et al. 2021; Zhu et al. 2015, Chen et al. 2021b), fruit coloration (Xiao et al. 2020, Ma et al. 2021), cold injury (Wang et al. 2018b), fruit cracking (Xue et al. 2020) among others. Additional studies focused on elucidating the association pattern of ncRNAs and their expression under different types of stress will have repercussions on expanding the information on their regulatory mechanism. Additional transcriptomic studies can be found in Table 3.4.

3.6 TROPICAL FRUITS PROTEOMICS AND METABOLOMICS ADVANCES

Proteins, serving as the fundamental structural and functional constituents of living organisms, constitute the proteome, reflecting the dynamic nature of biological entities and their responses to internal and external stimuli. Studying the proteome, such as that of tropical fruits, offers a comprehensive understanding facilitated by proteomics, a post-genomic approach enabling the analysis of complex protein mixtures. Mass spectrometry (MS) plays a pivotal role in proteomics, having undergone remarkable technological advancements over the past two decades, thus becoming integral to systems biology. Utilizing MS-based proteomics, researchers have efficiently identified, characterized, and quantified proteins, including those undergoing post-translational modifications. Proteomics approaches have been applied to investigate papaya ripening dynamics in response to ethylene and its inhibitor, 1-methylcyclopropane, shedding light on postharvest quality modulation (Huerta-Ocampo et al., 2012). Similarly, label-free

TABLE 3.4
Transcriptomic analysis of tropical fruits

Tropical fruit	Sample	Sequencing plataform	Number of Unigenes	Main findings	References
Garcinia mangostana	Rind and aril from mature mangosteen trees	Ion Proton System™	268,851	35,498 proteins were predicted. 21,316 proteins were annotated by UniProtKB	Matra et al. 2016
Musa acuminata	Unripe and ripe stages of banana	454 GSFLX-Titanium	112,233	Several members of gene families responsible for softening of banana, showed significantly higher expression in the ripe fruit compared to unripe fruit. Genes related to the cell wall softening were among the top up-regulated genes during banana fruit ripening. Genes related to aroma formation were identified. Several acyltransferases were identified that are likely involved in the synthesis aromatic volatiles and flavor components in the ripe sample.	Asif et al. 2014
Musa acuminata	Roots, leaves and fruits	Not mentioned	Not mentioned	10 MAPKK and 77 MAPKKK gene identified in the genome is involved in tissue development, fruit development and ripening, and response to abiotic stress	Wang et al. 2017
Musa sp	Pulp of different ripening stages of fruit	Illumina HiSeqTM 2000	Not mentioned	A great number of genes related to transcriptional regulation, signal transduction, cell wall modification, and secondary metabolism were up-regulated during pulp ripening.	Li et al. 2019

(continued)

TABLE 3.4 (Continued)
Transcriptomic analysis of tropical fruits

Tropical fruit	Sample	Sequencing plataform	Number of Unigenes	Main findings	References
Musa acuminata	Xylems of roots from banana resistant and susceptible cultivars. Both infected with *Fusarium oxysporum*	Illumina HiSeq™ 2000	~48 million high quality reads mapped to the Musa genome	Transcriptional reprogramming was more drastic in resistant banana cultivars, and defense-related genes were significantly upregulated.	Sun et al. 2019a
Musa balbisiana *Musa acuminata*	Root and corm tissues, foliar tissues, reproductive tissues and pseudo stem	Illumina GA II X	Not mentioned	1319 and 4288 putative lncRNAs were identified from *M. balbisiana* and *M. acuminata* respectively. Large lncRNAs were present only in *M. balbisiana*	Sampangi-Ramaiah et al. 2019
Citrus sinensis	Pulp of fruit	Illumina sequencing	19,440	The TCA cycle, and carotenoid biosynthesis were all differentially regulated during fruit development and ripening.	Yu et al. 2012
Myrica rubra.	Stems and leaves, buds, flowers, young fruit, fruit at breaker stage, red maturity stage, and dark red maturity stage	Illumina platform	41,239	The metabolic processes are enhanced and catalytic activity increases during fruit ripening. Expression of sucrose phosphate synthase (SPS) could be associated with the increase in sweetness, and glutamato descarboxilasa may participate in accumulation and loss of organic acids during fruit ripening. Genes related to anthocyanin biosynthesis and MrMYB1 played roles in accumulation of anthocyanin and the development of the deep dark-red color.	Feng et al. 2012

Ananas comosus	Ripe yellow pineapple fruit flesh	Illumina technology	28,728	Pineapple fruit undergoes multiple processes of development and stresses. Transcripts encoding enzymes related to different pathways were detected (fragrance synthesis in fruits, terpene biosynthesis straight chain ester biosynthesis from fatty acids, lignin biosynthesis, branched chain ester biosynthesis, citrate acid cycle and sucrose metabolism)	Ong et al. 2012
Mangifera indica	Pulp of fruit at five development stages	Novaseq 6000	21,349 genes expressed in at least one sample. 18,398 differentially expressed genes (DEGs) were identified through five stages	Ripening-related genes were identified, involved in sugar and carotenoid biosynthesis and softening-related metabolic processes. Abscisic acid (ABA) plays a key role in directly modulating mango fruit ripening through transcription factor MiHY5.	Wu et al. 2022
Mangifera indica	Fruit epidermal peel of ripe and overripe mango fruit	Illumina HiSeq 2500	107,744 unigenes	Biosynthesis of the cuticle component, cutin, is up-regulated during overripening. Pathways related to the biosynthesis of cuticle components were very active in the peel during mango fruit overripening. Unigenes involved in several metabolic processes related to mango fruit development were identified, such as the synthesis or degradation of lipids, cutin, secondary metabolites and cell wall polysaccharides, among others.	Tafolla-Arellano et al. 2017

(continued)

TABLE 3.4 (Continued)
Transcriptomic analysis of tropical fruits

Tropical fruit	Sample	Sequencing plataform	Number of Unigenes	Main findings	References
Annona muricata	Fruit under different postharvest storage conditions	Illumina Next-Seq 500	95.832 unigenes	Pectinesterase genes directly regulate the loss of firmness in fruits stored at 15 ± 2 °C	Palomino-Hermosillo et al. 2022
Annona muricata	Fast growing tissues (young leaves, shoots)	Illumina HiSeq	40,486 Assembled Transcripts	Not reported	Leebens-Mack et al. 2019
Citrus sinensis, Citrus sunki, Poncirus trifoliata and hybrids obtained from crosses between C. sunki and P. trifoliata	Leaves from inoculated and mock-inoculated plants, after 8 months of Candidatus *Liberibacter asiaticus* (CLas) infection.	Illumina HiSeq 2500	133,976 transcripts (reads were mapped on the *C. sinensis* genome)	several pathways were reprogrammed in response to in CLas *P. trifoliata* and resistant hybrids. Notably, a genetic tolerance mechanism was associated with the downregulation of gibberellin (GA) synthesis and the induction of cell wall strengthening. These defense mechanisms were triggered by a class of receptor-related genes and the induction of WRKY transcription factors.	Curtolo et al. 2020
Diospyros kaki	Persimmon fruit peels (freshly and softened)	Illumina HiSeq 2000	31,258 unigenes	Transcription involved in ethylene biosynthesis, perception and signaling was up-regulated. Transcription of basic chitinase, chitinase A, and UDP-glucose 6-dehydrogenase was up-regulated during softening. Chitinase and endochitinase activities would therefore have the physical effect of loosening cell walls.	Jung et al. 2017

Cocos nucifera	Spear leaves, young leaves and fruit flesh	Illumina Hiseq™ 2000	57,304 unigenes	39,109 (68.2%) were matched against the NCBI non-redundant protein database. The highest level of identity to of *Cocos nucifera* unigenes was *Oryza sativa* protein database, 347 unigenes involved in the biosynthesis and metabolism of fatty acid were identified.	Fan et al. 2013
Phoenix dactylifera	First true leaves from seedlings subjected to UV-B stress for 72 h and the control	Illumina HiSeq	26,091 unigenes	Genes related to the plants defense system against UV stress were most upregulated, these genes are involved in the absorption of ultraviolet radiation, synthesis of UV protectants and defense agents, which allows increased tolerance to UV-B stress.	Maher et al. 2021
Phoenix dactylifera	Date palm fruits at seven different developmental stages (F1-F7)	Roche/454 GS FLX	Range from 30 684 y 40 378	Cells are active for cell division in the F1 stage, the nutrient reservoir activity was higher in the first four stages. Beta-carotene hydroxylase plastocyanin and fruit ripening-related genes mainly expressed in the last three stages. The translation regulator activity was detected only in the last three developmental stages, and these genes may be associated with the regulation of fruit ripening.	Yin et al. 2012
Theobroma cacao	Various tissues and various conditions	Applied Biosystems BigDye V3.1 kits, and were resolved on ABI3730xl DNA Analyzers	From 169 to 3331	Developing an EST genomic resource database for cacao. Genes expressed in *Theobroma cacao* are more similar to *Vitis vinifera* proteins than to those of *Arabidopsis thaliana.* All major enzymes of the general flavonoid pathway, anthocyanin biosynthesis and terpene pathway were represented in EST resource. Genes related to resistance/defense mechanisms were also found.	Argout et al. 2008

proteomic analysis has elucidated the mechanisms underlying low-temperature-induced damage in hardy kiwifruit, providing insights crucial for quality enhancement and genetic manipulation. Furthermore, quantitative proteomic analysis in grape berries following melatonin application has revealed its role in oxidative stress tolerance and polyamine metabolism regulation, offering potential avenues for postharvest preservation strategies not only for grapes but also for other fruits (Zhang et al., 2021; Wang et al., 2022).

Tropical fruits face challenges in maintaining commercial value postharvest due to susceptibility to deterioration. Various strategies, including biotic and abiotic stimuli, controlled atmosphere, and cold storage, are employed to mitigate decay and preserve quality. Proteomics, especially utilizing MS-based techniques, has emerged as a valuable tool for comprehending ripening and physiological disorders in these fruits, promising insights into essential physiological processes (Fortes et al., 2017). The nutritional and organoleptic characteristics of tropical fruits are intricately linked to their biochemical composition, encompassing metabolites, minerals, proteins, starch, and cell walls, which are influenced by domestication, breeding, and cultural practices, as well as biotic and abiotic stress (Allwood et al., 2021).

Metabolites, being small molecules involved in metabolic pathways, serve as indicators of metabolic reprogramming during fruit development and postharvest, and are sensitive to biotic and abiotic stress conditions. Metabolomics, employing technologies like NMR and MS coupled with chromatographic techniques, enables the detection, quantification, and structural elucidation of metabolites, reflecting their diverse physical-chemical properties (Yanes, 2015). While NMR offers advantages such as non-destructiveness and reproducibility, MS provides higher sensitivity and specificity. Experimental design in metabolomics hinges on the targeted or untargeted approach, the former focusing on specific metabolites relevant to selected pathways, while the latter aims for broad coverage without a predefined hypothesis, guided by contextual information (Patti et al., 2012; Nagana Gowda and Raftery, 2017).

NMR-based metabolomics has been instrumental in characterizing soluble compounds in polar and semipolar extracts of various fruits like melon, kiwifruit, banana, grape berry, and persimmon, aiding in understanding their metabolic profiles during development and postharvest stages (Capitani et al., 2010; Moing et al., 2011; Ali et al., 2011; Maulidiani et al., 2018). These methodologies hold promise for further elucidating the intricate metabolic processes underlying fruit quality and response to stress, offering insights crucial for agricultural and food industries. Gas chromatography-mass spectrometry (GC-MS) is frequently utilized for the analysis of fruit polar metabolites post-derivatization or volatile organic compounds, employing electron impact (EI) or chemical ionization (CI) sources. GCxGC-MS enhances resolution, particularly beneficial for flavor analysis in fruits and fruit products (Allwood et al., 2021). Liquid chromatography-mass spectrometry (LC-MS) commonly employs reversed-phase (RP) C18 HPLC for nonpolar lipid species and certain fruit metabolites like flavonoids and alkaloids, while hydrophobic interaction liquid chromatography (HILIC) is suitable for polar metabolite separations. Coupling HILIC and RP methods can further increase resolution, with recent integration of ion mobility with MS showing promise for analyzing flavonoids, tannins, and phenolic acids (Allwood et al., 2021).

MS analyzers have undergone significant advancements in the past decade, facilitating accurate mass MS^n data collection on intact metabolites and their substructures, aiding in fruit metabolite identification (Allwood et al., 2019). Metabolomic analyses, including targeted and untargeted approaches using nuclear magnetic resonance (NMR) or MS, provide comprehensive insights into fruit development, quality, authenticity, and responses to postharvest treatments (D'Urso et al., 2018; Khakimov et al., 2016). Various fruits have been studied using GC–MS and LC-MS/MS metabolomic analyses to understand metabolic changes during ripening, postharvest senescence, and among different cultivars, highlighting correlations between metabolite profiles and fruit characteristics (Jiang et al., 2021). Furthermore, metabolomic methods, alongside molecular phylogenetic analyses, have been instrumental in revealing genetic relationships within fruit species, such as melons, demonstrating the utility of metabolomics in classification and phylogenomic studies (Moing et al., 2020).

3.7 ADVANCES IN MOLECULAR MARKER IDENTIFICATION AND GENOME EDITION IN TROPICAL FRUITS

The identification of molecular markers linked to important agronomic traits is essential to designing programs to improve crops more efficiently that allow us to face the challenges of getting closer to food security and sustainable agriculture. In this sense, multi-omics approaches have facilitated knowing the complete genomes and a great number of molecular markers, as well as key genes, proteins, and metabolites playing a role in different developmental, physiological, and biochemical phenomena of fruit tissues. In this section, we will describe the different experiments carried out using different omics approaches to identify the molecular markers, applications of genome editing tools to improve fruit crops, and databases based on omics information.

3.7.1 Molecular Markers

Various molecular markers have been developed and utilized to study genetic diversity and differentiation in several plant species. In Palmyra palm (*Borassus flabellifer*), efforts to differentiate between male and female inflorescences resulted in the development of 1,204 markers, though a sex-linked marker remained elusive (Pipatchartlearnwong et al., 2019). In bananas, including *Musa acuminata* and *Musa balbisiana*, genomic analysis has been conducted. Studies on *M. acuminata* revealed genomic homogeneity among subspecies (Rouard et al., 2018; D'Hont et al., 2012), with further analysis contributing to the construction of a pangenome for the *Musaceae* family (Rijzaani et al., 2022). Distinctness between *M. balbisiana* and *M. acuminata* cultivars was demonstrated using RAPD markers (Sales and Espino, 2004; Wahyudi et al., 2020), while ISSR markers aided in assessing genetic variation and phylogenetic relations (Padmesh et al., 2012; Creste et al., 2006; Hippolyte et al., 2012). Additionally, a genetic map with 489 markers was constructed for future reference (Hippolyte et al., 2010), and a database named BanSatDB was established (Arora et al., 2018).

Microsatellite markers proved useful in genotyping and genome diversity studies in *M. balbisiana* (Ravishankar et al., 2013; Rotchanapreeda et al., 2016). In breadfruit (*Artocarpus altilis*), SSRs from both nuclear and chloroplast genomes facilitated genetic studies and maternal lineage identification (De Bellis et al., 2016;). Similarly, SSR markers derived from an EST library of coconut leaves exhibited polymorphism and aided in genetic diversity studies (Preethi et al., 2020). Genomic data for *Coffea arabica* and *Coffea canephora* allowed the development of SSR markers for taxonomic purposes (Tran et al., 2018; Denoeud et al., 2014; Ferrao et al., 2015). Chloroplast genome sequencing in *Annona reticulata* and *Phoenix dactylifera* provided insights into phylogenetic relationships (Niu et al., 2020; Al-Dous et al., 2011). Moreover, sequencing efforts in *Hylocereus undatus* facilitated the identification of genes involved in betalain biosynthesis (Chen et al., 2021a; Xie et al., 2021). These studies collectively highlight the diverse applications of molecular markers in plant genetic research.

3.7.2 Genome Editing in Fruits

The genome editing tool is an efficient tool for introducing precise mutations into genomes and has become a better alternative to conventional breeding methods without extensive backcrossing and selection. This technique has great potential to improve fruit quality in tropical fruit crops (Mathiazhagan et al. 2021). In *Coffea canephora*, a search in the genome sequence allowed the identification of 5,338,568 possible gRNAs, which permitted the establishment of a genome editing protocol. The system was tested by successfully editing the phytoene desaturase gene using *A. tumefaciens* and somatic embryogenesis techniques. On the other hand, the technology of CRISPR/Cas9 genome editing was tested in *Citrus sinensis* embryogenic cell cultures. Two constructs including guides of RNA to target two regions of the phytoene desaturase gene under the control of the U6–26 promoter or the *CmYLCV* promoter were introduced by Agrobacterium infection. The transgenic embryos showed a stable and complete albino phenotype. An experiment using CRISPR/Cas9 to edit two alleles of the *CsWRK22* gene, was able to increase the resistance of the *C. sinensis* transgenic plants to citrus canker, which is a destructive disease in citrus (Wang et al. 2019a). In pomegranate (*Punica granatum*), the CRISPR/Cas9 technology was utilized to edit the genome. For this, a single guide RNA was designed to target *PgUGT84A23* and two guide RNA to target *PgUGT84A24* which are genes encoding UDP-dependent glycosyltransferase. The constructs were introduced by using *Agrobacterium rhizogenes*. It was recorded that the two genes edited lines have higher amounts of galloyglucose ethers and low concentrations of the isomers of punicalagin which is the most abundant hydrolysable tannins in pomegranate. This experiment clearly showed that the genome editing was successful (Chang et al. 2019).

3.7.3 Databases with Fruits Genomic Information

Efforts have been made to create databases that are comprised of omics data of tropical fruits, which will help bring the genomic information closer to breeders. Mathiazhagan et al. (2021) made an excellent compilation of databases that include genomic and transcriptomic information on important fruit crops. There is a web platform in which it is possible to access the complete *Coffea canephora* nuclear and chloroplast genome, genes, genetic markers, and transcriptomics, among other tools (https://coffee-genome-hub.southgreen.fr/, Dereeper et al. 2015). Similarly, it was created a web date palm genomic database named DRDB which is freely available for the public (http://drdb.big.ac.cn/home, He et al. 2017). In the *Citrus sinensis* annotation project (CAP), it is included the genomic and transcriptome data of the species. This information is freely available in http://citrus.hzau.edu.cn/ (Wang et al. 2014). While, in mango, there is a database freely available that includes 107,744 unigenes of mango exocarp (http://bioinfo.bti.cornell.edu/cgi-bin/mango/index.cgi, Tafolla-Arellano et al. 2017). Moreover, the mikan genome database includes genome annotation, diversity and CAPS molecular markers of the species *Citrus unshiu* and *Poncirus trifidiata,* which is of public access in https://mikan.dna.affrc.go.jp/ (Kawahara et al. 2020). In *Ananas comosus* known as PineElm_SSRdb, there is a database that is freely available for academic projects. This database includes 359,511 simple sequence repeats isolated from the genome, 45 isolated from chloroplasts, 249 SSR isolated from mitochondria, and 2832 isolated from expressed sequence tags (Chaudhary et al. 2016).

3.8 CONCLUSIONS AND FUTURE PERSPECTIVES

Tropical fruits are vital for human health, economy, and employment, yet face significant postharvest losses of 20–40% due to various stresses. Biotechnological interventions have been insufficient, highlighting the need for crop improvement programs to enhance stress resistance. Recent advances in omics sciences, including DNA sequencing and molecular biology, have revolutionized fruit research, leading to comprehensive analysis of genomes, transcriptomes, proteomes, and metabolomes. These studies have unveiled crucial genes, metabolic pathways, and regulatory mechanisms, enhancing our knowledge of different agronomic traits and stress resilience. Integrating multidisciplinary approaches is essential for a deeper understanding of fruit resistance mechanisms and efficient crop enhancement strategies.

ACKNOWLEDGMENTS

The authors acknowledge the national postgraduate scholarship program and the Researchers for Mexico program of CONAHCYT.

REFERENCES

Ajengui, A., Bertolini, E., Ligorio, A., Chebil, S., Ippolito, A., and S. M. Sanzani. 2018. Comparative transcriptome analysis of two citrus germplasms with contrasting susceptibility to *Phytophthora nicotianae* provides new insights into tolerance mechanisms. *Plant Cell Rep*. 37(3):483–99.

Al-Dous, E. K., George, B., Al-Mahmoud, M. E., et al. 2011. *De novo* genome sequencing and comparative genomics of date palm (*Phoenix dactylifera*). *Nat. Biotechnol.* 29(6):521–27.

Al-Mssallem, I. S., Hu, S., Zhang, X., et al. 2013. Genome sequence of the date palm *Phoenix dactylifera* L. *Nat. Commun.* 4(1):1–9.

Ali, K., Maltese, F., Fortes, A. M., Pais, M. S., Choi, Y. H. and Verpoorte, R. 2011. Monitoring biochemical changes during grape berry development in Portuguese cultivars by NMR spectroscopy. *Food Chem.* 124:1760–69.

Allwood, J. W., Gibon, Y., Osorio, S., et al. 2021. Developmental metabolomics to decipher and improve fleshy fruit quality. In *Advances in Botanical Research*, ed. P. Pétriacq and A. Bouchereaus, 3–34. London: Academic Press.

Allwood, J. W., Woznicki, T. L., Xu, Y., et al. 2019. Application of HPLC–PDA–MS metabolite profiling to investigate the effect of growth temperature and day length on blackcurrant fruit. *Metabolomics* 15:1–17.

Altendorf, S. 2017. Global prospects for major tropical fruits. The food outlook: biannual report on global food markets. The Food and Agriculture Organization of the United Nations (FAO) www.fao.org/fileadmin/templates/est/COMM_MARKETS_MONITORING/Tropical_Fruits/Documents/Tropical_Fruits_Special_Feature.pdf (accessed December 18, 2021).

Anaya-Dyck, J. M., Hernández-Oñate, M. Á., Tafolla-Arellano, J. C., Báez-Sañudo, R., Gutiérrez-Martínez, P., and M. E. Tiznado-Hernández. 2021. Soursop productive chain: An economic development option for Compostela, Nayarit. *Estud. Soc. Rev. Aliment. Contemp. Desarro. Reg.* 31(57):e211048.

Argout, X., Fouet, O., Wincker, P., et al. 2008. Towards the understanding of the cocoa transcriptome: Production and analysis of an exhaustive dataset of ESTs of *Theobroma cacao* L. generated from various tissues and under various conditions. *BMC Genom.* 9(1):1–19.

Argout, X., Martin, G., Droc, G., et al. 2017. The cacao Criollo genome v2.0: An improved version of the genome for genetic and functional genomic studies. *BMC Genom.* 18(1):1–9.

Argout, X., Salse, J., Aury, J. M., et al. 2011. The genome of *Theobroma cacao*. *Nat. Genet.* 43(2): 101–8.

Arora, V., Kapoor, N., Fatma, S., et al. 2018. BanSatDB, a whole-genome-based database of putative and experimentally validated microsatellite markers of three *Musa* species. *Crop J.* 6(6):642–50.

Asif, M. H., Lakhwani, D., Pathak, S., et al. 2014. Transcriptome analysis of ripe and unripe fruit tissue of banana identifies major metabolic networks involved in fruit ripening process. *BMC Plant Biol.* 14(1):1–15.

Baillo, E. H., Kimotho, R. N., Zhang, Z., and Xu, P. 2019. Transcription factors associated with abiotic and biotic stress tolerance and their potential for crops improvement. *Genes* 10(10):771.

Baweja, P. and Kumar, G. 2020. Abiotic stress in plants: An overview. In *Plant Stress Biology*, ed. B. Giri and M. P. Sharma, 1–15. Singapore: Springer.

Calanca, P. P. 2017. Effects of abiotic stress in crop production. In *Quantification of Climate Variability, Adaptation and Mitigation for Agricultural Sustainability*, ed. M. Ahmed and C. O. Stockle, 165–80. Switzerland: Springer Cham.

Capitani, D., Mannina, L., Proietti, N., et al. 2010. Monitoring of metabolic profiling and water status of Hayward kiwifruits by nuclear magnetic resonance. *Talanta* 82:1826–38.

Chang, L., Wu, S., and Tian, L. 2019. Effective genome editing and identification of a regiospecific gallic acid 4-O-glycosyltransferase in pomegranate (*Punica granatum* L.). *Hortic. Res.* 6:123.

Chaudhary, S., Mishra, B. K., Vivek, T., Magadum, S., and Yasin, J. K. 2016. PineElm_SSRdb: A microsatellite marker database identified from genomic, chloroplast, mitochondrial and EST sequences of pineapple (*Ananas comosus* (L.) Merrill). *Hereditas* 153(1):1–6.

Chen, J.-Y., Xie, F.-F., Cui, Y.-Z., et al. 2021a. A chromosome-scale genome sequence of pitaya (*Hylocereus undatus*) provides novel insights into the genome evolution and regulation of betalain biosynthesis. *Hortic. Res.* 8(1):164.

Chen, Y., Cheng, C., Feng, X., et al. 2021b. Integrated analysis of lncRNA and mRNA transcriptomes reveals the potential regulatory role of lncRNA in kiwifruit ripening and softening. *Sci. Rep.* 11(1):1–15.

Creste, S., Benatti, T. R., Orsi, M. R., Risterucci, A.-M., and Figueira, A. 2006. Isolation and characterization of microsatellite loci from a commercial cultivar of *Musa acuminata*. *Mol. Ecol. Notes* 6(2):303–6.

Cruz, A. F., Barka, G. D., and Blum, L. E. B. 2021. Cross-pathogenicity of *Colletotrichum* isolates on tropical fruit species in Brazil. *Arch. Phytopathol. Plant Prot.* 54(5–6): 295–307.

Curtolo, M., de Souza, P. I., Boava, L. P., et al. 2020. Wide-ranging transcriptomic analysis of *Poncirus trifoliata*, *Citrus sunki*, *Citrus sinensis* and contrasting hybrids reveals HLB tolerance mechanisms. *Sci. Rep.* 10(1):1–14.

D'hont, A., Denoeud, F., Aury, J. M., et al. 2012. The banana (*Musa acuminata*) genome and the evolution of monocotyledonous plants. *Nature* 488:213–17.

D'Urso, G., Piacente, S., Pizza, C., and Montoro, P. 2018. Metabolomics of healthy berry fruits. *Curr. Med. Chem.* 25:4888–902.

De Bellis, F., Malapa, R., Kagy, V., Lebegin, S., Billot, C., and Labouisse, J.-P. 2016. New development and validation of 50 SSR markers in breadfruit (*Artocarpus altilis*, Moraceae) by next-generation sequencing. *Appl. Plant Sci.* 4(8):1600021.

Denoeud, F., Carretero-Paulet, L., Dereeper, A., et al. 2014. The coffee genome provides insight into the convergent evolution of caffeine biosynthesis. *Science* 345(6201):1181–84.

Dereeper, A., Bocs, S., Rouard, M., et al. 2015. The coffee genome hub: a resource for coffee genomes. *Nucleic Acids Res.* 43(D1):D1028–35.

Ding, P. 2017. Tropical fruits. In *Encyclopedia of Applied Plant Sciences*, ed. B. Thomas, B. G. Murray, and D. J. Murphy, 431–34. Waltham: Elsevier – Academic Press.

Fan, H., Xiao, Y., Yang, Y., et al. 2013. RNA-Seq Analysis of *Cocos nucifera*: Transcriptome sequencing and *de novo* assembly for subsequent functional genomics approaches. *PLoS One* 8(3):e59997.

Fang, J., Miao, C., Chen, R., and Ming, R. 2016. Genome-wide comparative analysis of microsatellites in pineapple. *Trop. Plant Biol.* 9:117–35.

FAO STATS. 2019. The Food and Agriculture Organization of the United Nations (FAO). www.fao.org/faostat/ (accessed December 18, 2021).

Feng, C., Chen, M., Xu, C. J., et al. 2012. Transcriptomic analysis of Chinese bayberry (*Myrica rubra*) fruit development and ripening using RNA-Seq. *BMC Genom.* 13(1):1–15.

Feng, C., Feng, C., Lin, X., Liu, S., Li, Y., and Kang, M. A. 2021. Chromosome-level genome assembly provides insights into ascorbic acid accumulation and fruit softening in guava (*Psidium guajava*). *Plant Biotechnol. J.* 19:717–30.

Ferrao, L. F. V., Caixeta, E. T., Pena, G., et al. 2015. New EST-SSR markers of *Coffea arabica*: Transferability and application to studies of molecular characterization and genetic mapping. *Mol. Breed.* 35(1):31.

Fortes, A. M., Granell, A., Pezzotti, M., and Bouzayen, M. 2017. Editorial: Molecular and metabolic mechanisms associated with fleshy fruit quality. *Front. Plant Sci.* 8:1236.

Ge, Y., Zhang, T., Wu, B., et al. 2019. Genome-wide assessment of avocado germplasm determined from specific length amplified fragment sequencing and transcriptomes: Population structure, genetic diversity, identification, and application of race-specific markers. *Genes (Basel)* 10(3) (March):1–13.

Gepts, P. 2008. Tropical environments, biodiversity, and the origin of crops. In *Genomics of Tropical Crop Plants*, ed. P. H. Moore and R. Ming, 1–20. New York: Springer.

Haas, B. J. and Zody, M. C. 2010. Advancing RNA-seq analysis. *Nat. Biotechnol.* 28(5):421–23.

He, Z., Zhang, C., Liu, W., et al. 2017. DRDB: An online date palm genomic resource database. *Front. Plant Sci.* 8:1889.

Hippolyte, I., Bakry, F., Seguin, M., et al. 2010. A saturated SSR/DArT linkage map of Musa acuminata addressing genome rearrangements among bananas. *BMC Plant Biol.* 10(1):1–18.

Hippolyte, I., Jenny, C., Gardes, L., et al. 2012. Foundation characteristics of edible Musa triploids revealed from allelic distribution of SSR markers. *Ann. Bot.* 109(5):937–51.

Hrdlickova, R., Toloue, M., and Tian, B. 2017. RNA-Seq methods for transcriptome analysis. *Wiley Interdiscip. Rev. RNA* 8(1):e1364.

Hu, G., Feng, J., Xiang, X., et al. 2022. Two divergent haplotypes from a highly heterozygous lychee genome suggest independent domestication events for early and late-maturing cultivars. *Nat. Genet.* 54(1):73–83.

Huerta-Ocampo, J.Á., Mancilla-Margalli, N.A., Lino-López, G.J., et al. 2012. Ripening research in "maradol" papaya: a nutraceutical fruit. In *Hispanic Foods: Chemistry and Bioactive Compounds*, ed. M.H. Tunik, E. González de Mejía, 57–69. Washington: American Chemical Society.

Indiarto, R., Izzati, A. N., and Djali, M. 2020. Post-harvest handling technologies of tropical fruits: A review. *Int. J. Adv. Eng. Res. Sci.* 8(7):3951–57.

Jiang, N., Zhu, H., Liu, W., Fan, C., Jin, F., and Xiang, X. 2021. Metabolite differences of polyphenols in different litchi cultivars (*Litchi chinensis* Sonn.) based on extensive targeted metabonomics. *Molecules* 26(4):1181.

Jung, J., Choi, S. C., Jung, S., Cho, B. K., Ahn, G. H., and Ryu, S. B. 2017. A transcriptome approach toward understanding fruit softening in persimmon. *Front. Plant Sci.* 8:1556.

Kawahara, Y., Endo, T., Omura, M., et al. 2020. Mikan Genome Database (MiGD): Integrated database of genome annotation, genomic diversity, and CAPS marker information for mandarin molecular breeding. *Breed. Sci.* 70(2):200–11.

Khakimov, B., Mongi, R. J., Sørensen, K. M., Ndabikunze, B. K., Chove, B. E., and Engelsen, S. B. 2016. A comprehensive and comparative GC–MS metabolomics study of non-volatiles in Tanzanian grown mango, pineapple, jackfruit, baobab and tamarind fruits. *Food Chem.* 213:691–99.

Le Nguyen, K., Grondin, A., Courtois, B., and Gantet, P. 2019. Next-generation sequencing accelerates crop gene discovery. *Trends Plant Sci.* 24(3):263–74.

Leebens-Mack, J. H., Barker, M. S., Carpenter E. J., et al. 2019. One thousand plant transcriptomes and the phylogenomics of green plants. *Nature* 574:679–85.

Li, T., Wang, Y. H., Liu, J. X., Feng, K., Xu, Z. S., and Xiong, A. S. 2019a. Advances in genomic, transcriptomic, proteomic, and metabolomic approaches to study biotic stress in fruit crops. *Crit. Rev. Biotechnol.* 39(5):680–92.

Li, T., Yun, Z., Wu, Q., Qu, H., Duan, X., and Jiang, Y. 2019b. Combination of transcriptomic, proteomic and metabolomic analysis reveals the ripening mechanism of banana pulp. *Biomolecules* 9(10):523.

Ma, H., Yang, T., Li, Y., et al. 2021. The long noncoding RNA MdLNC499 bridges MdWRKY1 and MdERF109 function to regulate early-stage light-induced anthocyanin accumulation in apple fruit. *Plant Cell* 33(10):3309–30.

Maher, M., Ahmad, H., Nishawy, E., Li, Y., and Luo, J. 2021. Novel transcriptome study and detection of metabolic variations in UV-B-treated date palm (*Phoenix dactylifera* cv. Khalas). *Int. J. Mol. Sci.* 22(5):2564.

Mathiazhagan, M., Chidambara, B., Hunashikatti, L. R., and Ravishankar, K. V. 2021. Genomic approaches for improvement of tropical fruits: Fruit quality, shelf life and nutrient content. *Genes* 12(12):1881.

Matra, D. D., Kozaki, T., Ishii, K., Poerwanto, R., and Inoue, E. 2016. *De novo* transcriptome assembly of mangosteen (*Garcinia mangostana* L.) fruit. *Genom. Data* 10:35–37.

Maulidiani, M., Mediani, A., Abas, F., et al. 2018. 1H NMR and antioxidant profiles of polar and non-polar extracts of persimmon (*Diospyros kaki* L.) – Metabolomics study based on cultivars and origins. *Talanta* 184:277–86.

Ming, R., Hou, S., Feng, Y., et al. 2008. The draft genome of the transgenic tropical fruit tree papaya (*Carica papaya* Linnaeus). *Nature* 452:991–96.

Ming, R., VanBuren, R., Wai, C. M., et al. 2015. The pineapple genome and the evolution of CAM photosynthesis. *Nat. Genet.* 47:1435–42.

Mittal, A., Yadav, I. S., Arora, N. K., et al. 2020. RNA-sequencing based gene expression landscape of guava cv. Allahabad Safeda and comparative analysis to colored cultivars. *BMC Genom.* 21(1):1–19.

Moing, A., Aharoni, A., Biais, B., et al. 2011. Extensive metabolic cross-talk in melon fruit revealed by spatial and developmental combinatorial metabolomics. *New Phytol.* 190:683–96.

Moing, A., Allwood, J. W., Aharoni, A., et al. 2020. Comparative metabolomics and molecular phylogenetics of melon (*Cucumis melo*, Cucurbitaceae) biodiversity. *Metabolites* 10(3):121.

Moore, P. H. and Ming, R. 2008a. *Genomics of Tropical Crop Plants*. New York: Springer.

Moore, P. H. and Ming, R. 2008b. Papaya genome: A model for tropical fruit trees and beyond. *Trop. Plant Biol.* 1:179–80.

Motamayor, J.C., Mockaitis, K., Schmutz, J., et al. 2013. The genome sequence of the most widely cultivated cacao type and its use to identify candidate genes regulating pod color. *Genome Biol.* 14:r53.

Nagana Gowda, G. A. and Raftery, D. 2017. Recent advances in NMR-based metabolomics. *Anal. Chem.* 89:490–510.

Niu, Y.-f., Li, K.-X., and Liu, J. 2020. Complete chloroplast genome sequence and phylogenetic analysis of *Annona reticulata*. *Mitochondrial DNA B: Resour.* 5(3):3558–60.

OECD/FAO (2020), OCDE-FAO Perspectivas Agrícolas 2020-2029, OECD Publishing, Paris, https://doi.org/10.1787/a0848ac0-es.

Ong, W. D., Voo, L. Y. C., and Kumar, V. S. 2012. *De novo* assembly, characterization and functional annotation of pineapple fruit transcriptome through massively parallel sequencing. *PLoS One* 7(10):e46937.

Padmesh, P., Mukunthakumar, S., Vineesh, P. S., Skaria, R., Kumar, K. H., and Krishnan, P. N. 2012. Exploring wild genetic resources of *Musa acuminata* Colla distributed in the humid forests of southern Western Ghats of peninsular India using ISSR markers. *Plant Cell Rep.* 31(9):1591–601.

Palomino-Hermosillo, Y. A., Berumen-Varela, G., Ochoa-Jiménez, V. A., et al. 2022. Transcriptome analysis of soursop (*Annona muricata* L.) fruit under postharvest storage identifies genes families involved in ripening. *Plants* 11(14):1798.

Patti, G. J., Yanes, O., and Siuzdak, G. 2012. Metabolomics: The apogee of the omics trilogy. *Nat. Rev. Mol. Cell Biol.* 13:263–69.

Paull, R. E. and Duarte, O. 2011. *Tropical Fruits*. London: CAB International.

Pipatchartlearnwong, K., Juntawong, P., Wonnapinij, P., Apisitwanich, S., and Vuttipongchaikij, S. 2019. Towards sex identification of Asian Palmyra palm (*Borassus flabellifer* L.) by DNA fingerprinting, suppression subtractive hybridization and de novo transcriptome sequencing. *PeerJ* 7:e7268.

Preethi, P., Rahman, S., Naganeeswaran, S., et al. 2020. Development of EST-SSR markers for genetic diversity analysis in coconut (*Cocos nucifera* L.). *Mol. Biol. Rep.* 47(12):9385–97.

Rajapaksha, L., Gunathilake, D. M. C. C., and Pathirana, S. M. 2021. Reducing post-harvest losses in fruits and vegetables for ensuring food security–Case of Sri Lanka. *MOJ Food Process. Technols*. 9(1):7–16.

Rasmussen, S. K. 2020. Molecular genetics, genomics, and biotechnology in crop plant breeding. *Agronomy* 10(3):439.

Ravishankar, K. V., Raghavendra, K. P., Athani, V., et al. 2013. Development and characterisation of microsatellite markers for wild banana (*Musa balbisiana*). *J. Hortic. Sci. Biotechnol.* 88(5):605–9.

Rendón-Anaya, M., Ibarra-Laclette, E., Méndez-Bravo, A., et al. 2019. The avocado genome informs deep angiosperm phylogeny, highlights introgressive hybridization, and reveals pathogen-influenced gene space adaptation. *Proc. Natl. Acad. Sci U S A* 116:17081–89.

Rijzaani, H., Bayer, P. E., Rouard, M., Dolezel, J., Batley, J., and Edwards, D. 2022. The pangenome of banana highlights differences between genera and genomes. *Plant Genome* 15(1):e20100.

Rotchanapreeda, T., Wongniam, S., Swangpol, S. C., Chareonsap, P. P., Sukkaewmanee, N., and Somana, J. 2016. Development of SSR markers from *Musa balbisiana* for genetic diversity analysis among Thai bananas. *Plant Syst. Evol.* 302(7):739–61.

Rouard, M., Droc, G., Martin, G., et al. 2018. Three new genome assemblies support a rapid radiation in *Musa acuminata* (Wild Banana). *Genome Biol. Evol.* 10(12): 3129–40.

Sales, E. K. and Espino, R. R. C. 2004. RAPD markers for genetic analysis and classification of *Musa* B genome. *Philipp. Agric. Sci.* 87(2):165–73.

Sampangi-Ramaiah, M. H., Ravishankar, K. V., and Rekha, A. 2019. Long non-coding RNAs in banana: prediction, mapping and their comparative studies using *Musa balbisiana* and *Musa acuminata* transcriptome. *Trees* 33(2):359–69.

Sayago-Ayerdi, S., García-Martínez, D. L., Ramírez-Castillo, A. C., Ramírez-Concepción, H. R., and Viuda-Martos, M. 2021. Tropical fruits and their co-products as bioactive compounds and their health effects: A review. *Foods* 10(8):1952.

Schnell, R. J. and Priyadarshan, P. M. 2012. *Genomics of Tree Crops*. New York: Springer.

Scossa, F., Alseekh, S., and Fernie, A. R. 2021. Integrating multi-omics data for crop improvement. *J. Plant Physiol.* 257:153352.

Shalini, S., Singla, A., Goyal, M., et al. 2018. Omics in agriculture: Applications, challenges and future perspectives. In *Crop Improvement for Sustainability*, ed. P. Kumar, S. Kumar, S. Kumar, R. C. Yadav, 343–60. New Delhi: Daya Publishing House.

Singh, R. K., Prasad, A., Muthamilarasan, M., Parida, S. K., and Prasad, M. 2020. Breeding and biotechnological interventions for trait improvement: status and prospects. *Planta* 252(4):1–18.

Strijk, J. S., Hinsinger, D. D., Roeder, M. M., et al. 2021. Chromosome-level reference genome of the soursop (*Annona muricata*): A new resource for Magnoliid research and tropical pomology. *Mol. Ecol. Resour.* 21:1608–19.

Sun, J., Zhang, J., Fang, H., et al. 2019a. Comparative transcriptome analysis reveals resistance-related genes and pathways in *Musa acuminata* banana 'Guijiao 9' in response to Fusarium wilt. *Plant Physiol. Biochem.* 141:83–94.

Sun, X., Lin, L., and Sui., N. 2019b. Regulation mechanism of microRNA in plant response to abiotic stress and breeding. *Mol. Biol. Rep.* 46(1):1447–57.

Tafolla-Arellano, J. C., Zheng, Y., Sun, H., et al. 2017. Transcriptome analysis of mango (*Mangifera indica* L.) fruit epidermal peel to identify putative cuticle-associated genes. *Sci. Rep.* 7(1):1–13.

Teh, B. T., Lim, K., Yong, C. H., et al. 2017. The draft genome of tropical fruit durian (*Durio zibethinus*). *Nat. Genet.* 49(11):1633–41.

Thottathil, G. P., Jayasekaran, K., and Othman, A. S. 2016. Sequencing crop genomes: A gateway to improve tropical agriculture. *Trop. Life Sci. Res.* 27:93–114.

Tian, S., Jiang, L., Gao, Q., et al. 2017. Efficient CRISPR/Cas9-based gene knockout in watermelon. *Plant Cell Rep.* 36(3):399–406.

Tran, H. T. M., Ramaraj, T., Furtado, A., Lee, L. S., and Henry, R. J. 2018. Use of a draft genome of coffee (*Coffea arabica*) to identify SNPs associated with caffeine content. *Plant Biotechnol. J.* 16(10):1756–66.

Vázquez-Manjarrez, N., Ulaszewska, M., Garcia-Aloy, M., et al. 2020. Biomarkers of intake for tropical fruits. *Genes Nutr.* 15(1):1–21.

Vergara-Pulgar, C., Rothkegel, K., González-Agüero, M., et al. 2019. De novo assembly of *Persea americana* cv.'Hass' transcriptome during fruit development. *BMC Genom.* 20(1):1–14.

Wahyudi, D., Adnin, N. I., and Hapsari, L. 2020. Intraspecific variation of thirteen pisang ambon cultivars (*Musa Acuminata* CV. AAA) from east Java and central Java (Indonesia) based on random amplified polymorphic DNA (RAPD) marker. *Int. J. Conserv. Sci.* 11(4):1059–70.

Wang, J., Chen, D., Lei, Y., et al. 2014. *Citrus sinensis* annotation project (CAP): a comprehensive database for sweet orange genome. *PloS one*, *9*(1), e87723.

Wang, J., Feng, Y., Ding, X., Huo, J., and Nie, W. F. 2021. Identification of long non-coding RNAs associated with tomato fruit expansion and ripening by strand-specific paired-end RNA sequencing. *Horticulturae* 7(12):522.

Wang, L., Chen, S., Peng, A., Xie, Z., He, Y., and Zou, X. 2019a. CRISPR/Cas9-mediated editing of CsWRKY22 reduces susceptibility to *Xanthomonas citri* subsp. citri in Wanjincheng orange (*Citrus sinensis* (L.) Osbeck). *Plant Biotechnol. Rep.* 13(5):501–10.

Wang, L., Hu, W., Tie, W., et al. 2017. The MAPKKK and MAPKK gene families in banana: Identification, phylogeny and expression during development, ripening and abiotic stress. *Sci. Rep.* 7(1):1–15.

Wang, L., Yang, M., Dong, Y., et al. 2022. Melatonin confers enhanced polyamine metabolism and cell tolerance in *Vitis vinifera* against oxidative damage: Quantitative proteomic evidence. *Postharvest Biol Technol.* 184, 111756.

Wang, M., Zhao, W., Gao, L., and Zhao, L. 2018a. Genome-wide profiling of long non-coding RNAs from tomato and a comparison with mRNAs associated with the regulation of fruit ripening. *BMC Plant Biol.* 18(1):1–10.

Wang, P., Luo, Y., Huang, J., et al. 2020a. The genome evolution and domestication of tropical fruit mango. *Genome Biol.* 21(60) (March):1–17.

Wang, W. Q., Wang, J., Wu, Y. Y., Li, D. W., Allan, A. C., and Yin, X. R. 2020b. Genome-wide analysis of coding and non-coding RNA reveals a conserved miR164-NAC regulatory pathway for fruit ripening. *New Phytol.* 225(4):1618–34.

Wang, Y., Gao, L., Zhu, B., et al. 2018b. Integrative analysis of long non-coding RNA acting as ceRNAs involved in chilling injury in tomato fruit. *Gene* 667:25–33.

Wang, Z., Miao, H., Liu, J., et al. 2019b. *Musa balbisiana* genome reveals subgenome evolution and functional divergence. *Nat. Plants* 5:810–21.

Wu, S., Wu, D., Song, J., et al. 2022. Metabolomic and transcriptomic analyses reveal new insights into the role of abscisic acid in modulating mango fruit ripening. *Hortic. Res.* 9(May):uhac102.

Xia, Z., Huang, D., Zhang, S., et al. 2021. Chromosome-scale genome assembly provides insights into the evolution and flavor synthesis of passion fruit (*Passiflora edulis* Sims). *Hortic. Res.* 8(14):1–14.

Xiao, Y., Kang, B., Li, M., et al. 2020. Transcription of lncRNA ACoS-AS1 is essential to trans-splicing between SlPsy1 and ACoS-AS1 that causes yellow fruit in tomato. *RNA Biol.* 17(4):596–607.

Xiao, Y., Xu, P., Fan, H., et al. 2017. The genome draft of coconut (*Cocos nucifera*). *Gigascience* 6(11):1–11.

Xie, F., Hua, Q., Chen, C., et al. 2021. Genome-wide characterization of R2R3-MYB transcription factors in pitaya reveals a R2R3-MYB repressor *HuMYB1* involved in fruit ripening through regulation of betalain biosynthesis by repressing betalain biosynthesis-related genes. Cells 10(8):1949.

Xu, Z., Zhang, N., Fu, H., et al. 2021. Salt stress modulates the landscape of transcriptome and alternative splicing in date palm (*Phoenix dactylifera* L.). *Front. Plant Sci.* 12:807739.

Xue, L., Sun, M., Wu, Z., et al. 2020. LncRNA regulates tomato fruit cracking by coordinating gene expression via a hormone-redox-cell wall network. *BMC Plant Biol.* 20(1):1–15.

Yamamoto, T. 2016. Breeding, genetics, and genomics of fruit trees. *Breed. Sci.* 66(1):1–2.

Yanes, Ó. 2015. Metabolómica: la ciencia ómica más multidisciplinaria. *Soc. Española Bioquím. y Biol. Mol.* 186:7–10.

Yin, Y., Zhang, X., Fang, Y., et al. 2012. High-throughput sequencing-based gene profiling on multi-staged fruit development of date palm (*Phoenix dactylifera*, L.). *Plant Mol. Biol.* 78(6):617–26.

Yu, K., Xu, Q., Da, X., Guo, F., Ding, Y., and X. Deng. 2012. Transcriptome changes during fruit development and ripening of sweet orange (*Citrus sinensis*). *BMC Genom.* 13(1):1–13.

Zhang, W., Lin, J., Li, J., et al. 2021. Rambutan genome revealed gene networks for spine formation and aril development. *Plant J.* 108(4):1037–52.

Zhu, B., Yang, Y., Li, R., et al. 2015. RNA sequencing and functional analysis implicate the regulatory role of long non-coding RNAs in tomato fruit ripening. *J. Exp. Bot.* 66(15):4483–95.

4 Prospects of Association Mapping for Important Traits in Fruit Crops

Pooja, Madhurjit Singh Rathore, Paramjeet Sajwan, Lokesh Thakur, Jebi Sudan, Sajad Majeed Zargar, Deepak Sharma, and Vishal S Rana

4.1 INTRODUCTION

Plant traits crucial for fitness and agricultural value result from genetic, environmental, and gene-environment interactions. Genome-assisted breeding targets traits like disease resistance and nutrient efficiency, requiring genetic mapping of complex traits. Advances in sequencing, especially next-generation technologies, facilitate marker discovery in non-model crops. Genetic maps elucidate gene distances on chromosomes, with linkage mapping and association mapping (AM) identifying quantitative trait loci (QTLs). Linkage mapping utilizes known pedigrees, while AM leverages wild populations for trait analysis, collectively termed "association mapping" (Alvarez et al., 2015). Association mapping, in contrast, looks at how functional polymorphisms and nearby molecular markers are passed down in a group of genotypes whose ancestry is unknown. Since the unknown ancestry can go back thousands of years, the linked inheritance will only last for polymorphisms that are very closely linked. So, AM uses recombination that happened in the past, while linkage mapping only uses recombination that has happened since the mapping population was made. The goal of AM is to pinpoint genetic regions (loci, alleles) that contribute to observed phenotypic variation in a characteristic, in line with the overarching purpose of genetics to establish a correlation between genotype and phenotype (Botstein and Risch, 2003). This aids in identifying DNA sequence variations that underlie traits and/or selecting genotypes that exhibit the desired phenotype. A developing field in genetics is AM, often known as linkage disequilibrium (LD) mapping – a genetic technique with great potential for elucidating complicated traits. Association mapping could improve mapping resolution exploiting instances of recombination that occurred naturally in past populations. This has the potential to carry out gene-level mapping in species that are not used as models, which is currently impossible using linkage-based methods (Nordborg and Tavare, 2002). GWAS and LD-based AM have been used in a lot of studies on humans, livestock, and crop plants over the last 20 years. Most of the time, the

DOI: 10.1201/9781003343707-4

specific polymorphisms responsible for varying traits remain unknown so they can not be seen directly but can be found through statistical inference. Linkage disequilibrium (LD), which means that causative trait polymorphisms are not randomly linked to observed polymorphisms, is a necessary condition for finding these kinds of relationships. Genomic analysis, statistical genetics, molecular biology, and bioinformatics are all components of association genetics (Mariette et al., 2016). Together, these parts make up the basis for choosing, evaluating, and putting together genomic regions that are linked to trait variation. In addition to the aforementioned fields of study, expertise in population genetics and an in-depth understanding of phenotypic variability within the target species is essential. LD can happen for many different reasons, but the chief objective is to pinpoint the variations most closely associated with the trait-causing variants (Morris et al., 2013). Association genetics, then, also includes the methods used to figure out if these links are caused by things other than close physical proximity. Structured populations, such as homozygous inbred strains in plants with short generation times, are commonly used for QTL mapping, facilitating higher LD per base pair and enabling co-segregation of markers with QTL (Osborne et al., 2014). Conversely, unstructured populations exhibit lower LD due to numerous recombination events across many generations, leading to less consistent co-segregation with QTL (Font i Forcada et al., 2019). In QTL mapping, structured populations align with Mendelian theory, unlike unstructured populations lacking pedigrees and predetermined causative polymorphisms. Screening polymorphisms for mapping can arise from various sources, including whole-genome sequencing, phenotypically neutral sequence selection, or candidate gene selection. This chapter elucidates the LD-AM relationship, outlines AM procedures, and addresses potential methodological challenges.

4.2 NEED FOR AM IN FRUIT CROPS

Several complex variables of agricultural or evolutionary significance are controlled by many QTLs, their interaction, environment, and interaction between QTLs and environment, all of which contribute to the phenotypic variance of these characters. It is common practice to utilize linkage and AM to analyze complicated characters. Because of the small number of recombination events that occur during the generation of mapping populations and the cost of propagating and analyzing a large number of lines, linkage analysis in plants typically localizes QTLs to 10 to 20 cM intervals (Doerge, 2002; Holland, 2007). In spite of the hundreds of studies performed on different plant species over the past two decades (Holland, 2007; Kearsey and Farquhar, 1998), only a small fraction of the reported QTLs has been cloned or tagged at the gene level (Price, 2006).

4.2.1 Identification of Genetic Markers Responsible for Fruit Quality Traits

Fruit quality is an essential aspect of fruit crop breeding, and consumers expect fruits to have desirable characteristics, such as flavor, texture, and nutrient content.

Association mapping can be utilized to recognize genetic markers associated with fruit quality traits, enabling breeders to create new fruit varieties with enhanced fruit quality. For example, grapes and peach have been utilized for AM to recognize genetic markers for fruit quality (Cao et al., 2012; Font i Forcada et al., 2019).

4.2.2 Identification of Genetic Markers Associated with Disease Resistance

Fruit crops are susceptible to various diseases, which can significantly impact fruit productivity and quality, leading to economic losses for growers. Association mapping can be employed to identify genetic markers associated with disease resistance, which can be utilized in breeding programs to develop new fruit varieties with improved resistance to diseases. In strawberry, it is used to find genetic markers linked to resistance against fusarium wilt (Pincot et al., 2018). The AM has been used in apple to detect genetic markers linked for resistance against apple scab (McClure et al., 2018). Similarly, in peach, AM has been used to recognize genetic markers linked with resistance to peach brown rot (Fu et al., 2021).

4.2.3 Identification of Genetic Markers Linked with Climate Change Adaptation

Large effects from climate change are expected to impact fruit crop production, as altering temperatures, rainfall patterns, and extreme weather events can affect fruit yield, quality, and storage. Association mapping can be employed to identify genetic markers linked with traits that help fruit crops adapt to changing environmental conditions, such as drought tolerance, heat tolerance, and cold tolerance. In apple, it is used to locate genes associated with cold hardiness (Kumar et al., 2012; Chagne et al., 2016).

4.2.4 Efficient Breeding

Association mapping can speed up the breeding process by enabling breeders to identify desirable traits and genetic markers linked with those traits more efficiently and effectively than traditional breeding methods. As a result, breeders can develop new fruit varieties that meet consumer and grower needs in a shorter period (Portis et al., 2015).

4.3 APPROACHES USED FOR ASSOCIATION MAPPING (AM)

Polymorphisms in certain candidate genes that have been implicated in regulating phenotypic variance for specific traits can be related to one another and can be shown by a candidate-gene AM and genome-wide AM, also known by the term "genome scan," which looks at all of the genetic differences in the whole genome to find links between complex traits.

4.3.1 Genome-Wide Association Study (GWAS)

To detect all possible genomic areas involved in regulating the trait of interest, GWAS is frequently used. Several GWAS have been performed in higher plants. Yet, GWAS often returns statistically significant results at regions unrelated to the trait. LD between numerous factors, of which only some are physically linked to the trait, causes these indirect relationships. The problem can be exacerbated by epistasis and the population structure. The issue cannot be resolved by adding more markers or expanding the sample size. But a new multilocus mixed model (Segura et al., 2012) and a multitrait mixed model can help solve part of the problem (Korte et al., 2012). Markers for genotyping in GWAS are spread out across the entire genome, ideally in a uniform and dense pattern. This method allows for a comprehensive assessment of all the loci contributing to the regulation of entire traits displaying differences in the sample. For the selection of suitable markers, a genome-wide linkage map of markers from relevant species is required. Furthermore, much time and energy will be needed to ensure accurate phenotyping of the variable features. Last but not least, comparing hundreds of marker loci independently is necessary when using a high number of markers (Chagne et al., 2016). For this reason, to detect QTLs with moderate effect size, a large sample size (at least a thousand individuals) is required. Genome-wide QTL scanning with RILs and other F1-derived mapping populations is preferable to AM because it only requires the examination of a small number of markers and provides more statistical power to analyze the influence of a genomic area. Initially, the genome scanning strategy will likely only be applied to the more valued crops, such as bananas, grapes, apples, citrus fruits, and stone-fruits, due to the high cost of creating the sequence information on which the subsequent large-scale marker analysis must be based. Apples, for instance, would benefit from using the USDA Malus core assembly as their germplasm set of choice for AM research (Gross et al., 2012). This Malus core collection has been disseminated to numerous sites around the United States, assessed for multiple attributes, and continues to be evaluated. It contains a wide range of apple germplasm in addition to a huge number of accessions from the cultivated apple, *Malus x domestica.*

4.3.2 Candidate Gene (CG) Approach

Confirming the genes that govern the characteristic of interest is desirable when some data on the target trait genetics is available and CGs can be predicted using that data. After a GWAS, this can be used to find polymorphisms (Single nucleotide polymorphism (SNPs or haplotypes) linked to the characteristic of interest. Based on what is already known, a CG is one that is suspected of playing a role in the regulation of a characteristic of interest. This methodology provides a compromise between the marker-assisted selection and genome scan methods by allowing a much more limited application of markers in CG regions to give virtually equivalent statistical power of detection to the "large-scale" application of whole genome scanning (Singh and Singh, 2015). Comparative genomics, genome sequence annotation, transcript profiling, quantitative trait loci analysis, etc., are only some of the methods utilized to find potential genes involved. The accessibility of extensive

Express Sequence Tags (EST) sequence resources, such as those found in apples, grapes, and citrus fruits, considerably accelerates CG studies (Khan and Korban 2012). Candidate genes are frequently revealed through loss of function mutations in lab strains. However, it is unclear how these mutations connect to natural variation in the characteristics of wild populations (Gupta, 2014). Despite these obstacles, this approach has been utilized to find genes involved in the regulation of various features (i.e., morphological, phenotypic, and stress resistance qualities) (Ingvarsson and Street, 2011). Where genome-wide AM fails to reveal substantial marker-trait correlation after applying false discovery rate (FDR) modification, this method may be able to locate a QTL.

4.3.3 Populations Used in AM

Association mapping success hinges on the selection of suitable populations, sourced either from naturally occurring offspring or established families. While both biparental and multiparent populations can be utilized, the latter, notably doubled haploid, F3 families resulting from biparental crosses, are more commonly employed. Multiparent populations like nested association mapping and multiparent advanced generation inter crosses are gaining traction for their efficacy in AM, linkage mapping, and variety development. Plant AM relying heavily on carefully cultivated populations to examine genotype-phenotype-environment interactions. These populations typically comprise founder and recombinant populations, allowing for the exploration of diverse genetic and phenotypic traits through controlled crosses. Various plant populations, including natural, structured, and biparental mapping populations, serve as viable options for AM endeavors.

4.3.3.1 Natural Populations

Association mapping in natural populations is the research on the genetics of complicated features in natural populations. The populations reflect the range of variation within a given species since they were developed using only naturally occurring or indigenous material. The AM of fruit quality attributes, for instance, has been performed using apple samples collected from wild populations (Bus et al., 2011).

4.3.3.2 Structured Populations

To account for population stratification, AM in structured populations uses populations with established population structures. These populations are produced by mating individuals from different genetic backgrounds, and their offspring are subsequently genotyped and phenotyped to determine the relationships between genetic markers and observable characteristics. Association mapping of variables like berry size, sugar content, and acidity, for instance, have all been accomplished with the help of structured populations in grape breeding (Huang et al., 2012).

4.3.3.3 Bi-Parental and Breeding Populations

Using populations descended from crossing between two parents, bi-parental mapping populations conduct AM to pinpoint genetic markers associated with attributes.

Most QTL-IM/AM research can make use of either a biparental or multiparental mapping population, both of which are based on families. An association panel's varied genotypes can only be utilized for AM, not for identifying QTL-IM (Segura et al., 2012).

4.3.3.4 Multiparental Populations

It was first suggested in the 1990s that multiparental populations may be used for mapping purposes (Rebai & Goffinet, 1993). By this method, we were able to increase the size of a population from two parents to four by crossing them in a half-diallel. There was also a rise in the availability of statistical tools for studying families with more than two parents (Jannink & Wu, 2003; Rebai & Goffinet, 2000; Xu, 1998).

4.3.3.5 Cultivated Populations

These populations are developed from cultivated varieties and represent the genetic heterogeneity among the farmed germplasm. Association mapping of fruit attributes including acidity, sugar content, and flesh color has been performed, for instance, in peach using cultivated populations (Cao et al., 2019).

4.3.3.6 Association Panels

These populations are made up of a wide range of cultivars or accessions expected to reflect the genetic diversity of the species. For example, in strawberries, association panels have been utilized for AM of features such as fruit firmness, soluble solids content, and titratable acidity.

4.4 GENERAL STEPS IN AM ANALYSIS

The general steps in AM analysis include: 1) selection of population, 2) phenotyping, 3) genotyping for population structure analysis, 4) population structure and kinship analysis, 5) genotyping for LD analysis/decay, 6) linkage disequilibrium (LD) analyses, 7) identification of CG/s, 8) phenotyping, and 9) statistical analysis (see Figure 4.1).

4.4.1 Selection of Population

The careful selection of populations is crucial in AM processes. Rather than being rigidly categorized, a diverse population reflecting the genetic variation within the target species should be chosen (Alvarez et al., 2015). It is essential to consider variables such as germplasm variety, population structure, LD, and kinship, as they impact the effectiveness of statistical methods for detecting marker-phenotype associations (Breseghello and Sorrells, 2006; Stich and Melchinger, 2010). Researchers typically select large, randomly sampled populations from natural populations, germplasm collections, breeding lines, or populations resulting from multiparent crosses (Yu and Buckler, 2006). Population classification, based on population structure and family relatedness, influences the statistical approaches employed

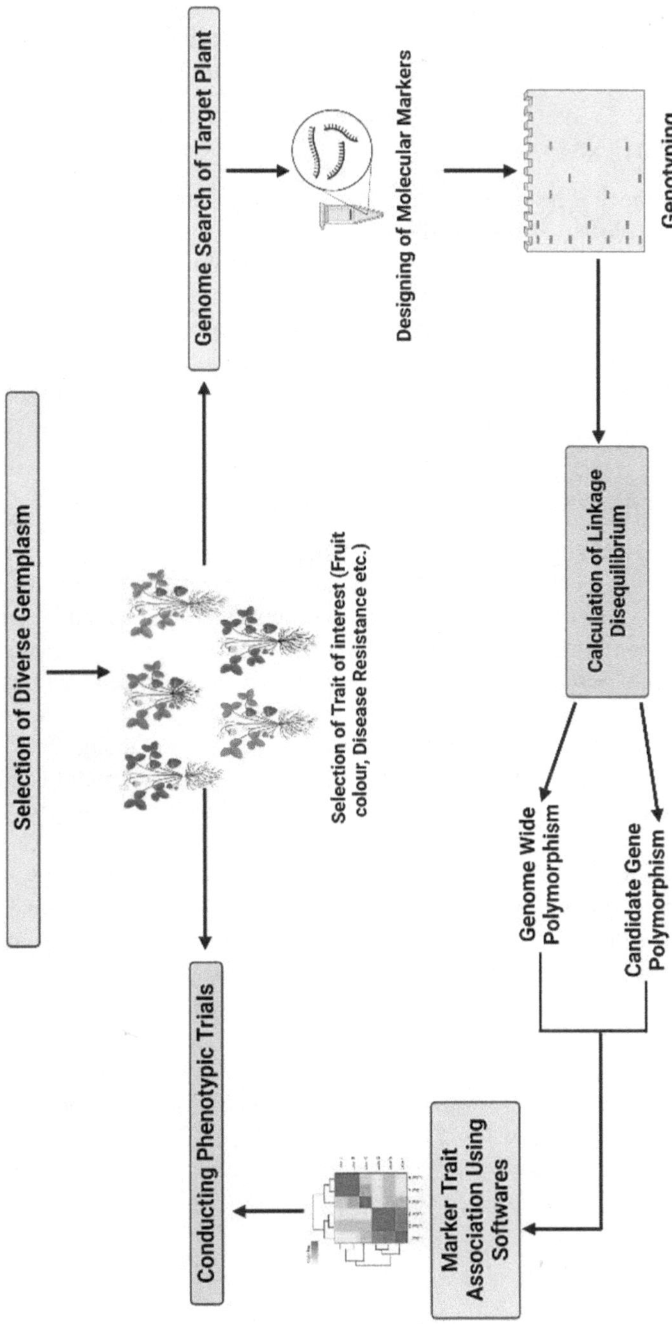

FIGURE 4.1 Steps involved in association mapping studies.

in association studies. Additionally, the origin of materials, including germplasm banks, naturally occurring populations, breeding lines, and artificial populations, aid in population classification for AM (Breseghello and Sorrells, 2006). The choice of population data sets depends on factors such as LD levels, genetic and phenotypic diversity, the significance of population structure, and family relatedness.

4.4.2 Phenotyping

Phenotyping is the process of analyzing the sample for several different characteristics of interest. The selected sample is allowed to mature in the field, where its morphological characteristics can be assessed. To reduce the potential impact of environmental factors on phenotyping, it is recommended that studies be replicated over many locations and time periods. An appropriate experimental design, such as a randomized block design, an augmented design, a nested design, etc., must be used to carry out the trials. For any mapping project to succeed, accurate and trustworthy phenotyping must be performed (Pieruschka and Schurr, 2019).

4.4.3 Genotyping for Population Structure Analysis

Genome-wide association studies in AM typically employ two main methodologies: GWAS itself and CG analysis. GWAS investigates broad genomic associations across populations, considering genotypes dispersed across genetic marker loci spanning all chromosomes (Rafalski, 2010). Conversely, CG analysis selects markers based on genomic position and prior QTL studies, focusing on genes relevant to the trait variance (Zhu et al., 2008). Neutral background markers utilized in association studies aid in elucidating genetic structures and grouping individuals (Pritchard and Rosenberg, 1999), overcoming limitations posed by population structure and kinship (Yu and Buckler, 2006), while also revealing inbreeding and kinship (Lynch and Ritland, 1999). DNA markers facilitate the identification of genetic loci linked to traits and can detect mutations like translocations and insertions (Hayward et al., 2015). High-throughput genotyping with active markers at 1 cM distance aids in marker-assisted selection (MAS) (Mohan et al., 1997). Despite limitations with AFLP and RAPD markers due to poor genome dispersion and polymorphism (Vos et al., 1995; Williams et al., 1990), SNPs and simple sequence repeats SSRs offer enhanced capabilities in establishing kinship matrices and population structures (Zhu et al., 2008). SNPs, with their modest mutation rates per generation compared to SSRs (Li et al., 2002), are preferred for evaluating population structures in crops due to their lower anticipated heterozygosity (Singh et al., 2001). SNPs prove valuable even in polyploid species and extensive sequencing efforts, aiding genome-wide association analyses (Syvanen, 2005; Collard and Mackill, 2007).

4.4.4 Population Structure and Kinship Analysis

The term "population structure" refers to the fact that members of a given population are not unified into a single group, but rather are fragmented into two or more groups with varying genetic compositions. So, it may be more likely to sample individuals with a certain characteristic trait more common in one or two of these categories than in the others, which could lead to inaccurate LD estimations. It is common for a phenotypic trait to display substantial association with non-expressive loci of the trait when there is a correlation between the phenotype and the population subset. When it comes to AM, one of the biggest limitations is the problem of non-random mating within a species' population structure because it causes shifts in allele frequencies (Ersoz et al., 2007). Justifications for familial relatedness and population structure are offered using methods like principal component analysis, mixed model approach, and structured association and genomic control, among others (Price et al., 2006; Yu and Buckler, 2006). In recent years, structured association has become the go-to method for addressing problems arising from the composition of study populations in the vast majority of association research. This is done by calculating and assigning individuals in population substructures using random, unlinked markers (Pritchard and Donnelly, 2001). With the use of the STRUCTURE programme (Pritchard et al., 1999), which employs a Bayesian approach to determine what percentage of a person's genome originated in each of several hypothesized populations, population structure can be determined. Based on their genomic classification, individuals are sorted into several groups. Principal component analysis, as noted by Price et al. (2006), is used to evaluate population structure since it is more efficient than STRUCTURE (Zhao et al., 2007).

4.4.5 Genotyping for LD Analysis/Decay

The resolution of AM investigations is assessed by examining the frequency with which alleles from different loci are associated with one another in a non-random fashion, a phenomenon known as LD (Oraguzie et al., 2007). The term "linkage disequilibrium" (LD) is used to describe the pattern of non-random connection of alleles across diverse genetic loci. The original definition of the phrase referred to a set of alleles that share a chromosome. LD occurs on a population-wide scale, although it is nonetheless commonly observed that LD is greatest between alleles that are physically adjacent to one another. That is why linkage might cause so-called disequilibrium by potentially decreasing the random relationship between alleles. It's It is worth noting that although displaying LD decays at a close distance necessitates a high number of markers, the projected resolution is extremely high. To discover LD between markers and the loci of importance, an extensive array of molecular markers is used to genotype the sample and achieve as high a level of genome coverage as possible. In addition, when LD spreads over great distances, mapping resolution suffers, even though only a few markers are needed. The level of LD identified across different populations of a species, as well as the distribution of LD in the relevant genomic sections of that species, would determine the number of markers needed to fully cover the whole genome. High LD is typically revealed at lower resolutions, and vice versa. The causes of LD are a wide range of characteristics, population size and structure, levels of inbreeding and genetic isolation between lineages, mutation and recombination rates, and the frequency with which new alleles are created (Mackay and Powell, 2007; Gupta et al., 2005).

4.4.6 LD Analysis

To determine which genes or QTLs are responsible for the characters of interest, a model-based examination of phenotype and genotype data is conducted. Assessments of population dynamics and kinship can be used as co-variates in the model to decrease false links among markers and QTLs of interest. These studies need substantial computational resources; thus, they are typically carried out with the use of specialized computer software. Differences between observed and expected frequencies of gamete haplotypes are quantified using LD (Soto-Cerda and Cloutier, 2012). The genetic distance between two polymorphic sites inside a locus or gene on a chromosome is plotted against their Pearson's squared value (product-moment) correlation coefficient to illustrate LD (Bradbury et al., 2007). Due to its reduced subjectivity, r2 is the method of choice for determining LD using pairwise measurements between markers (Soto-Cerda and Cloutier, 2012). For the simplest estimation of the number of markers needed for GWAS, knowledge of the LD decay rate over distances is essential. Before attempting to measure LD, it is significant to determine the total quantity of necessary markers for genome coverage for GWAS. In contrast to linkage-based mapping, LD can provide higher resolution when locating regions of the genome associated with a candidate's phenotype or disease (Mackay and Powell, 2007). Linkage disequilibrium is commonly used to measure genetic diversity and can be simply investigated to deduce the evolutionary background of the populations (Zhu et al., 2015; Slatkin, 2008). Factors that affect LD include natural selection, gene conversion, and fluctuating recombination rates, which are just a few of the many factors that shape a population's genetic makeup and mutation frequency (Ardlie et al., 2002). The degree of separation between two markers, as reported by Myles et al. (2009), is a function of the strength of their association. The resolution with which QTL may be mapped is proportional to the rate at which LD decays with increasing distance. Therefore, structural analysis of LD could serve as the first step in the design of AM research.

4.4.7 Identification of CGs

Candidate genes are those that have been shown to affect the development of characters with established biological activities, either directly or indirectly, and whose validity may be evaluated by analyzing the effects of causal gene differences in association analysis (Zhu and Zhao, 2007). In addition to a genome scan, location cloning of QTL governing significant variations in genes underlying desirable qualities is accomplished by CG analysis. To clarify, the causal genes are the QTL that point to a specific area of the genome as being important in influencing the character variants of interest. In AM, SNPs with causal polymorphism are typically given as a potential gene variation for genotyping because they have the greatest power for mapping QTL and LD (Ma et al., 2020). Therefore, resequencing of amplicons from many individuals with varying genetic backgrounds within a larger association population and particular genes is crucial to the process of identifying CG SNPs (Zhu et al., 2008). Promoter SNPs, exons, introns, and untranslated 5'/3' regions are all viable options for locating a potential gene, along with code sections having lower nucleotide diversity than the non-coding part (Zhu et al., 2008). The number of SNPs per unit of length needed to identify meaningful relationships is determined by a putative gene locus and is proportional to rate of LD degradation (Flint-Garcia et al., 2003; 2005). Many barriers to accurate CG detection have been found, and the digital candidate gene approach (DigiCGA) was developed to overcome them (Zhu and Zhao, 2007).

4.4.8 Phenotyping

For AM, a lot of different kinds of accessions are needed, which makes it hard to get the same kind of phenotypic data from different places and times. But for an uneven field, QTL x environment interactions, incomplete block design, and appropriate statistical approaches will boost mapping power (Eskridge, 2003). Trait phenotyping for G x interactions may be made more difficult by influences such as variations within and between years, habitats, and seasons (Atlin et al., 2011). In comparison to genotyping, phenotyping has received much less attention in AM (Zhu et al., 2008). It is just as challenging to obtain phenotypic data with enough replications over a range of sites and periods due to the huge and diverse accessions that are common in AM. Therefore, latex design (incomplete block design) should be used to efficiently arrange the experimental area because of its ability to boost the mapping capability (Piepho et al., 2006).

4.4.9 Statistical Analysis

The analysis of variance (ANOVA) is the quickest and most elementary statistical technique for comparing associations between quantitative variables (Yu and Buckler, 2006). However, the limitations of AM research, particularly those resulting from the quantitative transmission disequilibrium test for populations, were adapted so that they could be applied to inbred plant populations (Stich et al., 2010).

4.5 SOFTWARE PACKAGES

Evident from the above, it is the computational burden that would accompany any LD investigation. To achieve this goal, new data mining tools and other online assets are constantly being created. Various computational and statistical methods have been used in the study of AM. TASSEL, SAS, the R programme, STRUCTURE, SPAGeDi (Spatial Pattern Analysis of Genetic Diversity), EINGENSTRAT, MTD-REML (Multiple Trait Derivative-Free Restricted Maximum Likelihood), and ASR-REML (Residual Maximum Likelihood) are just a few of the available options (Table 4.1). Software such as STRAT, Bimbam, and GEN STAT 11 have also recently been integrated (Alvarez et al., 2015). Expression quantitative trait loci (eQTL) and genome- GWAS data have been used to test for pleiotropic interaction between gene expressive degree and intricate characteristics with the help of summary-data-based Mendelian randomization (SMR) and heterogeneity in dependent tools (HEIDI) tools (Zhu et al., 2015).

TABLE 4.1
List of softwares available for analysis of marker-trait analysis

S. No.	Software	Year of Release	Website
1.	TASSEL	2007	http://www.maizegenetics.net/tassel
2.	GAPIT	2011	http://www.maizegenetics.net/gapit
3.	EMMAX	2012	http://genetics.cs.ucla.edu/emmax/index.html
4.	GEMMA	2016	http://www.xzlab.org/software.html
5.	ANGSD	2015	http://www.popgen.dk/angsd/index.php/ANGSD
6.	PLINK	2014	http://pngu.mgh.harvard .edu/purcell/plink/
7.	Lrgpr	2013	http://lrgpr.r-forge.r-project.org/

4.6 MAJOR CONCERNS WITH AM IN PLANTS

In breeding and genetics, AM is a common method for locating genes and markers linked to useful agronomic characteristics. The reliability and consistency of the outcomes, however, are called into question by a number of issues linked to this methodology. Major issues with plant AM include the following.

4.6.1 Population Structure

Association mapping is vulnerable to changes in population structure, such as inbreeding, outcrossing, or hybridization, which can lead to either over- or under-representation of true associations due to cryptic relatedness (Ma et al., 2021). Confounding effects, which can either amplify or dampen association signals, are possible in structured populations. Because of this, it is crucial to consider population structure in AM research to rule out erroneous correlations (Yu and Buckler, 2006).

4.6.2 Linkage Disequilibrium

Linkage disequilibrium refers to the non-random connection of alleles at different loci and is the factor that can influence the efficacy and precision of AM. Linkage disequilibrium declines at varying rates depending on the species and its population history. It can be challenging to identify the causal variations when LD covers large geographic distances. Therefore, accurate AM findings require LD to be carefully considered (Huang and Han, 2014).

4.6.3 Phenotypic Variation

Association mapping only works if the phenotypic data is of high quality and trustworthy. Studies using AM are vulnerable to misidentification, which can lead to false positives because they cannot reliably distinct genetic and environmental influences on phenotypes. For instance, in fruit crops, the reliability of association studies can be compromised by inaccurate phenotypic measurement as a result of less-than-ideal environmental circumstances or improper field management (Waldchen et al., 2012). Variation in phenotype may result from exposure to different conditions, different stages of development, or even just plain bad data collection. Therefore, it is crucial to meticulously plan and carry out the phenotyping tests so as to reduce the influence of these potential confounds (Famoso et al., 2011).

4.6.4 Multiple Testing

An increased number of false positives may occur during AM because of the need to test many markers and attributes simultaneously. Methods like Bonferroni's correction, the false discovery rate, and permutation testing can be used to reduce the number of false positives in a study (Liu et al., 2008).

4.6.5 Sample Size

Association mapping's usefulness is highly sensitive to the size of the sample used. A lack of statistical power and the possibility of false negatives might result from an insufficient sample size. Therefore, a big enough sample size is necessary for identifying the real relationships with certainty (Wang et al., 2016). Due to the lack of statistical power, small sample sizes in AM research may prevent the discovery of real connections. Especially in complicated trait genetics investigations and in populations with limited genetic diversity or substantial, this is a serious cause of concern (Osborne et al., 2014).

4.7 USE OF AM IN FRUIT CROPS

Advancements in high-throughput genotyping have shifted focus from traditional QTL mapping to association or LD mapping, particularly benefiting economically valuable crops (Ranc et al., 2012). Initially, LD mapping methods are likely to be applied in lower-value crops such as banana, grape, citrus fruit, and apple (Foolad et al., 2012). These approaches prioritize marker distribution for MAS, conducting genome-wide scans, and identifying CGs for traits like disease resistance and fruit quality (Neale et al., 2007). However, QTL mapping in fruit trees faces challenges due to their long juvenility periods and difficulties in maintaining bi-parental crosses (Rikkerink et al., 2007). F1 progenies, commonly used in QTL mapping for perennial trees, result in poor resolution due to single meiotic events (Kumar et al., 2012). Fine-mapping within QTL regions is tedious and time-consuming, often yielding limited applicability across genetic backgrounds (Neale and Savolainen, 2004). These challenges underscore the need for genomic tools to expedite the identification of key genetic regions in trees (Savolainen et al., 2007). Association mapping offers promise for overcoming pedigree-based mapping limitations in perennial plants, particularly due to their large architectural frames and unstructured populations (Neale and Savolainen, 2004; Rikkerink et al., 2007) (see Table 4.2).

TABLE 4.2
List of important marker-trait studies in fruit plants

S.No.	Crop	Type of Association Mapping	Trait of Interest	Reference
1.	Grapes (*Vitis vinifera*)	Genome wide	Anthocyanin content, Fruit development, berry size	Fournier-Level et al. (2010), Houel et al. (2010)
2.	Apple (*Malus x domestica* borkh)	Pedigree based QTL mapping approach	Fruit firmness, weighted cortical intensity, browning, titratable acidity, fruit splitting, bitter pit	Cevik et al. (2010)
3.	Japanese pear (*Pyrus pyrifolia*)	Genome wide	Harvest time, resistance to black spot, firmness of fresh, fruit size, fruit shape in longitudinal selection, acid content, total soluble solid content, number of spurs	Iwata et al. (2013)
4.	Peach (*Prunus persica*)	Fine mapping approach	Flesh color around the stone, red pigment in the flesh, flesh texture, flesh adhesion, flesh firmness, chilling requirement, flowering time, fruit weight, ripening time, and fruit development period	Cao et al. (2012), Font i Forcada et al. (2019)
5.	Date palms (*Phoenix dactylifera*)	Genome wide approach	Sex determination, Fruit color, fruit color, Sugar composition, Abiotic stress tolerance	Rahman et al. (2022)
6.	Mango (*Mangifera indica*)	Marker-trait association	Fruit weight, TSS, pulp content and acidity	Padmakar et al. (2016)
7.	Banana (*Musa* spp.)	Genome- wide approach	Seedless trait	Sardos et al. (2016)
8.	Walnut (*Juglans regia* L.)	Marker-trait association by SNP markers	Kernel Weight, Kernel Size, Nut weight, Nut size	Arab et al. (2019)
9.	Almond (*Prunus amygdalus*)	Marker-trait association	Various chemical, physical and Kernal Traits	Font i Forcada et al. (2015)
10.	Kiwi (*Actinidia chinensis*)	Genome wide	Hairiness trait	Zheng et al. (2022)
11.	Guava (*Psidium guajava* L.)	Genome wide	Fruit skin, flesh color	Thakur et al. (2021)
12.	Plum (*Prunus armeniaca*)	Candidate gene approach	Resistance to plum pox virus, fruit firmness (FF), fruit flesh color (FFC), flowering time (FT), and soluble solid content (SSC)	Mariette et al. (2016), Ferik et al. (2022)
13.	Cherry (*Prunus avium*)	Marker-trait association	Fruit cracking, fruit firmness, total soluble solid (TSS) and fruit shape	Khadivi-Khub (2014)

4.8 LIMITATIONS OF AM

The ability of AM to identify a trait depends on the nature of the phenotype and its relationship with-associated marker locus (Alvarez et al., 2015). Even so, useful alleles can be discovered in the vast majority of germplasm collections, and these collections are often highly important potential sources of beneficial alleles (Rafalski, 2010). Similarly, when it comes to certain germplasms, the presence of unique individuals with varying growing conditions may be problematic. This is why it is important to take into account the phenotypic evaluation of genetic diversity in association studies (Myles et al., 2009). Several aspects, including selection history, population structure, kinship, etc. can skew AM data and lead to erroneous conclusions about the relationships between QTLs and markers. Although LD is the foundation of AM, it is possible that linkage is not the underlying mechanism for high levels of LD between a marker and a characteristic. Thus, it is preferable to employ a joint mapping technique that takes into account both LD and the linkage between markers and genes/QTLs. Even though AM provides a far more detailed picture of the genetic makeup of an organism, it still requires a massive marker set (hundreds of thousands, if not millions) to effectively cover the complete genome. The expenditure associated with genotyping would be significantly higher in the case of AM compared to linkage mapping. It is less certain that LD can be used to estimate the physical distance between loci because the rate of recombination varies across the genome. To identify a link between a marker and a trait, the allele frequency of the relevant gene or QTL must be high. In general, the phenotype of a trait is not significantly altered by the presence of low-frequency alleles in the population. Therefore, AM is not likely to discover low-frequency alleles, even when they have a big impact on the trait. As LD between a QTL and a marker of interest increases, so does AM's capacity to detect it. Detection of a low-effect QTL requires that it be in high LD with a marker (Van Inghelandt et al., 2011). The use of haplotypes and SNP-sets (instead of single SNPs) in GWAS has helped researchers prevent the errors of performing repeated tests and has led to more accurate identification of CGs, streamlining the process of conducting gene-set-based and gene-based association maps.

4.9 CONCLUSION

Association mapping application to plant breeding has enormous potential to advance agricultural progress. This is made even true by next-generation sequencing methods, which are both cheap and rapid, which will make GWA studies common and popular. Even though linkage mapping has been shown to be the gold standard for identifying and validating genes, AM continues to play a supportive role. When GWASs and next-generation mapping populations are used together, we will be better able to link phenotypes and genotypes. Genomic selection can use this information to speed up selection and breeding. Through these methods and the potential of enhanced genome modification technologies, comparative genomics has a chance to apply our knowledge to the enhancement of crops in the future. This has the potential to improve our knowledge of genome organization and polymorphism and will also yield a richness of markers for use in assessing germplasm for breeding purposes and in positional cloning research.

REFERENCES

Alvarez, M. F., Mosquera, T., and Blair, M.W. 2015. The use of association genetics approaches in plant breeding. *Reviews* 38, 17–68.

Arab, M. M., Marrano, A., Abdollahi-Arpanahi, R., Leslie, C. A., Askari, H., Neale, D. B. and Vahdati, K. 2019. Genome-wide patterns of population structure and association mapping of nut-related traits in Persian walnut populations from Iran using the Axiom J. regia 700K SNP array. *Scientific Reports* 9(1), 1–14.

Ardlie, K. G., Kruglyak, L. and Seielstad, M. 2002. Patterns of linkage disequilibrium in the human genome. *Nature Reviews Genetics*. 3(4), 299-309.

Atlin, G., Kleinknecht, K., Singh, K. and Piepho H. 2011. Managing genotype x environment interaction in plant breeding programs: a selection theory approach. *Journal of the Indian Society of Agricultural Statistics* 65, 237–247.

Botstein, D. and Risch, N. 2003. Discovering genotypes underlying human phenotypes: past successes for mendelian disease, future approaches for complex disease. *Nature Genetics* 33, 228–237.

Bradbury, P. J., Zhang, Z., Kroon, D. E., Casstevens, T. M., Ramdoss, Y. and Buckler, E. S. 2007. TASSEL: Software for association mapping of complex traits in diverse samples. *Bioinformatics* 23, 2633–2635.

Breseghello, F. and Sorrells, ME. 2006. Association mapping of kernel size and milling quality in wheat (*Triticum aestivum* L.) cultivars. *Genetics* 172, 1165–1177.

Bus, V. G., Chagne, D., Bassett, H. C., Bowatte, D. R., Calenge, F., Celton, J. M. and Schouten, H. J. 2011. Genome mapping of an apple scab, a powdery mildew and a woolly apple aphid resistance gene from open-pollinated Mildew Immune Selection. T*ree Genetics and Genomes* 7(3), 537–548.

Cao, K., Wang, L., Zhu, G., Fang, W., Chen, C. and Luo, J. 2012. Genetic diversity, linkage disequilibrium, and association mapping analyses of peach (Prunus persica) landraces in China. *Tree Genetics and Genomes* 8, 975–990.

Cao, K., Wu, X. and He, F. 2019. Genome-wide association study for fruit quality traits in peach using the Axiom® Peach Genotyping Array. *Horticulture Research* 6(1), 1–9.

Cevik, V., Ryder, C. D., Popovich, A., Manning, K., King, G. J. and Seymour, G. B. 2010. A FRUITFULL-like gene is associated with genetic variation for fruit flesh firmness in apple (*Malus domestica*Borkh.). *Tree Genetics and Genomes* 6, 271–279.

Chagne, D., Kirk, C., How, N., Whitworth, C., Fontic, C., Reig, G., ... and Iglesias, I. 2016. A functional genetic marker for apple red skin coloration across different environments. *Tree Genetics & Genomes*, 12, 1–9.

Collard, B. C. and Mackill, D. J. 2007. Marker-assisted selection: an approach for precision plant breeding in the twenty-first century. *Philosophical Transactions of the Royal Society B: Biological Sciences* 363, 557–572.

Doerge, R. W. 2002. Mapping and analysis of quantitative trait loci in experimental populations. *Nature Reviews Genetics*, 3, 43–45.

Ersoz, E. S., Yu, J. and Buckler, E. S. 2007. Applications of linkage disequilibrium and association mapping in crop plants. In: *Genomics-assisted Crop Improvement*. Springer, pp. 97–119.

Eskridge, K. 2003. Field design and the search for quantitative trait loci in plants. Available at: www.stat.colostate.edu/graybillconference2003/Abstracts/Eskridge.htm

Famoso, A. N., Zhao, K., Clark, R. T., Tung, C.-W., Wright, M. H., Bustamante, C. and McCouch, S. R. 2011. Genetic architecture of aluminum tolerance in rice (*Oryza sativa*) determined through genome-wide association analysis and QTL mapping. *PLoS Genetics*, 7(8), e1002221.

Ferik, F., Ates, D., Ercisli, S., Erdogan, A., Orhan, E. and Tanyolac, M. B. 2022. Genome-wide association links candidate genes to fruit firmness, fruit flesh color, flowering time, and soluble solid content in apricot (*Prunus armeniaca* L.). *Molecular Biology Reports*, 49(6), 5283–5291.

Flint-Garcia, S. A., Thornsberry, J. M. and Buckler, E. S. 2003. Structure of linkage disequilibrium in plants. *Annual Review of Plant Biology* 54 ,357–374.

Flint-Garcia, S. A., Thuillet, A. C., Yu, J., Pressoir, G., Romero, S. M., Mitchell, S. E. and Buckler, E. S. 2005. Maize association population: A high-resolution platform for quantitative trait locus dissection. *Plant Journal* 44(6), 1054–1064.

Font i Forcada, C., Guajardo, V., Chin-Wo, S. R. and Moreno, M. A. 2019. Association mapping analysis for fruit quality traits in *Prunus persica* using SNP markers. *Frontiers in Plant Science* 9, 2005.

Font i Forcada, C., Oraguzie, N., Reyes-Chin-Wo, S., Espiau, M. T., Sociasi Company, R. and Fernandez i Marti, A. 2015. Identification of genetic loci associated with quality traits in almond via association mapping. *PLoS One* 10(6), 146276.

Foolad, M. R. and Panthee, D. R. 2012. Marker-assisted selection in tomato breeding. *Critical Reviews in Plant Sciences* 31, 93–123.

Fournier-Level, A., Lacombe, T., Le Cunff, L., Boursiquot, J. M. and This, P. 2010. Evolution of the VvMybA gene family, the major determinant of berry colour in cultivated grapevine (*Vitis vinifera* L.). *Heredity* 104(4), 351–362.

Fu, W., da Silva Linge, C. and Gasic, K. 2021. Genome-wide association study of brown rot (Monilinia spp.) tolerance in peach. *Frontiers in Plant Science* 12, 635914.

Gross, B. L., Volk, G. M., Richards, C. M., Forsline, P. L., Fazio, G. and Chao, C. T. (2012). Identification of "duplicate" accessions within the USDA-ARS national plant germplasm system Malus collection. *Journal of the American Society for Horticultural Science* 137(5), 333–342.

Gupta, P. K., Kulwal, P. L. and Jaiswal, V. 2014. Association mapping in crop plants: Opportunities and challenges. *Advances in Genetics* 85, 109–147.

Gupta, P. K., Rustgi, S. and Kulwal, P. L. 2005. Linkage disequilibrium and association studies in higher plants: Present status and future prospects. *Plant Molecular Biology* 57: 461–485.

Hayward, A. C., Tollenaere, R., Dalton-Morgan, J. and Batley, J. 2015. Molecular marker applications in plants. In: *Plant Genotyping. Methods and protocols*. 13–27.

Holland, J. B. 2007. Genetic architecture of complex traits in plants. *Current Opinion in Plant Biology*, 10(2), 156–161.

Houel, C., Bounon, R., Chaib, J., Guichard, C., Peros, J. P., Bacilieri, R. and Adam-Blondon, A. F. 2010. Patterns of sequence polymorphism in the fleshless berry locus in cultivated and wild *Vitis vinifera* accessions. *BMC Plant Biology* 10(1), 1–15.

Huang, X. and Han, B. 2014. Natural variations and genome-wide association studies in crop plants. *Annual Review of Plant Biology* 65, 531–551.

Huang, Y. F., Doligez, A., Fournier-Level, A., Le Cunff, L., Bertrand, Y., Canaguier, A. and This, P. 2012. Dissecting genetic architecture of grape proanthocyanidin composition through quantitative trait locus mapping. *BMC Plant Biology* 12(1), 1–19.

Ingvarsson, P. K. and Street, N. R. 2011. Association genetics of complex traits in plants. *New Phytologist* 189, 909–922.

Iwata, H., Hayashi, T., Terakami, S., Takada, N., Sawamura, Y. and Yamamoto, T. 2013. Potential assessment of genome-wide association study and genomic selection in Japanese pear *Pyrus pyrifolia*. *Breeding Science* 63(1), 125–140.

Jannink, J. L., and Wu, X. L. 2003. Estimating allelic number and identity in state of QTLs in interconnected families. *Genetics Research*, 81(2), 133–144.

Kearsey, M. J. and Farquhar, A. G. 1998. QTL analysis in plants; where are we now? *Heredity* 80: 137–142.

Khadivi-Khub, A. 2014. Regression association analysis of fruit traits with molecular markers in cherries. *Plant Systematics and Evolution* 300(5), 1163–1173.

Khan, M. A., & Korban, S. S. (2012). Association mapping in forest trees and fruit crops. *Journal of Experimental Botany* 63(11), 4045–4060.

Korte, A., Vilhjalmsson, B. J., Segura, V., Platt, A., Long, Q. and Nordborg, M. 2012. A mixedmodel approach for genome-wide association studies of correlated traits in structured populations. *Nature Genetics* 44, 1066–1071.

Kumar, S., Bink, M. C., Volz, R. K., Bus, V. G. and Chagne, D. 2012. Towards genomic selection in apple (*Malus* × *domestica*Borkh.) breeding programmes: Prospects, challenges and strategies. *Tree Genetics & Genomes* 8, 1–14.

Li, Y. C., Korol, A. B., Fahima, T., Beiles, A. and Nevo, E. 2002. Microsatellites: Genomic distribution, putative functions and mutational mechanisms: a review. *Molecular Ecology* 11(12), 2453–2465.

Liu, N., Zhang, K. and Zhao, H. 2008. Haplotype-association analysis. Advances in Genetics 60, 335–405.

Lynch, M. and Ritland, K. 1999. Estimation of pairwise relatedness with molecular markers. *Genetics* 152, 1753–1766.

Ma, Y., Marzougui, A., Coyne, C. J., Sankaran, S., Main, D., Porter, L. D., ... and McGee, R. J. (2020). Dissecting the genetic architecture of Aphanomyces root rot resistance in lentil by QTL mapping and genome-wide association study. *International Journal of Molecular Sciences* 21(6), 2129.

Mackay, I. and Powell, W. 2007. Methods for linkage disequilibrium mapping in crops. *Trends in Plant Science* 12, 57–63.

Mariette, S., Wong Jun Tai, F., Roch, G., Barre, A., Chague, A., Decroocq, S. and Decroocq, V. 2016. Genome-wide association links candidate genes to resistance to Plum Pox Virus in apricot (*Prunus armeniaca*). *New Phytologist* 209(2), 773–784.

McClure, K. A., Gardner, K. M., Douglas, G. M., Song, J., Forney, C. F., DeLong, J. and Myles, S. 2018. A genome-wide association study of apple quality and scab resistance. *The Plant Genome* 11(1), 170075.

Mohan, M., Nair, S., Bhagwat, A., Krishna, T., Yano, M., Bhatia, C. and Sasaki, T. 1997. Genome mapping, molecular markers and marker-assisted selection in crop plants. *Molecular Breeding* 3, 87–103.

Morris, G. P., Ramu, P., Deshpande, S. P., Hash, C. T., Shah, T. and Upadhyaya, H. D. 2013. Population genomic and genome-wide association studies of agroclimatic traits in sorghum. *Proceedings of the National Academy of Sciences USA*, 110, 45.

Myles, S., Peiffer, J., Brown, P. J., Ersoz, E. S., Zhang, Z., Costich, D. E. and Buckler, E. S. 2009. Association mapping: Critical considerations shift from genotyping to experimental design. *The Plant Cell* 21, 2194–2202.

Neale, B., Ferreira, M., Medland, S. and Posthuma, D. 2007. *Statistical Genetics: Gene Mapping Through Linkage and Association*. Taylor & Francis.

Neale, D. B. and Savolainen, O. (2004). Association genetics of complex traits in conifers. *Trend in Plant Science* 9(7), 325–330.

Nordborg, M. and S. Tavare. 2002. Linkage disequilibrium: What history has to tell us. *TRENDS in Genetics*. 18, 83–90.

Oraguzie, N. C., Wilcox, P. L., Rikkerink, E. H. and De Silva, H. N. 2007. Linkage disequilibrium. In *Association mapping in plants* (pp. 11–39). New York, NY: Springer New York.

Osborne, T. S., Johnson, I. T. and Henderson, C. R. 2014. Challenges in plant genomics: Sample size and statistical power in genome-wide association studies. *The Plant Cell* 26(7), 2774–2780.

Padmakar, B., Dinesh, M. R., and Ravishankar, K. V. (2016). Marker-trait association for fruit characters in mango (*Mangifera Indica* L.) cultivars. *Journal of Horticultural Sciences*, *11*(2), 170–178.

Piepho, H. P., Williams, E. R. and Fleck, M. (2006). A note on the analysis of designed experiments with complex treatment structure. *HortScience*, *41*(2), 446–452.

Pieruschka, R. and Schurr, U. 2019. Plant phenotyping: past, present, and future. *Plant Phenomics*. https://doi.org/10.34133/2019/7507131.

Pincot, D. D., Poorten, T. J., Hardigan, M. A., Harshman, J. M., Acharya, C. B., Cole, G. S. and Knapp, S. J. 2018. Genome-wide association mapping uncovers Fw1, a dominant gene conferring resistance to Fusarium wilt in strawberry. *G3: Genes, Genomes, Genetics* 8(5), 1817–1828.

Portis, E., Cericola, F., Barchi, L., Toppino, L., Acciarri, N., Pulcini, L. and Rotino, G. L. 2015. Association mapping for fruit, plant and leaf morphology traits in eggplant. *PLoS One* 10(8), e0135200.

Price, A. H. 2006. Believe it or not, QTLs are accurate. *Trends in plant science*, *11*(5), 213–216.

Pritchard, J. K. and Donnelly, P. 2001. Case-control studies of association in structured or admixed populations. *Theoretical Population Biology* 60, 227–237.

Pritchard, J. K. and Rosenberg, N. A. 1999. Use of unlinked genetic markers to detect population stratification in association studies. *The American Journal of Human Genetics* 65, 220–228.

Rafalski, J. A. 2010. Association genetics in crop improvement. *Current Opinion in Plant Biology* 13, 174–180.

Rahman, H., Vikram, P., Hammami, Z. and Singh, R. K. 2022. Recent advances in date palm genomics: A comprehensive review. *Frontiers in Genetics*, 13, 959266.

Ranc, N., Munos, S., Xu, J., Le Paslier, M. C., Chauveau, A., Bounon, R. and Sabot, F. 2012. Genome-wide association mapping in tomato (*Solanum lycopersicum*) is possible using genome admixture of *Solanum lycopersicum* var. cerasiforme. *G3: Genes, Genomes, Genetics* 2(8), 853–864.

Rebaï, A. and Goffinet, B. (2000). More about quantitative trait locus mapping with diallel designs. *Genetics Research* 75(2), 243–247.

Rebai, A., and Goffinet, B. (1993). Power of tests for QTL detection using replicated progenies derived from a diallel cross. *Theoretical and Applied Genetics* 86, 1014–1022.

Rikkerink, E. H., Oraguzie, N. C., & Gardiner, S. E. (2007). Prospects of association mapping in perennial horticultural crops. *Association Mapping in Plants*, 249–269. New York, NY: Springer New York.

Sardos, J., Rouard, M., Hueber, Y., Cenci, A., Hyma, K. E., Van Den Houwe, I. and Roux, N. 2016. A genome-wide association study on the seedless phenotype in banana (Musa spp.) reveals the potential of a selected panel to detect candidate genes in a vegetatively propagated crop. *PLoS One* 11(5), e0154448.

Savolainen, O., Pyhajarvi, T., and Knurr, T. (2007). Gene flow and local adaptation in trees. Annual Review of Ecology, Evolution, and Systematics. 38, 595–619.

Segura, V., Vilhjalmsson, B. J., Platt, A., Korte, A., Seren, U. and Long, Q. 2012. An efficient multi-locus mixed-model approach for genome-wide association studies in structured populations. *Nature Genetics* 44, 825–830.

Singh, B. D. and Singh, A. K. 2015. *Marker-Assisted Plant Breeding: Principles and Practices*. National academic digital library Ethopia.

Singh, S. 2001. Pyramiding three bacterial blight resistance genes (xa5, xa13, and Xa21) using marker-assisted selection into indica rice cultivar PR106. *Theoretical and Applied Genetics* 102, 1011–1015.

Slatkin, M. 2008. Linkage disequilibrium—understanding the evolutionary past and mapping the medical future. Nature Reviews Genetics 9:477–485.

Soto-Cerda, B. J. and Cloutier, S. 2012. Association mapping in plant genomes. *Genetic Diversity in Plants* 29–54. 10-5772.

Stich, B. and Melchinger, A. E. 2010. An introduction to association mapping in plants. Centre for Agriculture and Biosciences International Reviews 5, 1–9.

Syvänen, A. C. 2005. Toward genome-wide SNP genotyping. *Nature Genetics* 37, S5.

Thakur, S., Yadav, I. S., Jindal, M., Sharma, P. K., Dhillon, G. S., Boora, R. S. and Mittal, A. 2021. Development of genome-wide functional markers using draft genome assembly of guava (*Psidium guajava* L.) cv. Allahabad safeda to expedite molecular breeding. *Frontiers in Plant Science* 12, 708332.

Van Inghelandt, D., Reif, J. C., Dhillon, B. S., Flament, P., and Melchinger, A. E. 2011. Extent and genome-wide distribution of linkage disequilibrium in commercial maize germplasm. *Theoretical and Applied Genetics* 123(1), 11–20.

Vos, P., Hogers, R., Bleeker, M., Reijans, M., Lee, T., Hornes, M., Friters, A., Pot, J., Paleman, J., Kuiper, M. and Zabeau, M. 1995. AFLP: A new technique for DNA fingerprinting. Nucleic Acids Research 23, 4407–4414.

Waldchen, M., Beckers, G. J. M. and Van Eeuwijk, F. A. 2012. Genome-wide association study in plants: Challenges and opportunities. *New Phytologist* 193(2), 333–346.

Wang, S. B., Feng, J. Y., Ren, W. L., Huang, B., Zhou, L., Wen, Y. J. and Zhang, Y. M. 2016. Improving power and accuracy of genome-wide association studies via a multi-locus mixed linear model methodology. *Scientific Reports* 6(1), 1–12.

Williams, J. G., Kubelik, A. R., Livak, K. J., Rafalski, J. A. and Tingey, S. V. 1990. DNA polymorphisms amplified by arbitrary primers are useful as genetic markers. *Nucleic Acids Research* 18, 6531–6535.

Xu, S. 1998. Mapping quantitative trait loci using multiple families of line crosses. *Genetics* 148(1), 517–524.

Yu, J. and Buckler, E. S. (2006). Genetic association mapping and genome organization of maize. *Current Opinion in Biotechnology* 17(2), 155–160.

Zhao, K., Aranzana, M. J., Kim, S., Lister, C., Shindo, C., Tang, C. and Dean, C. 2007. An Arabidopsis example of association mapping in structured samples. *PLoS Genetics* 3(1), 4.

Zheng, S., Wang, Y., Qu, D., Sun, W., Yu, Y. and Zhang, Y. 2022. Study on population structure of kiwifruit and GWAS for hairiness character. *Genetics. 821, 146276.*

Zhu, C., Gore, M., Buckler, E. S. and Yu, J. 2008. Status and prospects of association mapping in plants. *The Plant Genome* 1, 5–20.

Zhu, M. and Zhao, S. 2007. Candidate gene identification approach: Progress and challenges. *An International Journal of Biological Sciences* 3, 420.

Zhu, X., Xu, F., Zhao, S., Bo, W., Jiang, L., Pang, X. and Wu, R. 2015. Inferring the evolutionary history of outcrossing populations through computing a multiallelic linkage–linkage disequilibrium map. *Methods in Ecology and Evolution* 6, 1259–1269.

5 Improving Floral Attributes in Ornamental Crops by Biotechnological Approaches

Vipasha Verma, Mahinder Partap, Meenakshi Thakur, Shiv Rattan, Diksha Sharma, Deekshit HN, Anjali Chandel, Ujala, Priti, and Bhavya Bhargava

5.1 INTRODUCTION

Without a doubt, ornamental plants have redefined the horticultural industry. Various ornamental plants are presently broadly utilized in home gardening, professional landscaping, and cut flower production (Azadi et al. 2016; Mekapogu et al. 2022; Noman et al. 2017). The international floriculture industry has flourished in terms of volume, manufacturing value, specification, and commercialization and is estimated to be worth 100–150 billion Euros globally, with major markets in Europe, North America, and Japan (Chandler and Tanaka 2007; Mekapogu et al. 2022). Ornamental flowers such as roses, chrysanthemums, carnations, orchids, gerbera daisies, freesias, lilies, gladiolus, and ranunculus are the most prominent in the worldwide flower industry. Among the primary objectives of the ornamental plant industry is the introduction of new varieties with novel attributes, such as novel flower color, aroma, flower shape, plant architecture, improved stress resistance, and increased vase life. Even though conventional breeding methods are frequently employed in the development of new plant varieties, they do have their drawbacks, including a high degree of heterozygosity and subtractive one-point improvement (Shibata 2008). Since the last decade, biotechnological approaches such as genetic engineering and genome editing have gained widespread popularity over the past ten years as more practical means of overcoming the limitations of traditional breeding methods (Noman et al. 2016, 2017). Looking at the potential of biotechnology, interest in both public and private sectors has gradually increased in genetic engineering and its applications, which made experts lately concentrated on strengthening and upgrading qualitative

DOI: 10.1201/9781003343707-5

attributes for the flower industry (Parisi et al. 2016). The primary advantage of using genetic engineering in ornamental plants is that the novel genes from the related or different species' gene pools can be introduced for the improvement of desired traits (Chandler and Brugliera 2011; Kamthan et al. 2016). As an innovative technology, CRISPR-Cas9 mediated gene editing in ornamental plants will be of interest to both growers and consumers since it has the capacity to modify flower architecture, color, fragrance, and vase life of cut flowers by targeting genes of interest (Chandler and Sanchez 2012). Furthermore, multi-omics technologies have aided in the acceleration of molecular breeding and gene editing in floriculture crops (Mekapogu et al. 2022). This chapter has compiled the applications and recent achievements of genetic engineering and gene editing in ornamental plants.

5.2 SIGNIFICANCE OF FLORAL TRAIT MODIFICATION

Flowers or ornamental plants are typically quite colorful and showy. Floral traits such as shape, fragrance, and color are not only important for the plant's aesthetic appeal, but also for its very existence (Feng et al. 2010). Therefore, floral features are of special significance to plant geneticists. Due to high heterozygosity, chromosome number, gene pool, and sterility, ornamental plants often have sexual hybridization issues (Kim and Anderson 2006). Chrysanthemums are alloploid and have 36–75 chromosomes. Furthermore, *Anthurium sp.* have a life cycle of 3 years, which makes these species less amenable to improvement and the development of a new cultivar in such species may require approximately 8–10 years (Azadi et al. 2016). Likewise, the majority of carnation varieties are self-fertile and incapable of producing seeds (Nontaswatsri and Fukai 2010). Similarly, precocious petal pigmentation and enhanced sepals pigmentation in *Lisianthus spp.* is another matter of concern (Schwinn et al. 2014). The enormous genome size of ornamental plants, such as the lily, is a barrier to extracting genetic information (Du et al. 2015). Transgenic and gene editing technologies have the potential to generate novel floral phenotypes that do not naturally occur (Figure 5.1). Currently, efforts are being made to create genetically modified flowers with diverse colors and other important commercial attributes. This chapter thus focuses on biotechnological advances in modifying important floral traits that contribute to the generation of numerous novel ornamental plant varieties.

5.3 PLANT PIGMENTS AND FLOWER COLOR

Typically, conventional plant breeding methods are used to enhance the beauty and performance of ornamental plants. However, these methods are constrained by the gene pool and other characteristics of sexually related species (Da Silva Trentin et al. 2011). In the last few decades, biotechnology has given ornamental plants new and unique traits by using genes from other plant species (Noman et al. 2017). Florists and people who run businesses related to flowers are always looking for new flower color as novel-colored varieties are of utmost importance in the horticulture industry and determine purchase value. Anthocyanins, flavonoids, carotenoids, and betaines are the main pigments in flowers that make them look more vibrant and aesthetic

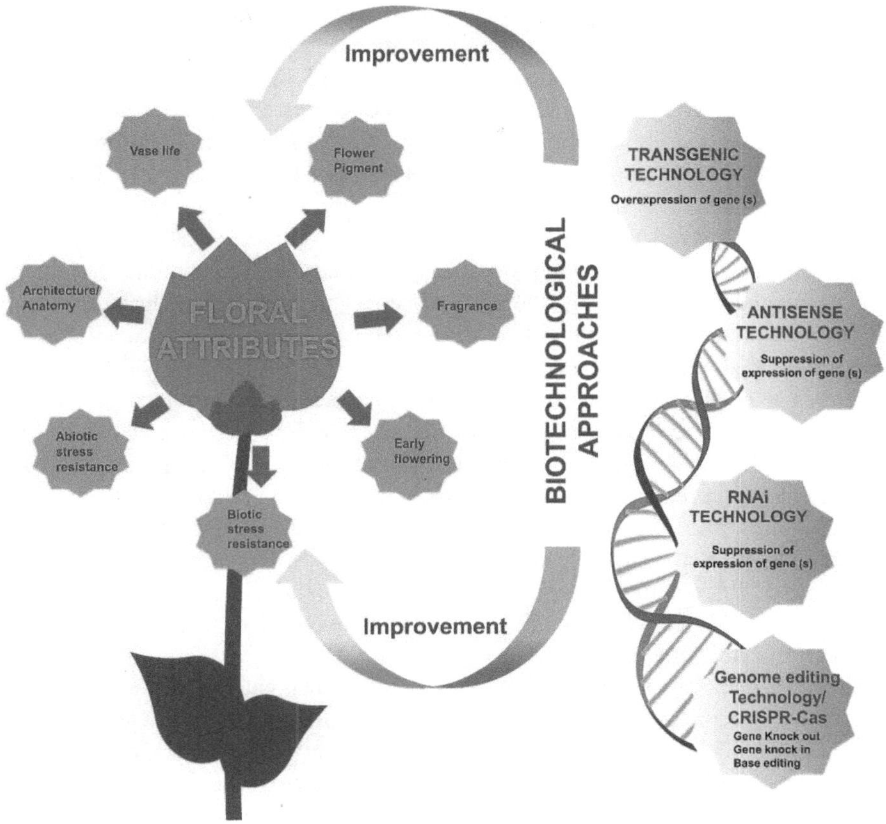

FIGURE 5.1 Floral traits improvement through biotechnology in ornamental crops.

(Veitch and Grayer 2008). Anthocyanidins-cyanidin, delphinidin, peonidin, petunidin, malvidin, and pelargonidin predominate in these pigments. Three anthocyanidins – delphinidin, cyanidin, and pelargonidin – are the most significant among cut flowers. In blue flowers, delphinidin and its derivatives are abundant, while pelargonidin acts as anthocyanidin base to give the bright red color (Tanaka et al. 2009). Floral color modifications resulting from genetic engineering are best studied in ornamental plants like petunia and torenia (Fukuchi-Mizutani et al. 2011; Zvi et al. 2012; Chen and Dubcovsky 2012). The color variations are closely tied to the regulation of specific genes that affect pigment precursor synthesis. Potential novel flower colors can be created using different ways of genetic engineering such as the introduction of genes encoding enzymes that are absent, diversion of existing pathways through the over-expression of genes, and/or the suppression of pathways through the down-regulation of genes. By inhibiting the expression of anthocyanin-related genes, transgenic plants have produced white blooms (Tanaka et al. 2008). Most of the important cut-flower species such as rose, chrysanthemum, and carnation lack *flavonoid 3′,5′-hydroxylase gene*, which is required for the synthesis of blue delphinidin-derived pigments (Brugliera et al. 2013). By inserting the *flavonoid 3',5'-hydroxylase gene* from petunia or pansy into the genomes of those cut flowers, scientists were able to create blue/purple colored carnations and roses (Table 5.1). Commercially accessible genetically modified (GM) carnations of the Moon® series have been on the market for the past 15 years (Tanaka et al. 2009), and a GM rose variety was introduced in 2009 (Katsumoto et al. 2007). Azadi et al. (2010) verified the presence of many colors in the calli and leaves of transgenic lily plants after transferring essential genes for the carotenoid biosynthesis pathway under the *35S CaMV* promoter. The over-expression of *GMYB10* in GM gerbera plants significantly increased accumulation of pigments and stimulated cyanidin biosynthesis (Laitinen et al. 2008). In Chrysanthemum, different colors are mostly caused by carotenoids and/or red malonylated cyanidin glucosides (Kishimoto et al. 2004). *CmCCD4a* is a gene unique to white chrysanthemum ray petals. One dominant gene prevents the accumulation of carotenoids in flower petals. The presence of *CmCCD4a* in white flowers is responsible for the conversion of synthetic carotenoids into a colorless compound (Yoshioka et al. 2012). Also, suppression technologies like antisense technology, RNAi technology or co-suppression have been found to be highly useful for creating the different colors of flowers (Tanaka et al. 2010). Two cyclamen cultivars were transformed with an antisense *CpF3'5'H c*onstruct by *Agrobacterium tumefaciens*. Flowers of the transgenic lines showed modified color and this correlated positively with the loss of endogenous F3'5'H transcript (Boase et al. 2010).

Flower scents enhance the aesthetic qualities of ornamental plants and have considerable economic relevance in cut flower and cosmetic industry. Many flower fragrance volatiles belong to the terpenoid, phenylpropanoid/benzenoid, and aromatic amino acid categories (Peled-Zehavi et al. 2015). The variety of specific metabolites generated by flowering plants is extensive (Zvi et al. 2012). Thus, an increase in floral-specific metabolites may aid in the detection, isolation, and identification of chemicals, as well as the enhancement of floral attributes like as aroma and color. Nonetheless, the discovery of master regulators that govern the generation of various volatile compounds and the accompanying metabolic pathways remains to be done. Prior to the emergence of high-throughput technology, the isolation of floral scent genes was limited to traditional biochemical methods. In recent years, Next-Generation Sequencing technologies have enabled the identification of novel fragrance genes in various ornamental species and suggests that transcription factors (TFs) have essential roles in regulating scent emission (Colquhoun and Clark 2011). In lavender (Lavandula), RNA seq led to the discovery of S-linalool synthase (LiS-LINS), which is significant since it gives a pleasant scent to lavender oils, hence increasing their market value (Adal et al. 2019). In addition to genes involved in terpenoid, phenylpropanoid pathway, few TFs responsible for controlling scent emissions have been uncovered (Verdonk et al. 2005). Emission of benezoids 1 (*EOBI*) in Petunia is due to a unique R2-R3 type floral TF that functions upstream of *ODO1* and downstream of *EOBII*. Several genes involved in the shikimate pathway and scent emission related genes were downregulated in response to *EOBI* silencing (Spitzer-Rimon et al. 2012). Du et al. (2019) categorized diverse flower scent kinds and discovered mono*TPS* polymorphisms and alternative splicing products by

TABLE 5.1
Applications of genetic engineering for enhancing floral attributes in various ornamental crops

Sr. No.	Source of gene	Target gene	Crop Transformed	Function/ Result	Reference
1.	*Agrobacterium rhizogenes*	*RolC*	Pelargoniums and Petunias	Dwarfed Pelargoniums and Petunias	Boase et al. (2004)
2.	*Agrobacterium tumefaciens*	*ipT*	Chrysanthemum	Increased branching and reduced internode length in Chrysanthemum	Khodakovskaya et al. (2009)
3.	*Arabidopsis thaliana*	*TCP3* chimeric repressor	Chrysanthemum	Production of fringed leaves	Narumi et al. (2011)
4.	*Arabidopsis thaliana*	*PAP1* transcription factor	Rose	Enhance production of anthocyanin, increased accumulation of eugenol (phenylpropanoid compound)	Zvi et al. (2012)
5.	*Bacillus thuringiensis*	*cry1Ab*	Chrysanthemum	Induces insect resistance	Shinoyama et al. (2008)
6.	Carnation/apple	*ACO* and *ACS*	Carnation	Improve the vase life of carnation	Zuker et al. (2001)
7.	Chrysanthemum	*Ls* *AP1*	Chrysanthemum	Less branching	Han et al. (2007)
8.	Chrysanthemum	*ERS1*	Chrysanthemum	Delay of petal senescence	Narumi et al. (2005)
9.	Chrysanthemum	*AP1*	Chrysanthemum	Accelerates time of flower in chrysanthemum	Jiang et al. (2010)
10.	Chrysanthemum	*CmMYB9a*	Chrysanthemum	Regulation of anthocyanin accumulation	Wang et al. (2022)
11.	Chrysanthemum	*CmDREB6*	Chrysanthemum	Enhanced heat tolerance	Du et al. (2018)
12.	Chrysanthemum	*CmbHLH2*+ *CmMYB6*	Chrysanthemum	Regulation of anthocyanin accumulation	Xiang et al. (2015)
13.	Chrysanthemum	*CmTFL1a*	Chrysanthemum	Delayed flowering	Haider et al. (2020)

(continued)

TABLE 5.1 (Continued)
Applications of genetic engineering for enhancing floral attributes in various ornamental crops

Sr. No.	Source of gene	Target gene	Crop Transformed	Function/ Result	Reference
14.	*Chrysanthemum nankingense*	*CnTCP4*	Chrysanthemum	Hypersensitivity to cold stress	Tian et al. (2022)
15.	*Clarkia breweri*	*Linalool synthase*	Carnation	Fragrance production	Lavy et al. (2002)
16.	*Clarkia breweri*	*Clarkia breweri gene (BEAT)*	Eustoma	Induces fragrances in petals	Aranovich et al. (2007)
17.	Gerbera	*GSQUA2*	Gerbera	Rapid vegetative alteration like dwarfism and vegetative axis elongation	Ruokolainen et al. (2010)
18.	Lotus	*CrtW*	Lotus	Overexpression alters the color of Petals from pale yellow to bright yellow or orange.	Suzuki et al. (2002)
19.	Orchid	*OMADS1*	Eustoma	Alteration in flowering time	Thiruvengadam and Yang (2009)
20.	Orchid/ Lily	*MADS-Box*	Lily	Ectopic expression changes the second round of petals into calyx in orchids and lilies	Thiruvengadam and Yang (2009)
21.	Rice	*Rice chitinase (chiII)*	Chrysanthemum	Creates Septoria leaf spot disease Resistance	Sen et al. (2013)
22.	Petunia	*DifF*	Lilium	Production of purple colored petal cells	Azadi et al. (2016)
23.	Petunia/Pansy	*F3' -5'H*	Petunia	Blue flowers in carnation (Petunia) and Roses (Pansy)	Katsumoto et al. (2007)
24.	*Xanthomonas oryzae cv. oryzae*	*hpaGxoo gene*	Chrysanthemum	Increases resistance to Alternaria leaf spot disease	Xu et al. (2010)

analysing volatile emissions from different 41 Lilium cultivars. Two monoterpene synthase genes (*LoTPS1* and *LoTPS3*), responsible for biosynthesis of ocimene and linalool, were isolated and functionally characterised in lilium "Siberia", a commercially significant cultivar valued by consumers for its fragrance and large, snow-white flowers (Hu et al. 2013, 2017). To make the petals of lisianthus flowers fragrant, the Clarkia breweri gene *BEAT* was introduced (Table 5.1). Various other genes have been identified that regulate floral scent biosynthesis in a variety of commercially important ornamental crops, including lily (*LhODO1*), rose (*AADC, COMT, MYB1*), carnation (linalool synthase), pelargonium (*TPS*), petunia (*BAHD, EOB1*), and dendrobium (*DoTPS10*). Although there have been notable successes in boosting plant defence and increasing the fragrance of flowers. But so far, research findings highlight the limitations caused by our lack of understanding of fragrance metabolic mechanisms and regulation.

5.5 EARLY FLOWERING

In conjunction with synthetic biology, flowers with altered colors have been produced. In addition to numerous associated advantages, the flower color modification is a significant and spectacular result of biotechnology. Despite of color, flowering period is also regarded as the primary determinant of success of flower crops. Flowers could be produced and distributed more quickly, for special occasions, or when there is a high market demand. (Jung and Muller 2009). The regulation mechanism of floral transition pathways has been elucidated by fundamental research on flowering time in model plants (Blazquez and Weigel 2000). Key flowering-related genes appear to be amenable to genetic modification, which appears to result in important and useful alteration in flowering process (Srikanth and Schmid 2011). Ornamental plant breeding goals include generating early flowering cultivars or plants that can flower during long days (Shulga et al. 2011). Several papers discussed in detail the successful introduction of genes for the production of flowers in a relatively short period of time (Litt and Irish 2003). Thus yet, only a small number of genes involved in flowering induction has been sequenced and utilized to regulate flowering in ornamental plants (Table 5.1). Among the known genes involved in flowering pathways, *FT* (Flowering locus T) is the principal integrator of numerous flowering genes that respond to a range of inputs (light, temperature, etc.) (Fornara et al. 2010). One *FT* orthologue gene was identified from the commercial cultivar 'white heaven' of *Lilium longiflorum* and three were obtained from the cultivar *Tulipa gesneriana* 'Dynasty (*LlFT and TgFT1, TgFT2, TgFT3*). Over-expression of *LlFT* in *L. longiflorum* also resulted in an early flowering, which is consistent with earlier findings (Leeggangers et al. 2018). Flowering was accelerated in 'Yuuka' by 20 days when the *CmBBX8* (*CmFTL* activator) gene was over-expressed, and was postponed by 15 days when CmBBX8 was silenced, relative to the wild type (Wang et al. 2020). *MADS* box genes play important role in floral development because they regulate flowering time and organ development (Thiruvengadam and Yang 2009). Overexpression of *AP1* during short days may induce bud formation 14 days earlier in transgenic chrysanthemum than in non-transgenic plants (Shulga et al. 2011). Agrobacterium mediated transformation in *Siningia* sp. demonstrated that exogenous LFY overexpression reduce flowering initiation duration (Zheng et al.

2001). The *CsFTL3* gene (Flowering Locus T like paralog) found to control flowering in *Chrysanthemum seticuspe* based on photoperiod. Overexpression of this gene in Chrysanthemum can cause flowering during long-day conditions (Oda et al. 2012). Eventually, for the quantitative impact of vernalization of bulb on flowering as well as to facilitate gene expression in the meristems, RNA sequencing technology became increasingly popular. Many plausible candidate genes involved in vernalization were discovered through studies of cold-treated lily bulbs (Huang et al. 2014; Villacorta-Martin et al. 2015). The *GSQUA2* gene introduction stimulated the flowering process in Gerbera (Ruokolainen et al. 2010). In *Lisianthus sp*., *OMADS1* transformation resulted in a considerable decrease of flowering time and enhanced number of flowers compared to non-transformed plants.

5.6 ANATOMY AND MORPHOLOGY OF FLOWERS

Whether indoors or outdoors, ornamental plants are grown for the purpose of decoration, and making human surroundings better. At flowering, the shoot meristem of flowering plants can undergo a sequence of programmable events that result in the formation of a floral meristem. Many floral meristem related genes have been discovered in several species, and their precise function is yet unknown (Chang et al. 2020). Overexpression of the tobacco phytochrome b1 gene in chrysanthemum produced tiny plants with wider branch angles (Zheng et al. 2001). Eventually, the height of chrysanthemum plants was reduced by introducing an Arabidopsis *GA-insensitive gene* (Petty et al. 2003). Owing to the ectopic expression of *CmYAB1* (a candidate abaxial gene), transgenic plants with flat petals and reduced petal curvature were formed, and pompon-shaped inflorescences were seen. The chrysanthemum inflorescence shaped like an anemone having various colored tubular flowers is one of the most well-known inflorescence varieties (Ding et al. 2019). In addition, Aida et al. (2008) discovered that an antisense *CAG* gene was introduced into *C. morifolium*. They discovered that suppression of the *CAG* gene transformed gynoecium and androecium into corolla-like tissues. Transformation of *DgLsL* gene in sense orientation led to the increase in chrysanthemum branch number, however, in contrast, branching was diminished in antisense transformants (Jiang et al. 2010). The silencing of the *DmCPD* and *DmGA20ox* genes is a viable method for developing dwarf chrysanthemum varieties (Xie et al. 2015). Thiruvengadam and Yang (2009) reported that transgenic plants containing the lily gene *35S::L MADS1-M* produced more flowers from leafy branches than their wild counterparts. Transgenic carnations containing the *rol C* gene with the 35S CaMV promoter exhibited increased flowering, stem organogenesis, and root organogenesis (Zuker et al. 2001; Casanova et al. 2004). Hence, the expression of these genes provides conclusive proof that they play a crucial role in promoting morphological and structural alterations in flowers.

5.7 BIOTIC STRESS TOLERANCE

The commercial value of cut flowers is determined by their phenotype and quality. Hence, any observable disease symptoms on a floriculture crop can have a significant effect on its quality, market value, and yield. Fungi cause the majority of flower crop diseases, followed by viruses and bacteria (Mathew 2019). Due to lack of number of resistance genes, breeding of ornamental plant for stress resistance/tolerance is a tedious process (Azadi et al. 2016). The introduction of genes encoding for chitinase, glucanase, osmotin, defensin, and other proteins has conferred resistance to bacterial and fungal infections in numerous ornamental plants (Chandler and Sanchez 2012; Teixeira da Silva et al. 2013). Genes overexpressing various antifungal proteins including class II *chitinase* and *type I RIP* (ribosome inhibiting protein) were introduced into Rose to acquire partial resistance to the diseases like black spot, powdery mildew, downy mildew, and rust (Dohm et al. 2002). The introduction of the rice *chitinase* gene enhanced resistance to powdery mildew disease in rose plants (Pourhosseini et al. 2013). Likewise, agrobacterium transformation of an antimicrobial protein gene*, Ace-AMP1* has also resulted in increased resistance to powdery mildew in *Rosa hybrida* cv. carefree beauty (Li et al. 2003). Transgenic *Rosa multiflora* expressing antisense *RhMLO1* showed improved resistance to powdery mildew (Qiu et al. 2015). Recently, two MLO members, *RgMLO6* and *RlMLO7*, were discovered as possible candidate genes that can produce resistance to powdery mildew in *Rosa species* (Xiang et al. 2019). A transcriptome analysis of rose petals infected with *B. cinerea* found that *RcERF099*, a gene encoding a member of the AP2/ERF TF family, regulates rose flower resistance to *Botrytis cinerea*. This study could help improve rose grey mould disease resistance (Li et al. 2020). Three N-methyltransferase genes (i.e., *CaXMT1, CaMYT1*, and *CaDXMT1)* were introduced to *Delphinium grandiflorum* cv. Shinba, resulted in higher level of *Botrytis cinerea* resistance (Kim and Xiao 2011). Similarly, *chiII* gene led to transgenic chrysanthemum plant-enhanced resistance to septoria leaf spot disease (Sen et al. 2013). Recently, several SNP markers and other transcripts involved in Botrytis resistance pathways have been identified for additional linkage mapping (Fu et al. 2016). In transgenic chrysanthemum, for example, overexpression of the *hpaGxoo* gene increased resistance to *Alternaria tenuissima* (Xu et al. 2011). *Agrobacterium*-mediated transformation of the *LLA* gene (*Lycoris Longituba agglutinin*) in transgenic chrysanthemum plants increased aphid resistance (He et al. 2009). Transgenic lines with increased resistance to Cucumber mosaic virus were produced by introducing a defective replicase or *coat protein subgroup II gene* from the CMV into gladiolus using Agrobacterium (Kamo et al. 2010). Moreover, overexpression of the *chloroperoxidase* gene from *Pseudomonas pyrrocinia* and the exochitinase and endochitinase genes from *Fusarium venetanum* induced resistance in gladoilus against *Fusarium oxysporum* f. sp. gladioli (Kamo et al. 2016).

5.8 ABIOTIC STRESS RESISTANCE

Floriculture crops are economically important in horticulture. Nevertheless, abiotic factors like soil salinity, nutrient acidity, nutrient imbalance, and water deficiency are significant constraints that drastically reduce plant yield and other attributes (Ali et al. 2016; Noman and Aqeel 2017; Zafar et al. 2016). To survive stress, physiological, biochemical, and molecular pathways become triggered. Throughout the past few decades, floriculture crop stress adaptability genetics has been extensively studied. Physiological and omics methods have identified abiotic stress resistance genes and regulatory networks. Moreover, various methods, including conventional plant breeding and genetic engineering are used to create abiotic stress-tolerant varieties (Noman et al. 2017). *Rosa chinensis* Jacq. genome transformed with the Medicago *MtDREB1C* gene performed well in freezing temperatures, demonstrating an increase in cold tolerance (Chen et al. 2010). It was proposed that transforming the *AtCBF3* gene could enhance petunia's frost tolerance (Warner 2010). Generally, TFs regulate structural gene expression in response to physical and environmental stimuli and are thus involved in conferring abiotic stress tolerance. Among *WRKY* TFs (TFs), *CmWRKY17* from *Chrysanthemum morifolium*, increased salt tolerance in chrysanthemum by downregulating stress response genes like *AtRD29*, *AtDREB2B*, *AtSOS1*, *AtSOS2*, and *AtSOS3* (Li et al. 2015). Similarly, overexpression of *CmWRKY1* in chrysanthemum cultivar "Jinba" improved drought stress resistance by increasing expression of stress-responsive genes such as *DREB1A, ABI1, ABI2, MYB2*, and *RAB18* of ABA signalling pathway (Fan et al. 2016). *Chrysanthemum morifolium BBX24* zinc finger TF gene affected flowering time and drought and cold stress resistance (Yang et al. 2014). As heat shock proteins play important role in imparting heat stress/ temperature stress in plants. Thus, *CgHSP70* overexpression in chrysanthemum increased heat tolerance. Compared to control, the transgenic plants had higher proline content, higher peroxidase activity, and reduced malondialdehyde (MDA) level. Transgenic plants recovered faster after heat stress, drought, or salinity, suggesting a beneficial effect of *CgHSP70* (Song et al. 2014). The introduction of the *CBF3* gene from *Arabidopsis thaliana* into *Petunia hybrida* (petunia) resulted in increased frost tolerance (Warner 2010). Overexpression of a TF *(RhNAC3)* identified from *Rosa hybrida* resulted in a similar consequence of abiotic stress/drought resistance in rose (Jiang et al. 2014). RNAi technology also demonstrated its potential for generating abiotic stress resistant ornamental crops. *PhERF2-RNAi* petunia lines showed that *PhERF2* (ethylene-responsive element binding factor) negatively regulates waterlogging tolerance (Yin et al. 2019). Both drought stress and ABA treatments reduced *CmBBX19* expression. Researchers have established *CmBBX19* overexpression (*CmBBX19OX*) and *CmBBX19* suppression (*CmBBX19RNAi*) lines and discovered that the latter showed improved drought tolerance in comparison to the wild-type (WT) controls, while the former showed diminished tolerance (Xu et al. 2020). *Chrysanthemum morifolium CmNF-YB8-RNAi* lines were more drought-resistant than overexpression lines. RNAi lines modify stomatal opening rate and leaf epidermal cuticle thickness to reduce transpiration (Wang et al. 2022). Environmental stress tolerance can be enhanced by generating transgenic cultivars, along with an expanded product line, thanks to genetic engineering's unprecedented potential for improvement in plant characteristics mainly in ornamental plants.

5.9 VASE LIFE

Extending vase of cut flower while maintaining their aroma and flavour life is an important component in persuading customers to repurchase cut flowers (Kamthan et al. 2016). After harvest, cut flowers are typically dispersed without their roots, which makes it extremely challenging to preserve their quality (Mekapogu et al. 2022). Various biotechnological approaches have been employed to achieve the goal of extending the vase life (Matas et al. 2009). In the ageing process of flowers, ethylene can be a pivotal factor in particular plant species. By manipulating genes involved in ethylene biosynthesis or signalling, researchers have been able to extend the postharvest life of flowers in various species that normally experience ethylene-dependent flower senescence. Transcripts for the several phases of the ethylene biosynthesis pathway in carnation have been found using transcriptome analysis (Tanase et al. 2012). Using a floral transcriptome, ethylene-responsive transcripts, which are likely essential regulators of ethylene-influenced cell growth, were also found in *Rosa hybrida* (Pei et al. 2013). Yet, in recent years, numerous studies have concentrated on the understanding of ethylene biosynthesis. Carnation plants that are transgenic and have a delayed onset of petal senescence were produced by silencing the *ACO* gene, which is responsible for inhibiting ethylene biosynthesis (Savin et al. 1995). By maintaining ethylene resistance or inhibiting ethylene biosynthesis genes, it is possible to extend the shelf life of a product. By altering the ethylene receptor gene in Oncidium and Odontoglossum, increased vase life has been observed (Raffeiner et al. 2009). Different research groups have conducted antisense experiments with *ACS* or *ACO* genes to enhance the shelf life of flowers of ornamental plants (Bapat et al. 2010; Litz and Padilla 2012). Agrobacterium-mediated transformation of *PSAG12-ipt* into the Rose cultivar "Linda" increased the level of *ipt* expression in transgenic plants when the plants were exposed to darkness and a high level of exogenous ethylene. Before senescence, cut flower leaves turned yellow and ethylene also accelerated yellowing. This reduced flower beauty, quality, and vase life. Thus, transgenic chrysanthemums with reduced ethylene sensitivity should have longer vase life (Satoh et al. 2006). In addition, the endogenous production and distribution of plant growth regulators such as IAA, iPA, and ABA during flower induction and initiation may enable us to better plan the harvest timing for optimal flower uniformity and ornamental quality (Jiang et al. 2010; Teixeira da Silva et al. 2013). Reduced ethylene sensitivity is directly proportional to improved vase life, therefore, chrysanthemum transformed with a mutated ethylene receptor gene *mDG-ERS1 (etr1-4)* reduced leaf senescence, which has been proved very useful (Satoh and Yokoya 2008). Similarly, compared to non-transformed plants, petunias constitutively expressing a mutant *etr1-1* (ethylene receptor-1) transgene from Arabidopsis were resistant to ethylene and had larger flowers with a longer vase life (Clevenger et al. 2004). Yet, they have a deleterious effect on a number of morpho- physiological characteristics. Using a flower-specific fructose-bisphosphatase 1 (*FBP1*) promoter to drive the expression of the same transgene, it was possible to create normal plants with longer-lasting flowers in kalancho, Campanula, and *Dianthus caryophyllus* (Olsen et al. 2015). A recent study of Japanese morning glory has revealed that a NAC TF, EPHEMERAL1 (*EPH1*), is a key regulator in ethylene-independent petal senescence, despite the fact that little is known about the regulatory mechanisms of petal senescence in flowers that show ethylene-independent senescence. Suppressing *EPH1* expression markedly delays petal senescence in morning glory.

5.10 GENOME EDITING IN ORNAMENTAL CROPS

Traditional breeding methods like hybridization, mutagenesis, polyploidization, and double haploid induction have historically been employed for plant trait enhancement but suffer limitations such as narrow gene pool, high heterozygosity, lengthy lifecycle, and seed production challenges. The advent of CRISPR/Cas9 genome editing offers a precise and efficient approach to induce targeted DNA modifications, potentially overcoming these drawbacks (Hilscher et al., 2017; Ahn et al., 2020). This efficient and versatile genome editing technology can be used to improve the plant canopy architecture, colors, enhance shelf life as well as tolerance to abiotic and biotic stress in plants (Ahn et al., 2020; Jin et al., 2023). Genome editing in ornamental plants paves way for generating of new generation ornamental plants with desired floral attributes. Since *Agrobacterium*- mediated transformation is difficult in some ornamental crops, it is usually necessary for devising strategies to transfer desired genes of useful traits (Erpen-Dalla Corte et al. 2019). Many studies on ornamental plants have aimed to increase floricultural traits such as flower color, size, form, scent, disease resistance, vase life, etc. However, because it is difficult to apply genome-editing systems to ornamental plants whose genome sequence information is unknown, only a limited number of research utilizing CRISPR-Cas genome-editing systems have been published in ornamental plants (Table 5.2). *Petunia hybrida* was the first ornamental plant genome to be edited using the CRISPR-Cas9 method. The successful production of mutations in the phytoene desaturase (*PhPDS*) gene of petunia producing the predicted albino phenotype was reported in petunia (Zhang et al. 2016). Recently, a crucial gene encoding the ethylene biosynthesis enzyme 1-aminocyclopropylate-1-carboxylate oxidase1 (*PhACO1*) was modified using CRISPR-Cas9 in the Petunia cultivar "Mirage Rose." Transgenic Petunia plants produced less ethylene and their petals aged more slowly. Mutant Japanese morning glory (Ipomoea nil "Violet") plants created by the CRISPR-Cas9 method that targeted the EPHEMERAL1 (*EPH1*) gene, encoding the master regulator of petal senescence as a NAC transcription, exhibited delayed petal senescence (Shibuya et al. 2018). The gene editing of the dihydroflavonol-4-reductase (*DFR*) gene in *Ipomoea nil* resulted in the first floral color modification in higher plants by CRISPR-Cas9-mediated mutagenesis (Watanabe et al. 2017; 2018) used CRISPR-Cas9-mediated mutagenesis for the carotenoid cleavage dioxygenase 4 (*CCD4*) gene to create mutant *Ipomoea nil* plants with pale-yellow petals. Mutant Lilium species of the albino, pale-yellow, and albino-green chimeric types were produced with various patterns of mutation, including base insertion, deletion, and substitution, by targeting the phytoene desaturase (*PDS*) gene encoding a key enzyme of carotenoid synthesis for induced mutagenesis (Yan et al. 2019). Furthermore, Nishihara et al. (2018) observed a color change in *Torenia fournieri* from pale blue to white via CRISPR-Cas9-mediated mutation on the flavone 3-hydrolase (*F3'H*) gene, which encodes a key enzyme in the flavonoid biosynthetic pathway. In addition, ornamental plant species with higher polyploidy, such as roses (tetraploid) and chrysanthemums (hexaploid), require a very effective editing platform for the future enhancement of these plants. Thus, more accurate information regarding whole-genome sequencing is needed.

TABLE 5.2
CRISPR-Cas based genome editing in ornamental plant species

Sr. No.	Plant species	Target gene	Gene function	Traits	References
1.	Chrysanthemum *(Chrysanthemum Morifolium)*	*CpYGFP*	Express yellowish-green fluorescent protein (marker)	Disrupts fluorescence protein	Kishi-Kaboshi et al. (2017)
2.	Petunia (*Petunia hybrida*)	*PDS*	Key enzyme in carotenoid biosynthesis	Albino phenotype	Zhang et al. (2016)
3.	Petunia (*P. Hybrid; Madness*)	*NR*	Facilitate nitrogen uptake & nitrate metabolism	Deficiency in nitrate assimilation	Subburaj et al. (2016)
4.	Orchid *(Dendrobium officinale)*	*C3H, C4H, 4CL, CCR, IRX*	Lignocelluloses biosynthesis pathway	Reduced ligno cellulose biosynthesis	Kui et al. (2017)
5.	Wishbone flower (*Torenia fournieri)*	*TfRAD1*	Regulate petal shape & corolla pigmentation	Abnormal flowers	Sui et al. (2017)
6.	Petunia *(Petunia inflate)*	*PiSSK1*	SCF's subunit, degrade non-self S-RNases non-self S-RNases	Self-incompatibility	Sun and Kao (2018)
7.	Japanese morning glory *(Ipomoea nil*)	*DFR-B*	Anthocyanin biosynthesis enzyme	Anthocyanin-less stems, leaves and flowers	Watanabe et al. (2017)
8.	Wishbone Flower (*Torenia fournieri* L.)	*F3H*	Flavonoid biosynthesis	Pale blue (almost white) flowers	Nishihara et al. (2018)
9.	Japanese morning glory (*I. nil*)	*EPH1*	Key regulator of petal senescence	Delayed petal senescence	Shibuya et al. (2018)
10.	Japanese morning glory (*I. nil*)	*CCD*	Carotenoid degradation	White petals to turn pale	Watanabe et al. (2018)
11.	Lilium (*Lilium longiflorum*)	*LlPDS*	Key enzyme of carotenoid synthesis	Yellow	Yan et al (2019)

(*continued*)

TABLE 5.2 (Continued)
Applications of genetic engineering for enhancing floral attributes in various ornamental crops

Sr. No.	Plant species	Target gene	Gene function	Traits	References
12.	Japanese Gentian (*Gentiana scabra*)	*Gt5GT, Gt3'GT, Gt5/3'AT*	Encodes anthocyanin (vivid blue flowers)	Chimeric mutations that are completely albino, albino green & pale yellow	Tasaki et al. (2020)
13.	Japanese Gentian (*G. scabra*)	*GST1*	Anthocyanin accumulation in petals	Pale red violet, dull pink, and pale mauve flowers	Chib et al. (2020)
14.	Moon Orchid (*Phalaenopsis amabilis*)	*PDS3*	Regulate enzyme in carotenoid synthesis	Reduced anthocyanin in petals, white and pale blue flower	Semiarti et al. (2020)
15.	Moth Orchids (*P. equestris)*	*MADS*	Encodes DBP for flower initiation & development	Albino leaf	Tong et al. (2020)
16.	Petunia (*Petunia hybrid;* Mirage Rose)	*ACO1*	Ethylene production linked to flower senescence	MADS-null mutants (Long juvenile period)	Xu et al. (2020)
17.	Petunia (*Petunia hybrid;* Mirage Rose)	*ACO3, ACO4*	Ethylene production linked to flower senescence	Reduced ethylene production reduced and enhanced vase life	Xu et al. (2021)
18.	Petunia (*P. hybrid;* Madness Midnight*)*	*F3'H*	Anthocyanidin and flavonol synthesis	Reduced ethylene production and enhanced vase life	Yu et al. (2021)
19.	Petunia (*P. hybrid;* Mitchell Diploid)	*DPL*	Regulate corolla tube venation	Pale purplish-pink flower color	Zhang et al. (2021)
20.	Poinsettia (*Euphorbia pulcherrima)*	*F3'H*	Cyanidin-type anthocyanins formation	Absence of the vein associated anthocyanin pattern	Nitarska et al. (2021)
21.	*Petunia hybrid* (Mirage Rose)	*ACO1, ACO3, ACO4*	Seed germination	Flower color changed from vivid red to vivid reddish-orange	Naing et al. (2021)

5.11 CONCLUSION

The economic value of ornamental plants is on the rise globally due to the potential of gene modification, offering benefits such as diverse floral appearances, novel colors, and enhanced fragrances. Consumer acceptance of GM flowers, such as roses and carnations, has been positive, with no observed adverse effects during vegetative propagation. Exploitable characteristics of ornamental plants, particularly in secondary metabolite production, hold promise for pharmaceutical applications. Advances in molecular markers and genomic sequencing are expected to lead to the development of new transgenic varieties. Cutting-edge techniques like ZFNs, TALENs, and CRISPR offer avenues for targeted genomic modifications, facilitating innovation in the floriculture industry. However, economic and regulatory challenges hinder the commercialization of GM ornamentals. Addressing regulatory barriers and relaxing requirements for non-food plants could facilitate the commercial release of genetically modified ornamental products.

REFERENCES

Adal, A. M., Sarker, L. S., Malli, R. P., Liang, P., & Mahmoud, S. S. (2019). RNA-Seq in the discovery of a sparsely expressed scent-determining monoterpene synthase in lavender (Lavandula). *Planta*, 249, 271–290.

Ahn, C.H., Ramya, M., An, H.R., Park, P.M., Kim, Y.J., Lee, S.Y. and Jang, S., 2020. Progress and challenges in the improvement of ornamental plants by genome editing. Plants, 9(6), p.687.

Aida, R., Komano, M., Saito, M., Nakase, K., & Murai, K. (2008). Chrysanthemum flower shape modification by suppression of chrysanthemum-*AGAMOUS* gene. *Plant Biotechnology*, 25(1), 55–59.

Ali, Q., Haider, M. Z., Iftikhar, W., Jamil, S., Tariq Javed, M., Noman, A., ... & Perveen, R. (2016). Drought tolerance potential of *Vigna mungo* L. lines as deciphered by modulated growth, antioxidant defense, and nutrient acquisition patterns. *Brazilian Journal of Botany*, 39, 801–812.

Aranovich, D., Lewinsohn, E., & Zaccai, M. (2007). Post-harvest enhancement of aroma in transgenic lisianthus (*Eustoma grandiflorum*) using the Clarkia breweri benzyl alcohol acetyl transferase (BEAT) gene. *Postharvest biology and technology*. 43, 255–260. doi:10.1016/j.postharvbio.2006.09.001

Azadi, P., Bagheri, H., Nalousi, A. M., Nazari, F., & Chandler, S. F. (2016). Current status and biotechnological advances in genetic engineering of ornamental plants. *Biotechnology Advances*, 34(6), 1073–1090.

Azadi, P., Otang, N. V., Chin, D. P., Nakamura, I., Fujisawa, M., Harada, H., ... & Mii, M. (2010). Metabolic engineering of Lilium× formolongi using multiple genes of the carotenoid biosynthesis pathway. Plant Biotechnology Reports, 4, 269–280.

Bapat, P. N., Shankar Rao, D. S., Krishna Prasad, S., & Yelamaggad, C. V. (2010). High-pressure dielectric investigations of nanocolloidalaerosil– Nematic liquid crystal composites. *The Journal of Physical Chemistry B*, 114(40), 12825–12832.

Blazquez, M. A., & Weigel, D. (2000). Integration of floral inductive signals in Arabidopsis. *Nature*, 404(6780), 889–892.

Boase, M.R., Winefield, C.S., Lill, T.A. and Bendall, M.J., 2004. Transgenic regal pelargoniums that express the rol c gene from Agrobacterium rhizogenes exhibit a dwarf floral and vegetative phenotype. *In Vitro Cellular & Developmental Biology-Plant*, 40, 46–50.

Boase, M.R., Lewis, D.H., Davies, K.M., Marshall, G.B., Patel, D., Schwinn, K.E. and Deroles, S.C., 2010. Isolation and antisense suppression of flavonoid 3', 5'-hydroxylase modifies flower pigments and colour in cyclamen. BMC plant biology, 10, pp.1–12.

Brugliera, F., Tao, G.Q., Tems, U., Kalc, G., Mouradova, E., Price, K., Stevenson, K., Nakamura, N., Stacey, I., Katsumoto, Y. and Tanaka, Y., 2013. Violet/blue chrysanthemums—metabolic engineering of the anthocyanin biosynthetic pathway results in novel petal colors. Plant and Cell Physiology, 54(10), pp.1696–1710.

Casanova, D., Cirera, J., Llunell, M., Alemany, P., Avnir, D., & Alvarez, S. (2004). Minimal distortion pathways in polyhedral rearrangements. *Journal of the American Chemical Society*, 126(6), 1755–1763.

Chandler, S., & Tanaka, Y. (2007). Genetic modification in floriculture. *Critical Reviews in Plant Sciences*, 26(4), 169–197.

Chandler, S. F., & Brugliera, F. (2011). Genetic modification in floriculture. *Biotechnology Letters*, 33(2), 207–214.

Chandler, S. F., & Sanchez, C. (2012). Genetic modification; the development of transgenic ornamental plant varieties. *Plant Biotechnology Journal*, 10(8), 891–903.

Chang, C. J., Chen, J. K., Lin, K. S., Wei, Y. H., Chao, P. Y., & Huang, C. Y. (2020). Enhanced visible-light-driven photocatalytic degradation by metal wire-mesh supported Ag/flower-like Bi2WO6 photocatalysts. *Journal of Alloys and Compounds*, 813, 152186.

Chen, A., & Dubcovsky, J. (2012). Wheat TILLING mutants show that the vernalization gene VRN1 down-regulates the flowering repressor VRN2 in leaves but is not essential for flowering. *PLoS Genetics*, 8(12), e1003134.

Chen, H., Lai, Z., Shi, J., Xiao, Y., Chen, Z., & Xu, X. (2010). Roles of Arabidopsis WRKY18, WRKY40 and WRKY60 transcription factors in plant responses to abscisic acid and abiotic stress. *BMC Plant Biology*, 10(1), 1–15.

Chib, S., Thangaraj, A., Kaul, S., Dhar, M. K., & Kaul, T. (2020). Development of a system for efficient callus production, somatic embryogenesis and gene editing using *CRISPR/Cas9* in Saffron (*Crocus sativus* L.). *Plant Methods*, 16(1), 1–10.

Clevenger, D. J., Barrett, J. E., Klee, H. J., & Clark, D. G. (2004). Factors affecting seed production in transgenic ethylene-insensitive petunias. *Journal of the American Society for Horticultural Science*, 129(3), 401–406.

Colquhoun, T. A., & Clark, D. G. (2011). Unraveling the regulation of floral fragrance biosynthesis. *Plant Signaling & Behavior*, 6(3), 378–381.

da Silva Trentin, D., Giordani, R. B., Zimmer, K. R., Da Silva, A. G., Da Silva, M. V., dos Santos Correia, M. T., ... & Macedo, A. J. (2011). Potential of medicinal plants from the Brazilian semi-arid region (Caatinga) against *Staphylococcus epidermidis* planktonic and biofilm lifestyles. *Journal of Ethnopharmacology*, 137(1), 327–335.

Ding, L., Zhao, K., Zhang, X., Song, A., Su, J., Hu, Y., Zhao, ... & Chen F. (December 2019). Comprehensive characterization of a floral mutant reveals the mechanism of hooked petal morphogenesis in *Chrysanthemum morifolium*. *Plant Biotechnology Journal.* 17(12), 2325–2340.

Dohm, A., Ludwig, C., Schilling, D., Debener, T. (2002). Transformation of roses with genes for antifungal proteins to reduce their susceptibility to fungal diseases. *Acta horticulturae.* 572, 105–111

Du, X., Li, W., Sheng, L., Deng, Y., Wang, Y., Zhang, W., ... & Chen, S. (2018). Over-expression of chrysanthemum *CmDREB6* enhanced tolerance of chrysanthemum to heat stress. *BMC Plant Biology*, 18(1), 1–10.

Du, Y., Mao, L., Queenborough, S. A., Freckleton, R. P., Chen, B., & Ma, K. (2015). Phylogenetic constraints and trait correlates of flowering phenology in the angiosperm flora of C hina. *Global Ecology and Biogeography*, 24(8), 928–938.

Du, F., Wang, T., Fan, J.M., Liu, Z.Z., Zong, J.X., Fan, W.X., Han, Y.H. and Grierson, D., 2019. Volatile composition and classification of Lilium flower aroma types and identification, polymorphisms, and alternative splicing of their monoterpene synthase genes. Horticulture research, 6.

Erpen-Dalla Corte, L., Mahmoud, L. M., Moraes, T. S., Mou, Z., Grosser, J. W., & Dutt, M. (2019). Development of improved fruit, vegetable, and ornamental crops using the *CRISPR/Cas9* genome editing technique. *Plants*, 8(12), 601.

Fan, Q., Song, A., Jiang, J., Zhang, T., Sun, H., Wang, Y., ... & Chen, F. (2016). *CmWRKY1* enhances the dehydration tolerance of chrysanthemum through the regulation of ABA-associated genes. *PLoS One*, 11(3), e0150572.

Feng, F., Jun, Y. B., Liu, X., & Li, L. (2010). An adjustable approach to fuzzy soft set based decision making. *Journal of Computational and Applied Mathematics*, 234(1), 10–20.

Fornara, F., de Montaigu, A., & Coupland, G. (2010). SnapShot: Control of flowering in Arabidopsis. *Cell*, 141(3), 550.

Fu, Y. B., Peterson, G. W., & Dong, Y. (2016). Increasing genome sampling and improving SNP genotyping for genotyping-by-sequencing with new combinations of restriction enzymes. *G3: Genes, Genomes, Genetics*, 6(4), 845–856.

Fukuchi-Mizutani, M., Akagi, M., Ishiguro, K., Katsumoto, Y., Fukui, Y., Togami, J., ... & Tanaka, Y. (2011). Biochemical and molecular characterization of anthocyanidin/flavonol 3-glucosylation pathways in *Rosa× hybrida*. *Plant Biotechnology*, 28(2), 239–244.

Haider, S., Gao, Y., & Gao, Y. (2020). Standardized genetic transformation protocol for Chrysanthemum cv.'Jinba'with *TERMINAL FLOWER 1* homolog *CmTFL1a*. *Genes*, 11(8), 860.

Han, B. H., Suh, E. J., Lee, S. Y., Shin, H. K., & Lim, Y. P. (2007). Selection of nonbranching lines induced by introducing Ls-like cDNA into Chrysanthemum (*Dendranthema × grandiflorum* (Ramat.) Kitamura) Shuho-no-chikara. *ScientiaHorticulturae*. 115, 70–75. doi:10.1016/j.scienta.2007.07.012

He, J., Chen, F., Chen, S., Fang, W., Miao, H., & Luo, H. (2009). Transformation of*Lycorislongituba agglutinin* gene to cut chrysanthemum and identification of aphid resistance in the transgenic plants. *Acta Botanica Boreali-OccidentaliaSinica*, 29(11), 2318–2325.

Hilscher, J., Bürstmayr, H., & Stoger, E. (2017). Targeted modification of plant genomes for precision crop breeding. *Biotechnology Journal* 12, 1600173.

Hu, Z., Tang, B., Wu, Q., Zheng, J., Leng, P., & Zhang, K. (2017). Transcriptome sequencing analysis reveals a difference in monoterpene biosynthesis between scented Lilium 'Siberia'and unscented Lilium 'Novano'. *Frontiers in Plant Science*, 8, 1351.

Hu, Z., Zhang, H., Leng, P., Zhao, J., Wang, W., & Wang, S. (2013). The emission of floral scent from Lilium 'siberia'in response to light intensity and temperature. *Acta Physiologiae Plantarum*, 35, 1691–1700.

Huang, J., Liu, X., Wang, J., & Lü, Y. (2014). Transcriptomic analysis of Asiatic lily in the process of vernalization via RNA-seq. *Molecular Biology Reports*, 41, 3839–3852.

Jiang, B., Miao, H., Chen, S., Zhang, S., Chen, F., and Fang, W. (2010). The lateral suppressor-like gene,DgLsL, alternated the axillary branching in transgenic chrysanthemum (*Chrysanthemum × morifolium*) by modulating IAA and GA content. *Plant Molecular Biology Reporter*, 28, 144–151. doi:10.1007/s11105-009-0130-3

Jiang, X., Zhang, C., Lü, P., Jiang, G., Liu, X., Dai, F., & Gao, J. (2014). Rh NAC 3, a stress-associated NAC transcription factor, has a role in dehydration tolerance through regulating osmotic stress-related genes in rose petals. *Plant Biotechnology Journal*, 12(1), 38–48.

Jin, C., Dong, L., Wei, C., Wani, M.A., Yang, C., Li, S. and Li, F., 2023. Creating novel ornamentals via new strategies in the era of genome editing. Frontiers in Plant Science, 14, p.1142866.

Jung, C., & Müller, A. E. (2009). Flowering time control and applications in plant breeding. *Trends in Plant Science*, 14(10), 563–573.

Kamo, K., Jordan, R., Guaragna, M. A., Hsu, H. T., & Ueng, P. (2010). Resistance to Cucumber mosaic virus in Gladiolus plants transformed with either a defective replicase or coat protein subgroup II gene from Cucumber mosaic virus. *Plant Cell Reports*, 29, 695–704.

Kamo, K., Lakshman, D., Pandey, R., Guaragna, M. A., Okubara, P., Rajasekaran, K., ... & Jordan, R. (2016). Resistance to *Fusarium oxysporum* f. sp. gladioli in transgenic Gladiolus plants expressing either a bacterial chloroperoxidase or fungal chitinase genes. *Plant Cell, Tissue and Organ Culture (PCTOC)*, 124, 541–553.

Kamthan, A., Chaudhuri, A., Kamthan, M., & Datta, A. (2016). Genetically modified (GM) crops: Milestones and new advances in crop improvement. *Theoretical and Applied Genetics*, 129, 1639–1655.

Katsumoto, Y., Fukuchi-Mizutani, M., Fukui, Y., Brugliera, F., Holton, T.A., Karan, M., Nakamura, N., Yonekura-Sakakibara, K., Togami, J., Pigeaire, A. and Tao, G.Q., 2007. Engineering of the rose flavonoid biosynthetic pathway successfully generated blue-hued flowers accumulating delphinidin. *Plant and cell physiology*, 48(11), pp.1589–1600.

Khodakovskaya, M.,Vanková, R., Malbeck, J., Li, A., Li, Y., & McAvoy, R. (2009). Enhancement of flowering and branching phenotype in chrysanthemum by expression of ipt under the control of a 0.821kb fragment of the LEACO1 gene promoter. *Plant Cell Reports* 28, 1351–1362. doi:10.1007/s00299-009-0735-x

Kim, D. C., & Anderson, N. O. (2006). Comparative analysis of laboratory freezing methods to establish cold tolerance of detached rhizomes and intact crowns in garden chrysanthemums (*Dendranthema* × *grandiflora Tzvelv.*). *Scientia Horticulturae*, 109(4), 345–352.

Kim, Y. K., & Xiao, C. L. (2011). Stability and fitness of pyraclostrobin-and boscalid-resistant phenotypes in field isolates of *Botrytis cinerea* from apple. *Phytopathology*, 101(11), 1385–1391.

Kishi-Kaboshi, M., Aida, R., & Sasaki, K. (2017). Generation of gene-edited *Chrysanthemum morifolium* using multicopy transgenes as targets and markers. *Plant and Cell Physiology*, 58(2), 216–226.

Kishimoto, S., Maoka, T., Nakayama, M., & Ohmiya, A. (2004). Carotenoid composition in petals of chrysanthemum (*Dendranthema grandiflorum* (Ramat.) Kitamura). *Phytochemistry*, 65(20), 2781–2787.

Kui, L., Chen, H., Zhang, W., He, S., Xiong, Z., Zhang, Y., ... & Cai, J. (2017). Building a genetic manipulation tool box for orchid biology: Identification of constitutive promoters and application of *CRISPR/Cas9* in the orchid, *Dendrobium officinale*. *Frontiers in Plant Science*, 7, 2036.

Laitinen, R. A., Ainasoja, M., Broholm, S. K., Teeri, T. H., & Elomaa, P. (2008). Identification of target genes for a MYB-type anthocyanin regulator in Gerbera hybrida. *Journal of Experimental Botany*, 59(13), 3691–3703.

Lavy, M., Zuker, A., Lewinsohn, E., Larkov, O., Ravid, U., Vainstein, A. and Weiss, D., 2002. Linalool and linalool oxide production in transgenic carnation flowers expressing the Clarkia breweri linalool synthase gene. *Molecular Breeding*, 9, pp.103–111.

Leeggangers, H. A., Rosilio-Brami, T., Bigas-Nadal, J., Rubin, N., Van Dijk, A. D., Nunez de Caceres Gonzalez, F. F., ... & Zaccai, M. (2018). Tulipa gesneriana and *Lilium longiflorum* PEBP genes and their putative roles in flowering time control. *Plant and Cell Physiology*, 59(1), 90–106.

Li, D., Liu, X., Shu, L., Zhang, H., Zhang, S., Song, Y., & Zhang, Z. (2020). Global analysis of the AP2/ERF gene family in rose (*Rosa chinensis*) genome unveils the role of RcERF099 in Botrytis resistance. *BMC Plant Biology*. 20, 1–15.

Li, X., Gasic, K., Cammue, B., Broekaert, W., & Korban, S. S. (2003). Transgenic rose lines harboring an antimicrobial protein gene, Ace-AMP1, demonstrate enhanced resistance to powdery mildew (*Sphaerotheca pannosa*). *Planta*, 218, 226–232.

Li, X., Luo, L., Yang, J., Li, B., & Yuan, H. (2015). Mechanisms for solubilization of various insoluble phosphates and activation of immobilized phosphates in different soils by an efficient and salinity-tolerant *Aspergillus niger* strain An2. *Applied Biochemistry and Biotechnology*, 175, 2755–2768.

Litt, A., & Irish, V. F. (2003). Duplication and diversification in the *APETALA1/FRUITFULL* floral homeotic gene lineage: Implications for the evolution of floral development. *Genetics*, 165(2), 821–833.

Litz RE, Padilla G (2012) Genetic transformation of fruit trees. In: Priyadarshan PM, Schnell RJ (eds) Genomics of tree crops. Springer, Berlin, pp 117–153

Matas, A. J., Gapper, N. E., Chung, M. Y., Giovannoni, J. J., & Rose, J. K. (2009). Biology and genetic engineering of fruit maturation for enhanced quality and shelf-life. *Current Opinion in Biotechnology*, 20(2), 197–203.

Mathew, I., Shimelis, H., Shayanowako, A. I. T., Laing, M., & Chaplot, V. (2019). Genome-wide association study of drought tolerance and biomass allocation in wheat. *PLoS One*, 14(12), e0225383.

Mekapogu, M., Kwon, O. K., Song, H. Y., & Jung, J. A. (2022). Towards the improvement of ornamental attributes in chrysanthemum: Recent progress in biotechnological advances. *International Journal of Molecular Sciences*, 23(20), 12284.

Naing, A.H., Xu, J. & Kim, C.K.2021. Editing of 1-aminocyclopropane-1-carboxylate oxidase genes negatively affects petunia seed germination. *Plant Cell Reports* 41, 209–220

Narumi, T., Aida, R., Koyama, T., Yamaguchi, H., Sasaki, K., Shikata, M., Nakayama, M., Ohme-Takagi, M. and Ohtsubo, N., 2011. Arabidopsis chimeric TCP3 repressor produces novel floral traits in Torenia fournieri and Chrysanthemum morifolium. *Plant Biotechnology*, 28(2), pp.131–140.

Narumi, T., Aida, R., Ohmiya, A., & Satoh, S. (2005). Transformation of chrysanthemum with mutated ethylene receptor genes: mDG-ERS1 transgenes conferring reduced ethylene sensitivity and characterization of the transformants. *Postharvest Biology Technology*. 37, 101–110. doi:10.1016/j.postharvbio.2005.04.008

Nishihara, M., Higuchi, A., Watanabe, A., & Tasaki, K. (2018). Application of the *CRISPR/Cas9* system for modification of flower color in *Torenia fournieri*. *BMC Plant Biology*, 18(1), 1–9.

Nitarska, D., Boehm, R., Debener, T., Lucaciu, R. C., & Halbwirth, H. (2021). First genome edited poinsettias: Targeted mutagenesis of flavonoid 3′-hydroxylase using *CRISPR/Cas9* results in a colour shift. *Plant Cell, Tissue and Organ Culture (PCTOC)*, 147(1), 49–60.

Noman, A., Aqeel, M., Deng, J., Khalid, N., Sanaullah, T., & Shuilin, H. (2017). Biotechnological advancements for improving floral attributes in ornamental plants. *Frontiers in Plant Science*, 8, 530.

Noman, A., Aqeel, M., & He, S. (2016 November 21). CRISPR-Cas9: Tool for qualitative and quantitative plant genome editing. *Frontiers in Plant Science*,7 ,1740.

Nontaswatsri, C., & Fukai, S. (2010). Genetic transformation of carnation (*Dianthus caryophylus* L.). *Protocols for in vitro Propagation of Ornamental Plants, Methods in Molecular Biology* 589 87–96.

Oda, A., Narumi, T., Li, T., Kando, T., Higuchi, Y., Sumitomo, K., ... & Hisamatsu, T. (2012). *CsFTL3*, a chrysanthemum *FLOWERING LOCUS T*-like gene, is a key regulator of photoperiodic flowering in chrysanthemums. *Journal of Experimental Botany*, 63(3), 1461–1477.

Olsen, A., Lütken, H., Hegelund, J. N., & Müller, R. (2015). Ethylene resistance in flowering ornamental plants–improvements and future perspectives. *Horticulture Research*, 2. 15038

Parisi, C., Tillie, P., & Rodríguez-Cerezo, E. (2016). The global pipeline of GM crops out to 2020. *Nature Biotechnology*, 34(1), 31–36.

Pei, H., Ma, N., Chen, J., Zheng, Y., Tian, J., Li, J., Zhang, S., Fei, Z. and Gao, J., 2013. Integrative analysis of miRNA and mRNA profiles in response to ethylene in rose petals during flower opening. PloS one, 8(5), p.e64290.

Peled-Zehavi, H., Oliva, M., Xie, Q., Tzin, V., Oren-Shamir, M., Aharoni, A. and Galili, G., 2015. Metabolic engineering of the phenylpropanoid and its primary, precursor pathway to enhance the flavor of fruits and the aroma of flowers. Bioengineering, 2(4), pp.204–212.

Petty, L. M., Harberd, N. P., Carré, I. A., Thomas, B., & Jackson, S. D. (2003). Expression of the Arabidopsis gai gene under its own promoter causes a reduction in plant height in chrysanthemum by attenuation of the gibberellin response. *Plant Science*, 164(2), 175–182.

Pourhosseini, L., Kermani, M. J., Habashi, A. A., & Khalighi, A. (2013). Efficiency of direct and indirect shoot organogenesis in different genotypes of *Rosa hybrida. Plant Cell, Tissue and Organ Culture (PCTOC)*, 112, 101–108.

Qiu, X., Wang, Q., Zhang, H., Jian, H., Zhou, N., Ji, C., ... & Tang, K. 2015. Antisense *RhMLO1* gene transformation enhances resistance to the powdery mildew pathogen in *Rosa multiflora. Plant Molecular Biology Reports* 33, 1659–1665.

Raffeiner, B., Serek, M., & Winkelmann, T. (2009). *Agrobacterium tumefaciens*-mediated transformation of Oncidium and Odontoglossum orchid species with the ethylene receptor mutant gene etr1-1. *Plant Cell, Tissue and Organ Culture (PCTOC)*, 98, 125–134.

Ruokolainen, S., Ng, Y. P., Albert, V. A., Elomaa, P., & Teeri, T. H. (2010). Large scale interaction analysis predicts that the Gerbera hybrida floral E function is provided both by general and specialized proteins. *BMC Plant Biology* 10, 129. doi:10.1186/1471-2229-10-129

Sato, T., Takahoko, M., & Okamoto, H. (2006). HuC: Kaede, a useful tool to label neural morphologies in networks in vivo. *Genesis*, 44(3), 136–142.

Sato, Y., & Yokoya, S. (2008). Enhanced tolerance to drought stress in transgenic rice plants overexpressing a small heat-shock protein, sHSP17. 7. *Plant Cell Reports*, 27, 329–334.

Savin, K. W., Baudinette, S. C., Graham, M. W., Michael, M. Z., Nugent, G. D., Lu, C. Y., ... & Cornish, E. C. (1995). Antisense ACC oxidase RNA delays carnation petal senescence. *HortScience*, 30(5), 970–972.

Schwinn, K. E., Boase, M. R., Bradley, J. M., Lewis, D. H., Deroles, S. C., Martin, C. R., & Davies, K. M. (2014). MYB and bHLH transcription factor transgenes increase anthocyanin pigmentation in petunia and lisianthus plants, and the petunia phenotypes are strongly enhanced under field conditions. *Frontiers in Plant Science*, 5, 603.

Semiarti, E., Nopitasari, S., Setiawati, Y., Lawrie, M. D., Purwantoro, A., Widada, J., ... & Asano, Y. (2020). Application of *CRISPR/Cas9* genome editing system for molecular breeding of orchids. *Indonesian Journal of Biotechnology*, 25(1), 61–68.

Sen, S., Kumar, S., Ghani, M., & Thakur, M. (2013). Agrobacterium mediated genetic transformation of chrysanthemum (*Dendranthema grandiflora* Tzvelev) with rice chitinase gene for improved resistance against Septoria obesa. *Plant Pathology Journal*. 12, 1–10. doi: 10.3923/ppj.2013.1.10

Shibata, M. (2008). Importance of genetic transformation in ornamental plant breeding. *Plant Biotechnology*, 25(1), 3–8.

Shibuya, K., Watanabe, K., & Ono, M. (2018). *CRISPR/Cas9*-mediated mutagenesis of the *EPHEMERAL1* locus that regulates petal senescence in Japanese morning glory. *Plant Physiology and Biochemistry*, 131, 53–57.

Shinoyama, H., Mochizuki, A., Nomura, Y., & Kamada, H. (2008). Environmental risk assessment of genetically modified chrysanthemums containing a modified *cry1Ab* gene from Bacillus thuringiensis. *Plant Biotechnology* 25, 17–29. doi:10.5511/plantbiotechnology.25.17

Shulga, O. A., Mitiouchkina, T. Y., Shchennikova, A. V., Skryabin, K. G., & Dolgov, S. V. (2011). Overexpression of AP1-like genes from Asteraceae induces early-flowering in transgenic Chrysanthemum plants. *In Vitro Cellular & Developmental Biology-Plant*, 47, 553–560.

Song, A., Zhu, X., Chen, F., Gao, H., Jiang, J., & Chen, S. (2014). A chrysanthemum heat shock protein confers tolerance to abiotic stress. *International Journal of Molecular Sciences*, 15(3), 5063–5078.

Spitzer-Rimon, B., Farhi, M., Albo, B., Cna'ani, A., Ben Zvi, M. M., Masci, T., ... & Vainstein, A. (2012). The R2R3-MYB-like regulatory factor EOBI, acting downstream of EOBII, regulates scent production by activating *ODO1* and structural scent-related genes in petunia. *The Plant Cell*, 24(12), 5089–5105.

Srikanth, A., & Schmid, M. (2011). Regulation of flowering time: All roads lead to Rome. *Cellular and Molecular Life Sciences*, 68, 2013–2037.

Su, S., Xiao, W., Guo, W., Yao, X., Xiao, J., Ye, Z., ... & Luo, D. (2017). The *CYCLOIDEA–RADIALIS* module regulates petal shape and pigmentation, leading to bilateral corolla symmetry in *Torenia fournieri* (Linderniaceae). *New Phytologist*, 215(4), 1582–1593.

Subburaj, S., Chung, S. J., Lee, C., Ryu, S. M., Kim, D. H., Kim, J. S., ... & Lee, G. J. (2016). Site-directed mutagenesis in *Petunia* × *hybrida* protoplast system using direct delivery of purified recombinant *Cas9* ribonucleoproteins. *Plant Cell Reports*, 35, 1535–1544.

Sun, L., & Kao, T. H. (2018). CRISPR/Cas9-mediated knockout of PiSSK1 reveals essential role of S-locus F-box protein-containing SCF complexes in recognition of non-self S-RNases during cross-compatible pollination in self-incompatible *Petunia inflata*. *Plant Reproduction*, 31, 129–143.

Suzuki, K., Mizutani, M., Fukui, Y., Ueyama, Y., Katsumoto, Y., Miyazaki, K., Ohkawa, H., Kusumi, T. and Yanaka, Y., 2002. Flower color modification of Torenia by engineering gene expression of cytochromes P450 involved in flavonoid biosynthesis. *In Vitro Cell and Development Biology* 38, 23–28.

Tanaka, Y., Brugliera, F., & Chandler, S. (2009). Recent progress of flower colour modification by biotechnology. *International Journal of Molecular Sciences*, 10(12), 5350–5369.

Tanaka, Y., Brugliera, F., Kalc, G., Senior, M., Dyson, B., Nakamura, N., Katsumoto, Y. and Chandler, S., 2010. Flower color modification by engineering of the flavonoid biosynthetic pathway: practical perspectives. Bioscience, biotechnology, and biochemistry, 74(9), pp.1760–1769.

Tanaka, Y., Sasaki, N., & Ohmiya, A. (2008). Biosynthesis of plant pigments: Anthocyanins, betalains and carotenoids. *The Plant Journal*, 54(4), 733–749.

Tanase, K., Nishitani, C., Hirakawa, H., Isobe, S., Tabata, S., Ohmiya, A., & Onozaki, T. (2012). Transcriptome analysis of carnation (*Dianthus caryophyllus* L.) based on next-generation sequencing technology. *BMC Genomics*, 13(1), 1–11.

Tasaki, K., Yoshida, M., Nakajima, M., Higuchi, A., Watanabe, A., & Nishihara, M. (2020). Molecular characterization of an anthocyanin-related glutathione S-transferase gene in Japanese gentian with the *CRISPR/Cas9* system. *BMC Plant Biology*, 20, 1–14.

Teixeira da Silva, J. A., Dobránszki, J., & Ross, S. (2013). Phloroglucinol in plant tissue culture. *In Vitro Cellular & Developmental Biology-Plant*, 49, 1–16.

Thiruvengadam, M., & Yang, C. H. (2009). Ectopic expression of two MADS box genes from orchid (*Oncidium Gower Ramsey*) and lily (*Lilium longiflorum*) alters flower transition and formation in *Eustoma grandiflorum*. *Plant Cell Reports*, 28, 1463–1473.

Tian, C., Zhai, L., Zhu, W., Qi, X., Yu, Z., Wang, H., ... & Chen, S. (2022). Characterization of the TCP gene family in Chrysanthemum nankingense and the role of CnTCP4 in cold tolerance. *Plants*, 11(7), 936.

Veitch, N. C., & Grayer, R. J. (2008). Flavonoids and their glycosides, including anthocyanins. *Natural Product Reports*, 25(3), 555–611.

Verdonk, J. C., Haring, M. A., Van Tunen, A. J., & Schuurink, R. C. (2005). *ODORANT*1 regulates fragrance biosynthesis in *petunia* flowers. *The Plant Cell*, 17(5), 1612–1624.

Villacorta-Martin, C., Núñez de Cáceres González, F. F., de Haan, J., Huijben, K., Passarinho, P., Lugassi-Ben Hamo, M., & Zaccai, M. (2015). Whole transcriptome profiling of the vernalization process in *Lilium longiflorum* (cultivar White Heaven) bulbs. *BMC Genomics*, 16, 1–16.

Wang, H., Wang, H., Ge, Q., & Dai, J. (2020). The interactive effects of chilling, photoperiod, and forcing temperature on flowering phenology of temperate woody plants. *Frontiers in Plant Science*, 11, 443.

Wang, Y., Zhou, L.J., Wang, Y., Geng, Z., Liu, S., Chen, C., Chen, S., Jiang, J. and Chen, F., 2022. CmMYB9a activates floral coloration by positively regulating anthocyanin biosynthesis in chrysanthemum. Plant Molecular Biology, 108, 1–13.

Wang, T., Wei, Q., Wang, Z., Liu, W., Zhao, X., Ma, C., Gao, J., Xu, Y. and Hong, B., 2022. CmNF-YB8 affects drought resistance in chrysanthemum by altering stomatal status and leaf cuticle thickness. Journal of Integrative Plant Biology, 64(3), pp.741–755.

Warner, R., 2010. Genetic Approaches to Improve Cold Tolerance of Petunia, American Floral Endowment. United States of America. Retrieved from https://policycommons.net/artifacts/1721837/genetic-approaches-to-improve-cold-tolerance-of-petunia/2453492/ on 11 Jul 2024. CID: 20.500.12592/6mtf98.

Watanabe, K., Kobayashi, A., Endo, M., Sage-Ono, K., Toki, S., & Ono, M. (2017). *CRISPR/Cas9*-mediated mutagenesis of the dihydroflavonol-4-reductase-B (DFR-B) locus in the Japanese morning *glory Ipomoea* (Pharbitis) nil. *Scientific Reports*, 7(1), 10028.

Watanabe, K., Oda-Yamamizo, C., Sage-Ono, K., Ohmiya, A., & Ono, M. (2018). Alteration of flower colour in *Ipomoea* nil through *CRISPR/Cas9*-mediated mutagenesis of carotenoid cleavage dioxygenase 4. *Transgenic Research*, 27, 25–38.

Xiang, G., Zhang, H., Jian, H., Yan, H., Wang, Q., Zhou, N., ... & Qiu, X. (2019). De Novo assembly and characterization of the transcriptome of susceptible and resistant rose species in response to powdery mildew. *Scientia Horticulturae*, 257, 108653.

Xiang, L. L., Liu, X. F., Li, X., Yin, X. R., Grierson, D., Li, F., & Chen, K. S. (2015). A novel bHLH transcription factor involved in regulating anthocyanin biosynthesis in Chrysanthemums (*Chrysanthemum morifolium Ramat.*). *PLoS One*, 10(11), e0143892.

Xie, Q., Chen, G., Liu, Q., Zhu, Z., & Hu, Z. (2015). Dual silencing of *DmCPD* and *DmGA20ox* genes generates a novel miniature and delayed-flowering *Dendranthema morifolium* variety. *Molecular Breeding*, 35, 1–13.

Xu, G., Chen, S., and Chen, F. (2010). Transgenic chrysanthemum plants expressing a *harpinXoo* gene demonstrate induced resistance to Alternaria leaf spot and accelerated development. *Russian. Journal of Plant Physiology* 57, 548–553. doi: 10.1134/S1021443710040138

Xu, J., Kang, B.C., Naing, A.H., Bae, S.J., Kim, J.S., Kim, H. and Kim, C.K., 2020. *CRISPR/Cas9*-mediated editing of 1-aminocyclopropane-1-carboxylate oxidase1 enhances Petunia flower longevity. *Plant biotechnology journal* 18(1), 287–297.

Xu, J., Kang, B. C., Naing, A. H., Bae, S. J., Kim, J. S., Kim, H., & Kim, C. K. (2020). CRISPR/Cas9-mediated editing of 1-aminocyclopropane-1-carboxylate oxidase1 enhances Petunia flower longevity. *Plant Biotechnology Journal*, 18(1), 287–297.

Xu, G., Chen, S. and Chen, F., 2010. Transgenic chrysanthemum plants expressing a harpin Xoo gene demonstrate induced resistance to Alternaria leaf spot and accelerated development. *Russian Journal of Plant Physiology*, 57, pp.548–553.

Xu, J., Naing, A. H., Bunch, H., Jeong, J., Kim, H., & Kim, C. K. (2021). Enhancement of the flower longevity of petunia by *CRISPR/Cas9*-mediated targeted editing of ethylene biosynthesis genes. *Postharvest Biology and Technology*, 174, 111460.

Yan, R., Wang, Z., Ren, Y., Li, H., Liu, N., & Sun, H. (2019). Establishment of efficient genetic transformation systems and application of *CRISPR/Cas9* genome editing technology in *Lilium pumilum* DC. Fisch. and *Lilium longiflorum* White Heaven. *International Journal of Molecular Sciences*, 20(12), 2920.

Yang, L., Wu, K., Gao, P., Liu, X., Li, G., & Wu, Z. (2014). *GsLRPK*, a novel cold-activated leucine-rich repeat receptor-like protein kinase from Glycine soja, is a positive regulator to cold stress tolerance. *Plant Science*, 215, 19–28.

Yin, D., Sun, D., Han, Z., Ni, D., Norris, A. and Jiang, C.Z., 2019. PhERF2, an ethylene-responsive element binding factor, plays an essential role in waterlogging tolerance of petunia. Horticulture Research, 6.

Yoshioka, S., Aida, R., Yamamizo, C., Shibata, M., & Ohmiya, A. (2012). The carotenoid cleavage dioxygenase 4 (*CmCCD4a*) gene family encodes a key regulator of petal color mutation in chrysanthemum. *Euphytica*, 184, 377–387.

Yu, J., Tu, L., Subburaj, S., Bae, S., & Lee, G. J. (2021). Simultaneous targeting of duplicated genes in *Petunia* protoplasts for flower color modification via *CRISPR-Cas9* ribonucleoproteins. *Plant Cell Reports*, 40, 1037–1045.

Zafar, N., Athar, M., Iqbal, M. Z., & Shafiq, M. (2016). Effect of diesel generator exhaust pollutants on growth of *Vinca rosea* and *Ruellia tuberosa*. *Journal of Applied Sciences and Environmental Management*, 20(4), 1191–1197.

Zhang, B., Xu, X., Huang, R., Yang, S., Li, M., & Guo, Y. (2021). CRISPR/Cas9-mediated targeted mutation reveals a role for AN4 rather than DPL in regulating venation formation in the corolla tube of Petunia hybrida. *Horticulture Research*, 8, 116

Zhang, B., Yang, X., Yang, C., Li, M., & Guo, Y. (2016). Exploiting the *CRISPR/Cas9* system for targeted genome mutagenesis in *petunia*. *Scientific Reports*, 6(1), 20315.

Zheng, Z. L., Yang, Z., Jang, J. C., & Metzger, J. D. (2001). Modification of plant architecture in chrysanthemum by ectopic expression of the *tobacco phytochrome B1* gene. *Journal of the American Society for Horticultural Science*, 126(1), 19–26.

Zuker, A., Tzfira, T., Scovel, G., Ovadis, M., Shklarman, E., Itzhaki, H., & Vainstein, A. (2001). *RolC*-transgenic carnation with improved horticultural traits: Quantitative and qualitative analyses of greenhouse-grown plants. *Journal of the American Society for Horticultural Science*, 126(1), 13–18.

Zvi, M. M. B., Shklarman, E., Masci, T., Kalev, H., Debener, T., Shafir, S., ... & Vainstein, A. (2012). PAP1 transcription factor enhances production of phenylpropanoid and terpenoid scent compounds in rose flowers. *New Phytologist*, 195(2), 335–345.

6 *In Vitro* Production of Some Underutilized and Unconventional Horticultural Crops

Sapna Tyagi and Amit Sharma

6.1 INTRODUCTION

Agriculture in today's context is one of the most important sources of renewable wealth in the world. There are many plant species still lying unexplored and underexploited. Fruit crops are rich sources of vitamins, minerals, micronutrients, soluble dietary fiber, and various phytochemicals and have medicinal potential. For ages, several underutilized indigenous fruit crops have been utilized in various indigenous medicinal systems like Ayurveda, Unani and Homoeopathy. The numbers of underutilized crops in India are not well-known or well-documented (Kour *et al.,* 2018). India is a tropical country, where many minor fruit tree species are very popular and their fruits are always in high demand as seasonal delicacies (Vividha *et al*., 2019). However, most of these fruit trees are not commercially cultivated due to the poor viability of their seeds and remain underutilized for their medicinal and nutritional value. These fruit trees provide significant support to the livelihood of rural and tribal people as they generate additional household income. Underutilized crops may be termed as the crops that are valuable but neither grown on a commercial or large scale nor are traded widely around the world. These crops are less well-known plant species in terms of research and marketing, yet it can be demonstrated that they adapt well to harsh and stressful environments. Underutilized horticultural crops are nutrient-dense and well-suited to minimal input farming. Increased investment in research and development of these crops will undoubtedly have a significant impact on food security and nutrition. India is homeland of many important yet underutilized fruit crops such as bael (*Aegle marmelos*), ber (*Zizyphus mauritiana*), Indian gooseberry (*Emblica officinalis*), Jamun or black plum (*Syzygium cumini*), jackfruit (*Artocarpus hetrophyllus*), Khejri (*Prosopis cineraria*), wood apple (*Limonia acidissima*), tamarind (*Tamarindus indica*) and many others. Most of these species are in a state of neglect, while a trivial number of them are restricted to their geographical location and cultivated by local communities for specific household uses only. With the dearth of scientific knowledge about them, their current use is linked to their economic potential. These

DOI: 10.1201/9781003343707-6

species were immensely constructive by surviving harsh agro-climatic conditions and can be established on degraded lands, which are presently being underutilized either due to poor soil fertility or moisture scarcity. Due to unsustainable market pressures and rapid urbanization, the majority of these species have come to near extinction. A holistic approach is hence proposed that includes both *in-situ* and *ex-situ* conservation strategies, as well as re-governance of the market chain. Reinforcement of their domestication through standardization of cultivation practices, facilitation for the supply of planting material and increasing the demands by exploring their uses, and creation of awareness among consumers (Kamboj, 2020). Therefore, there has been focused attention by researchers on exploiting underutilized horticultural species for multifarious uses. Conventional methods of propagation are slow, season bound and economically unviable. Micropropagation techniques can play a crucial role in the rapid and large-scale production of desirable planting material for these indigenous underutilized fruits. Thus, it is necessary to develop high-yielding varieties, production and protection technologies. The protocols for micropropagation of the majority of indigenous underutilized fruit crops are not available and there is a need to refine or scale protocols for mass multiplication in order to encourage economic development and security of small and marginal farmers. In this chapter, an attempt has been made to summarize work done on *in vitro* production of some underutilized and unconventional fruit crops; i.e., ber (*Zyziphus mauritiana*), kaith (*Limonia acidissima*), khejri ((*Prosopis cineraria*), and jamun (*Syzygiumcumini*).

6.2 BER/JUJUBE (*ZYZIPHUS MAURITIANA* LAMK.)

Ber, also known as Indian jujube, Indian plum, Chinese apple, Chinese date, and jujube, are species of Zizyphus belonging to the family Rhamnaceae. It is a native of Indo-China and India. There are almost 130–140 species in the *Zizyphus* genus (Islam & Simmons, 2006), out of which two major domesticated commercial species cultivated over vast areas of the world are *Z. mauritiana* Lamk. and *Z. jujube* Mill. Other major species available in India are *Z. nummularia, Z. rotundifolia, Z. xylopyrus, Z. oenopila, Z. vulgaris, Z. sativa, Z. rugosa, Z. fumiculosa,* and *Z. oxyphylla.* Almost 125 ber varieties are cultivated in India, predominantly Mundia, Gola, Umran, Parbani, Karaka, Banarasi, Kaithli, Mehrun, Elaichi, Sanam, and Sebeti (Hussain, 2021). India stands as the leading global producer of ber due to its extensive cultivation and adaptable growing conditions, thriving particularly in semi-arid and dry regions of tropical and subtropical climates. Regarded as a "poor man's apple," ber boasts high nutritional value and therapeutic benefits, demanding minimal cultivation efforts (Diengngan and Hassan, 2015). Its greenish-yellow to reddish-brown fruits offer spongy, sweet, and sour pulp rich in vitamin C, surpassing citrus and apples in content. Additionally, they are abundant in vitamins A and B complex, along with essential minerals like iron, phosphorus, and calcium, promoting bone strength and aiding in various health conditions. Ber fruits are valued for their medicinal properties, acting as blood purifiers and effective against pathogens, as evidenced by secondary metabolites such as saponins and flavonoids. Despite its genetic diversity, achieving true-to-type planting material remains a challenge, necessitating rapid propagation methods like in vitro techniques to meet commercial demand effectively.

The proliferation of *in vitro* shoots is greatly influenced by the culture medium used. So, it is important to standardize culture conditions and medium for a particular cultivar to promote commercial-scale cultivation. Protocols for *in vitro* culture of *Z. jujube* are successfully obtained for various cultivars via leaf explants (Feng *et al.*, 2010; Qing Rong *et al.*, 2011; Chun Hua *et al.*, 2012; Ma *et al.*, 2012, Ye *et al.*, 2012), embryos (Kim *et al.*, 2006; Wang *et al.*, 2009), cotyledons (Kim *et al.*, 2006; Wang *et al.*, 2011), nodal explants (Goyal *et al.*, 2006; Soliman and Hegazi, 2013), the effect of some crucial factors on *in vitro* propagation of Chinese Jujube from different types of explants. Efforts have been made by various workers to establish protocols for *in vitro* culture of Indian jujubes through shoot tip and nodal stem segments (Goyal & Arya, 1985; Mathur *et al.*,1995; Sudhersan *et al.*, 2001; Sudhersan & Hussain, 2003). Rathore *et al.* (1992) reported efficient cloning from a nodal segment of *Z. nummularia* established on MS (Murashige and Skoog) medium containing BAP 5.0 mg l^{-1} + IAA 0.05 mg l^{-1} and BAP 7.5 mg l^{-1} + IAA 0.1 mg l^{-1} in case of *Z. mauritiana*. Rooting was induced on a filter paper bridge in White's liquid medium containing 25 mg l^{-1} IBA for 48 h and then on agar-gelled hormone-free White's medium. Mathur *et al.* (1995) obtained *in vitro* plantlets from stem explants of *Ziziphus mauritiana* grown on modified Murashige and Skoog's medium containing 3800 mg/L potassium nitrate, 2475 mg/L ammonium nitrate, 11 μM BA and 0.5 μM IAA, and rooting on pretreatment with 50 μM IAA for 24 h followed by transfer to auxin-free White's medium. Sudhersan *et al.* ((2001) reported *in vitro* culture of *Z. mauritiana* cultivar Umran, but failed to obtain adventitious root induction even on higher concentrations of IBA-fed MS basal medium. Abass (2010) reported that shoot growth at the starting stage required MS full-strength media (growth regulators free), and for shoot multiplication, MS medium supplemented with 0.5 mg/L BA or 3.5 mg/L IBA or 3.5 mg/LKin and 3.5 mg/L IBA+3.5 mg/L Kin proved to be best. Shoot length was found to be significantly affected by the presence of 0.5 mg/L BA or 3.5 mg/L IBA+3.5 mg/L Kin. Bottlenecks during *in vitro* rooting were cracked and it was reported that micropropogated shoots not showing rooting on medium supplemented with various concentrations of auxin got rooted readily on peat substrate. Sudhersan *et al.* (2001) established shoot tip of *Ziziphus mauritiana* cultivar Umran *in vitro* on MS medium supplemented with 100 mg/l glutamine, 125 mg/l myo-inositol, and 0.1 mg/l benzyladenine (BA) and proliferated multiple shoots on MS medium supplemented with 0.01–0.1 mg/l BA and obtained seven times multiplication on hormone-free MS medium. Thirty percent % rooting after 30 days was observed on MS medium containing 10 mg/l IBA. Wang *et al.* (2009) successfully proliferated regenerated shoots of Chinese Jujube Dongzao on MS medium augmented with 1 mg/l BAP. Higher rate of shoot multiplication was found after six month-long subculturings in *Z. jujuba* cv. Huizao (Abbas *et al.*, 2014). The constituents of medium, composition, and concentration of plant hormones are the key factors influencing adventitious shoot induction (Feng *et al.*, 2010).

Ibrahim *et al.* (2011) demonstrated efficient cloning of *Ziziphus mauritiana* Lamk. cv. Zaytoni through somatic embryogenesis from shoot tip-derived callus and developed somatic embryos on half-strength MS medium supplemented with BA 1.0 mg/L. Embryos were germinated and matured on half-strength MS medium augmented with 0.1 mg/L BA and 0.1 mg/L NAA. Earlier, several authors used shoot tips and nodal stem segments for *in vitro* propagation of Indian jujube of various cultivars, but the major problem was low rooting efficiency (Goyal and Arya, 1985; Mathur *et al.,* 1995; Sudhersan and Hussain, 2003). Similarly, Ibrahim *et al.* (2012) reported first use of indirect organogenesis for *in vitro* propagation of Indian jujube and achieved *in vitro* plant regeneration using indirect and direct organogenesis in *Ziziphus sativa* L. (wild jujube) and *Z. jujuba* Mill. (Chinese jujube). Callus initiation occurred on MS medium containing 5.0 mg/L BA + 10.0 mg/L IBA within 60 days and organogenic callus was obtained on MS medium supplemented with BA at 5.0 mg/L. Regenerated shoots were 100% rooted on half-strength MS medium fed with 0.2 mg/L NAA. Soliman and Hegazi (2013) worked on *Z. jujuba* cultivars Comethry, Balahy, and Balady and found that MS medium containing 0.05 mg/L β-naphthalene-acetic acid (NAA) + 2 mg/L N6-(2-isopentenyl) adenine (2iP) was best for the *in vitro* establishment of Comethry and Balady cv., but for Balahy cv. MS medium containing 0.05 mg/L NAA + 1 mg/L thidiazuron (TDZ) showed 100% growth. Eighty percent rooting was observed on MS medium containing 2 mg/L indole-3-butyric acid (IBA) for cultivars Comethry and Balahy, whereas Balady cv. rooted well on MS medium containing 2 mg/L IBA + 0.5 mg/L NAA. Rooted plantlets were survived well in pots containing garden soil, sand and peat moss (1:1:1 v:v:v) in the greenhouse conditions. Abbas *et al.* (2014) successfully achieved *in vitro* plant regeneration in *Ziziphus mauritiana* Lamk., cv. Zaytoni, through callus obtained from the shoot tip cultured on MS medium augmented separately with 10.0 mg/L IBA, 15.0 mg/L NAA, and 5.0 mg/L 1 BA and obtained organogenic callus from primary callus established on MS medium containing 5.0 mg/L BA, which showed adventitious shooting when cultured on MS medium supplemented with BA at 1.0 mg/L and NAA at 0.1 mg/L. No rooting was observed when axillary shoots were planted on full-strength MS medium containing various concentrations of IBA or NAA; however, 45% rooting was obtained, when the axillary shoots were cultured on half-strength MS containing 0.2 mg/L NAA.

Hansika *et al.* (2017) found Captan 50% (1.2 g/L) as the most effective fungicide for surface sterilization of Jujube shoot tips and nodal segments. The highest percentage of elongated bud (96.66%) was recorded on MS (Murrashige and Skoog's) medium containing 1.5 mg/L BAP and the lowest significant rate was recorded in TDZ 0.2 mg/L (3.33%) as TDZ produced stunted shoots, whereas higher concentrations of TDZ can induce callusing, adventitious shooting, or somatic embryogenesis (Lu, 1993). Rooting was not recorded in both shoot tips and *in vitro* generated shoots on media supplemented with IBA. Lu *et a*l. (2018) established a highly efficient *in vitro* propagation and regeneration system for the "Teapot" jujube via shoot organogenesis by employing response surface methodology (RSM) and reported that more than 75% regeneration was observed by employing one-week pre-culture darkness and on wood plant medium (WPM) supplemented with 0.25 mg/L GA3, 0.5 mg/L BAP and 0.1 mg/L IAA. One hundred percent rooting was found on the modified 1/2 Murashige and Skoog (MS) medium supplemented with 0.4 mg/L IBA.

6.3 JAMUN/INDIAN BLACKBERRY [*SYZYGIUM CUMINI* (L.)]

Jamun, also known as Jambul, Black Plum, Java Plum, Indian Blackberry, and Jamblang, belongs to species Syzygium of the family Myrtaceae. It is also known as *Syzygium cumini*, *Syzygium jambolanum* and *Eugenia cumini* (Swami *et al.,* 2012). It is a significant minor and underutilized treefruit of the Indian subcontinent and grows naturally in clayey loamy soil in tropical as well as in sub-tropical zones of the Indo-gangetic plain. Jamun is a significant restorative plant utilized for its edible fruit and medicinal value. Jamun fruits and seeds are sweet, acrid, and sour and possess extremely high nutritional value. It is reported to be very rich in tannins, flavonoids, essential oils, anthocyanins, other phenolic constituents (Sharma *et al.*, 2003; Reynertson *et al.*, 2008), and iron minerals, sugar, protein, etc. The seeds contain alkaloid jambosine and glycoside jambolin or antimellin, the gallic and ellagic acid content that lowers blood glucose levels, used for treatment and control of diabetes mellitus (Sagrawat *et al.,* 2006). Diarrhea, ulcers, and inflammation are also successfully treated with it. It is also reported to be a blood pressure regulator (Rai *et al.,* 2011). Different parts of Jamun were also reported for its antioxidant, anti-inflammatory, neuropsycho-pharmacological, anti-microbial, anti-bacterial, anti-HIV, and antifungal, nitric oxide scavenging, free radical scavenging, anti-diarrheal, antifertility, anorexi-genic, gastroprotective and anti-ulcerogenic, and radio-protective activities (Navneet and Singh, 2018). It is also reported to have antineoplastic (Barh and Viswanathan, 2008), radioprotective (Jagetia and Baliga, 2002), and chemopreventive properties (Jagetia *et al.,* 2008); it is considered to be useful in the prevention and treatment of cancer. The high anthocyanin content of fruit pulp might be a valuable resource for the food processing industry as natural colorants. Fruits are utilized for invigorating well-being drinks, making jams, squashes, and good quality wine. Great quality vinegar is additionally produced from slightly unripe fruit juice is diuretic, stomachic, and carminative (Kamboj *et al.,* 2020). Due to unorganized farming, complete information about the total area under its cultivation in India is apparently not available. It suffers from feeble seed viability and low germination rate in its natural habitat (Dent, 1948). Thus, vegetative propagation is an alternative, but not feasible for obtaining a large quantity of planting material within a short time interval.

Micropropagation is an advanced vegetative propagation technique for producing a large number of genetically uniform and pathogen-free transplants in a limited time and space. In this context, it is necessary to standardize an efficient method for rapid clonal propagation of its elite genotypes (Remashree *et al.,* 2007). Genetically uniform plants, which can serve as the source for less variable medicinal formulations, can be regenerated using *in vitro* systems. Due to the recalcitrant nature of this tree and intrinsic issues such as phenolic exudation, explant browning, etc., attempts to regenerate plantlets from the explants of adult trees have been unsuccessful. However, from nodal and shoot tip explants derived from *in vitro* seedlings, plantlets may be micropropagated. Studies on many fruit crops have revealed that the growth of explants under laboratory conditions is affected by many factors like genotype, media composition, growth regulators, and growth conditions. Yadav *et al.* (1990) obtained multiple shoots from nodal and shoot tip segments of 10–15 days-old seedlings of *Syzygium cumini* cultured on modified Murashige & Skoog (MS) medium supplemented with BA (0.23–8.90 μM) singly or in combination with NAA, IAA, and IBA. Rathore *et al.* (2004) cultured rejuvenated shoot sprouts and observed multiple shoots regenerated on MS medium fed with 9.0 μM BAP and growth additives. Shoots were then proliferated by repeated subculturings of original explants on an amended medium fed with 4.5 μM BAP and 2.25 μM kinetin. Rooting was observed on ½ MS supplemented with 0.1% activated charcoal and 10 μM IBA or 15 μM NAA. The rooted plantlets were hardened and then transferred to a pot. One hundred percent *ex vitro* rooting was observed in the cloned shoots. Jain and Babbar (2003) studied the shoot proliferation potential of epicotyl explant of jamun and found 4% sucrose to be the best for caulogenic response and shoot developing potential among various carbon sources tested. Shoot elongation was observed on MS medium supplemented with 1.0 mg/L BA and casein hydrolysate (1.5 g/L) or glutamine (200 mg/L) and profuse shooting was observed on MS medium with 1 mg/L BA. These explants did not require an additional supply of reduced nitrogen for further normal development. Rooting was promoted on Knop's medium supplemented with 2% sucrose and 1 mg/L IAA and *in vitro* plants were acclimatized and successfully transferred to the field.

Remashree *et al.* (2007) developed an efficient protocol for micropropagation and *ex vitro* rooting of *Syzigium cumini* and induced multiple shoots from mature nodal explants using Woody Plant Medium (WPM) supplemented with combinations of 8.8 μM BA, 9.3 μM Kinetin, and 5.37 μM NAA. Rooting was induced in WPM supplemented with 4.9 μM IBA and *ex vitro* rooting was also obtained successfully. Plantlets established in the field were found morphologically identical to the donor plant with 80% survival rate. However, Jejurker *et al.* (2016) observed callusing on MS basal medium + 6 BAP with additional supplement of 0.60 mg/l IBA and 0.60 mg/l kinetin, whereas MS basal medium supplemented with IBA & kinetin (0.02–0.5 mg/L) resulted in shoot development. Although IAA resulted in some growth response in explant, but did not induce callusing (Singh *et al.*, 2019). However, Yadav *et al.* (2014) induced callus from the seedling explants on MS medium containing 2,4-D (0.05–1.0 mg/L) with BA (0.5–2.0 mg/L) which responded shooting on MS medium fed with BAP (0.5–2.0 mg/L) alone or with NAA (0.05–0.2 mg/L). 85% shooting (25 shoots/culture) was observed on medium fed with 1.0 mg/L of BA in combination with 0.05 mg/L of NAA. Adventitious roots initiated when subcultured on MS medium containing IBA or NAA (1.0 mg/L).

Choudhri *et al.* (2013) achieved axillary bud proliferation in jamun nodal explants excised from *in vitro* seedlings. Best shoot proliferation was observed on half-strength MS supplemented with 2 mg/L BAP, 3% sucrose, and 3% activated charcoal. The shootlets thus obtained were successfully rooted on 1/4th strength MS fed with 3% sucrose and 2.5 mg/L IBA. Hardening was done with 32.5% survival rate on vermiculite for six weeks and then field transferred. Yadav *et al.* (2014) reported that the cotyledonary nodes obtained from seeds and immature seeds were best for callogenesis. The success of the whole process of tissue culture technology depends on vital hardening. According to Prabhuling and Carmelita (2018), improper hardening leads to the failure of whole technology and the industry itself, so protocol for acclimatization of *in vitro* propagated plantlets of Jamun cv. AJG-85 was developed and reported that cocopeat alone and mixture of cocopeat+perlite+vermiculite (1:1:1 v/v) was found superior for primary hardening for four weeks under polytunnels. In the case of secondary hardening, perlite+red soil (1:1 v/v) and vermiculite+red soil (1:1

v/v) under shadehouse for eight weeks gave superior and effective results. Jejurker *et al.* (2016) observed callusing on MS medium fed with 6- BAP along with (0.60 mg/L – 0.60 mg/L) IBA and kinetin and shoot development on 0.02–0.5 mg/L of IBA & kinetin. The medium with 6-BAP in combination with IBA & kinetin induced faster growth than alone. Auxin in higher concentration inhibits the explant growth. No response was found in the medium with IAA. Prabhuling *et al.* (2017) raised *in vitro* shoots from shoot tip explants of jamun cv. AJG-85 and found 0.50% activated charcoal better in controlling phenolic browning. Half-strength WPM supplemented with BAP (2 mg/l) + NAA (0.1 mg/l) gave the best results among all the media tested. Best shoot elongation was found in the medium fed with GA3 (2 mg/L), whereas Naaz *et al.* (2018) regenerated shoots via *de novo* organogenesis-induced callus raised from a cotyledonary node on MS medium supplemented with 2, 4-D (7.5 μM). Ninety three percent shoot induction was observed on MS medium augmented with 5.0 μM BA + 0.25 μM NAA along with 10 mg/L $AgNO_3$. The profuse shooting was recorded after 4 weeks of transfer. 80% rooting was obtained on the medium fed with IBA (200 μM) and acclimatized successfully on sterilized soilrite after 4 weeks. Recently, Vividha *et al.* (2019) developed an efficient method for rapid shoot proliferation from the nodal segment of Jamun (cv. Rajamun) on Lloyd and McCown WPM augmented with 8 μM/L BAP and Kn. WPM with 13μM BAP + 8.5μM Kn + 5μM NAA responded to high-frequency shoot multiplication after 4 weeks of the first subculture and maximum shoot proliferation at 6 weeks after the second subculturing with significant shoot length elongation at 10 weeks.

6.4 KAITH/WOOD APPLE (*LIMONIA ACIDISSIMA* L.)

Limonia acidissima L. (syn. *Feronia limonia*), also known as wood apple, Elephant Apple, Monkey Fruit, Kaitha, Kath Bel, or Kabeet, is the only species of its genus, in the family Rutaceae. It is one of the underappreciated and underutilized fruit trees, and its other components have powerful traditional uses as well. Its fruits have a broad spectrum of traditionally established therapeutic properties (Tiwari & Gupta, 1959) and are widely used in several Ayurvedic preparations like Panch kapittha and Kapitthaashtaka churna (Jain *et al.*, 2011). Until processing methods were discovered in the middle of the 1950s, the fruit was considered a “poor man’s food” in India. The fruits are considered as sour, sweet, aromatic, astringent, refrigerant, aphrodisiac, alexipharmic, and useful in whooping cough (Mukherjee and Wahile, 2006). According to Yunani medication, the fruits are tonic to heart, liver and lungs. Pulp is a good antidote for snakebites (Gupta *et al.*, 2009). The pulp of unripe fruit is used in the treatment of diarrhea and dysentery. According to Ayurveda, the fruit may be used to cure a variety of conditions, including cancer, asthma, wounds, diarrhoea, dysentery, cardiac dysfunction, hepatitis, sore throat, ophthalmia, leucorrhoea, and more. From wood apple fruits, Saima *et al.* (2000) extracted an acidic heteropolysaccharide that significantly inhibited the development of Ehrlich ascites carcinoma cells *in vivo.* According to Seeja *et al.* (2005), indigenous people substitute wood apple fruits for bael (*Aegle marmelos* Correa) and vice versa. The plant has great potential for exploitation in view of the value of its pharmaceutical products, use as food, fodder, medicine, energy and for industrial purposes. It can be cultivated even on the barrel land, and can be a profitable business for farmers as well. A perusal of literature showed that the lack of adequate planting material is a major constraint preventing the commercial expansion of this fruit tree. All the plants that exist in different parts of India are of seedling origin and thus offer great scope for the selection of suitable genotypes of high yield potentiality with good fruit quality and other desirable characteristics. Purohit and Tak (1992) tried to micropropopgate *L. accidissima* and observed shoot development with callus formation at the basal end of the explant but not enough shoot proliferation. Handique and Bhattacharjee (2000) also observed similar findings. Hiregoudar *et al.* (2003) observed 100% *in vitro* shoot regeneration from leaf and nodal explants excised from seedling grown *in vitro* on MS medium supplemented with 5 μM BAP and obtained 20–32 shoots after repeated subculturings for five months. These shoots were then rooted on a half-strength MS medium fed with 1 μM NAA and successfully transferred to the field. On the contrary, Hiregoudar *et al.*(2005) also recorded an average of 12 and 8 shoots from hypocotyl and internodal explants, respectively, on the medium containing 2 μM BA alone. Contrary to this, Vyas *et al.* (2005) observed best *in vitro* shooting response from hypocotyls on MS medium augmented with 2.22 μM BA and 2.32 μM Kn and multiplication on MS medium fed with 2.22 μM BA, 6.96 μM Kn, and 0.05 μM NAA. Sixty percent rooting was obtained on one-quarter strength MS medium supplemented with 7.36 μMIBA. Vyas and Purohit (2006) reported that a controlled enhanced CO_2 environment provided to *in vitro* shoots accelerates the shoot growth and shoot multiplication in comparison to ambient air in the growth room. Even in the absence of sucrose in medium, shoots proliferated subjected to 10.0g m^{-3} CO_2. Sucrose and CO_2 together produced enhanced maximum growth. However, Islam *et al.* (2010) observed the best shoot growth and proliferation on MS medium augmented with 0.2 mg/L BAP and the best rooting response on NAA (0.5 mg/L) with 80% rooting within seven weeks in *F. limonia*. This might be a result of the cultural and environmental circumstances influencing the genotypic variance of explants (Ramakrishna and Shasthree, 2015) and indicates that either IAA or IBA or NAA may be used for rooting. It is well known that for root induction, auxin is necessary, but its continued presence in the medium inhibits the development of adventitious roots (Teixeira da Silva *et al.*, 2019). Saini and Tyagi (2019) regenerated *in vitro* shoots through callus developed from hypocotyl and direct regeneration from internodal explants on MS medium supplemented with 0.5 mg/L kinetin and 0.5 mg/L BAP; whereas poor shooting response was observed on MS Medium supplemented with either BAP or Kinetin. No response was observed in the case of cotyledons. One hundred percent shoot proliferation and a higher number of shoots (27.3+1.18) per explants were obtained after two rounds of subculturing at the interval of 8 weeks of inoculation.

6.5 KHEJR/SHAMI [*PROSOPIS CINERARIA* (L.) DRUCE]

Khejri or Jand/Shami is a member of the *Prosopis* species in the Fabaceae family. It is a desert tree considered the lifeline tree of the Indian Thar desert and also known as the wonder tree, nature's gift, the king of desert, and the golden tree. It is a multipurpose tree as it provides a vegetable pod, flour, fodder, timber, fuel, medicine, gum, and resin (Pasiecznik *et al.,* 2004). It is highly drought-resistant and adapted to harsh conditions and hence planted for forest restoration in arid landscapes. It is a high litter accumulating tree and increases soil fertility by fixing atmospheric nitrogen and increases soil fertility and as a result, increases the amount of organic matter, soluble calcium, and available phosphorus and reducing soil pH (Man and Shankarnarayan, 1980; Yadav *et al.,* 2008). Moreover, the Khejri tree is considered a productivity booster in inter-cropping and companion cropping systems, thus it is highly compatible with agri-horticultural crops (Rani *et al.*, 2013; Pareek *et al.*, 2015) in arid and semi-arid regions. The unripe pods ("Sangri") are consumed as vegetable in green and dried form and also as pickles in many parts of the Thar Desert. The pods are milled to flour which serves as protein-rich nutritious food supplement in these famine-prone desert margins. All the organs of Khejri have medicinal properties and are used as a folk remedy for various ailments.

Haroon *et al.* (2018) documented various uses of bark as a refrigerant, anthelmintic and in dysentery, asthma, leucoderma, rheumatism, bronchitis, leprosy, piles, wandering of the mind and muscle tremors. The flowers mixed with sugar are used to prevent miscarriages (Orwa *et al.*, 2009), diabetes, and for treating Lewis lung carcinoma. Leaves are used to treat mouth ulcers. Pod and seed are used for preventing protein malnutrition and calcium and iron deficiency in the blood, asthma, piles, leprosy, etc. (Samadia, 2021). Furthermore, this promising multipurpose tree not only provides shade but is also used as fuel wood, timber, charcoal, animal fodder, and human food as timber and domestic fuel wood (Kumar *et al.,* 2009). Even though khejri is a multipurpose tree but studies on its improvement are scarce. Generally, *Prosopis cineraria* is propagated by seeds and established well with 80–90% germination (Mahoney, 1990). Jordan and Balboa (1985) reported its propagation by air layering and root suckers. But it seems that *in vitro* propagation is more challenging in *P. cineraria* in comparison to other species (Pasiecznik *et al.,* 2004). As explants derived from mature and older plants do not regenerate appropriately in culture, *in vitro* expertise has not been effectively achieved from explants excised from mature and selected individuals of *Prosopis* tree (Shekhawat *et al.,* 1993). Micropropagation offers not only a means for the conservation of elite and rare tree germplasm but also for bringing about their genetic improvement. Though considered a problematic and recalcitrant species to *in vitro* regeneration, during the past few years, a few reports with limited success have appeared employing both juvenile and mature explants (Batchelor *et al.,* 1989; Shekhawat *et al.,* 1993; Kumar and Singh, 2009).

According to Shekhawat *et al.* (1993), *in vitro* clonal propagation of *P. cineraria* is extremely affected by genotype, age of tree, nature of explant and size, season, the orientation of explant on medium, plant growth regulators, and some addendums like ascorbic acid and citric acid, adenine sulphate, L-arginine, glutamine and ammonium citrate), incubation conditions, and subculturing period. They obtained maximum number of shoots from the nodal shoot segment from pruned thorny adult trees on MS medium fortified with 0.1 mg/L IAA + 2.5 mg/L BAP + additives and incubated at higher temperature (31+-2°C) and mixed (fluorescent and incandescent) light for 12 h per day, whereas explants from thornless trees produced 6–8 shoots per explant on MS medium containing 0.1 mg/L IAA+5.0 mg/L BAP + additives. It was observed that nodal shoots raised from root and stub sprouts produced multiple shoots. Root segments differentiated into multiple shoots on MS medium containing 0.5 mg/L IBA+2.5 mg/L BAP. Differentiated shoots multiplied best on MS medium containing 0.1 mg/L naphthalene acetic acid (NAA) + 1.0 mg/L BAP + additives. Multiple shoots were obtained after 6 subculturings on fresh medium. Rooting was obtained by palpitating raised shoots with 100 mg/L IBA for 4 h and then culturing on half-strength hormone-free MS medium. Initial dark incubation for 5 days at high temperature (33±2°C) was found essential for root induction from shoots and induced 63% rooting within two weeks. Kumar and Singh (2009) developed an efficient protocol for *in vitro* propagation of khejri from nodal explants excised from mature plants of *Prosopis cineraria*. Multiple shoots were induced on MS medium fortified with various cytokinins and auxins individually and in various combinations. Best results were obtained on MS medium fed with 5.0 mg/L BAP and 1.0 mg/L IAA among all others tested. Rooting was observed on half-strength MS medium augmented with 3.0 and 5.0 mg/L IBA and hardened in pots containing sterilized sand and vermiculites (1:1). Plantlets were successfully acclimatized to the field with a 60% survival rate after four weeks. Kumar *et al.* (2017) investigated three different basal media (i.e., MS, SH, and B5) to observe the impact of medium composition and the nature of explants on *in vitro* responses in plant tissue culture and found MS medium to be the best. The importance of MS medium and its significance for *in vitro* response has been advocated by many earlier reports (Hossain *et al.*, 2008; Ibrahim *et al.,* 2008; Anabzhagan *et al.*, 2010). Different explants of *P. cinerariaviz* leaf, apical bud and binodal segments were also tried for induction of auxillary buds. Binodal segments were recorded as best explants which showed the best shoot induction response. The induction of callus was observed to be best over 2,4-D (2.5 mg/L) with Kn (0.5 mg/L) from the binodal explant. Organogenesis was induced on filter bridge dipped in liquid MS medium fed with BAP (2.0 mg/L) with IAA (0.1 mg/L). MS medium fortified with BAP (2.0 mg/L) + αNAA (0.1 mg/L) and 1% activated charcoal enhanced the shooting response. These results were in agreement with previous findings (Debnath, 2008 and Ali *et al.,* 2010), which reported BAP as the best hormone for organogenesis.

In this study, the impact of calcium was observed with a significant effect on the multiplication of shoots, and it was found that MS (BAP 2.0 mg/L) + (α-NAA 0.1 mg/L) + ($CaCl_2$.2H2O 2.0 gm/L) concentration showed the highest number of shoots. It was found that media with higher CaCl2.2H2O content was much better for plant *in vitro* regeneration (Kumar *et al.*, 2017). Khatak *et al.* (2018) analyzed the impact of plant growth regulators on bud break and shoot proliferation and explored that sterilization of explants is a crucial step in plant tissue culture and achieved 80% sterilization by subjecting the explants to 0.3% mercuric chloride for 3 min without prior treatment of ethanol and antifungal supplements. However, treatment with 0.1% HgCl2 for 3 minutes followed by a pretreatment of ethanol (rinse) and Bavistin (45 min) resulted in effective explant sterilization. The axillary nodes were excised from *in vivo* grown young seedlings, sterilized and cultured on MS medium supplemented with different concentrations of cytokinins and auxins. Eighteen different media combinations for micropropagation were tried and bud break was observed in 15 to 20 days at 0.5 and 1.0 mg/L of BAP and 0.5 mg/L NAA. In treatment with NAA (1.5 mg/L) and kinetin at two different concentrations (2.0 and 3.0 mg/L), the shoot length doubled and the initiation process of shoot proliferation was also shortened to 30 days. Haroon *et al.* (2018) developed efficient and reproducible micropropagation practice by forcing softwood shoots from stem segments of the mature tree on three different media (i.e., sand, peat moss, and sawdust medium). It was revealed that the maximum number of sprouting with a higher number of shoots and maximum length which also has maximum number of leaves and nodes were observed in peat moss followed by sand media. A similar technique was also employed for recalcitrant woody species by Preece and Read (2003). Results also showed that the stem segments under the dark conditions in all media tested have better outcomes when compared with forced large segments under natural conditions. This protocol provides uniform shoots for field transfer from sprouted logs without any need for fresh explants. It is very difficult to germinate seeds of *Prosopis* (Rivera *et al.*, 2020). Narang and Shukla (2022) successfully germinated the seeds by various pre-treatments. Scarification of seeds is done followed by soaking in distilled water for 4–12 hr at room temperature. It is observed that 100% germination was achieved when seeds were treated with concentrated sulphuric acid, soaked in distilled water for 8 hr and then established on Knop's medium. Cotyledonary node explants excised from 10-day-old *in vitro* grown seedlings produced the maximum number of shoots on MS medium supplemented with 2.5 mg/L BAP after 60 d of culture. These adventive shoots were differentiated in the axil of cotyledons. Eighty nine percent rooting was achieved when shoots were subcultured on MS medium supplemented with 3 mg/LIBA within 20–25 days. The plants developed *in vitro* were gradually transferred to the soil. In this investigation within 60 days, approximately nine shoots per axil of the cotyledonary node were developed in contrast to earlier reports of Arya and Shekhawat (1986); hence in this study, double yield was achieved in half duration. Thus, not only the incubation time was reduced to nearly half but also the yield was doubled. The plantlets were transferred to the field successfully. Similarly, Pareek *et al.* (2012) reported bud break and shoot proliferation in cotyledonary nodal segments excised from 15 to 20 days old *in vitro*-grown seedlings and achieved the best shoot proliferation on MS medium supplemented with 0.05 mg/L IBA and 2.0 mg/L BAP. *In vitro* shoots were rooted on MS medium supplemented with 2.0 mg/L IBA and transferred to pots under field conditions.

6.6 CONCLUSION

Underutilized horticultural fruits and vegetables are nutritionally rich and adapted to low-input agriculture. These crops are often neglected due to their low commercial value or limited availability, but they have unique flavors and nutritional properties that make them attractive to consumers. There is an urgent need to rapidly multiply important underutilized fruit crops in order to increase commercial production, but there are certain constraints in the commercial exploitation of these underutilized natural crops as propagation strategies are not well developed in the absence of desirable seeds and planting material. Awareness and mindfulness among the cultivating network about the healthful and therapeutic product is also not present. Thus, systems for their exploitation need to be developed. For the production and distribution of potential selections/varieties, strategies need to be developed, especially at the national and regional levels so that small agricultural communities can generate more money by increasing output, supplying local needs, and fostering home markets. *In vitro* propagation of underutilized horticultural fruit crops offers a sustainable and efficient method for the production of disease-free, genetically uniform plants that can help to increase the availability and diversity of these unique fruits in the market. This can help to promote their cultivation on a larger scale and lead to increased productivity and profitability for farmers and provide consumers with high-quality, nutritious fruits. With the increasing demand for healthy and nutritious food, the future prospects of *in vitro* propagation are promising, and it is expected to play a significant role in the horticultural industry in the coming years.

REFERENCES

Abass, N.H. 2010. Rapid in vitro multiplication and ex vitro rooting of Zizyphus mauritiana. *Journal of Applied Sciences Research,* 6(1), 63–69.

Abbas, M., Ibrahim, A.-H. and Jasim, A. 2014. Micropropagation of Indian jujube (Ziziphus mauritiana Lam.cv. Zytoni) through shoot tip culture. Aquaqculture, *Aquaqrium, Conservation and Legislation* – International Journal of Bioflux Society, 6, 11–15.

Ali, M.S., Azhar, I., Ahmad, F., Ahmad, V. and Usmanghani, K. 2010. Antimicrobial screening of Mimoaceous plants. *Pharmaceutical Biology*, 39, 43–46.

Anabzhagan, M., Kalpana, M., Rajendran, R., Ntaarajan, V. and Dhanavewl, D. 2010. In vitro production of Stevia rebaudiana Bertoni. *Emirates Journal of Food and Agriculture*, 222, 216–222.

Arya, H.C. and Shekhawat, N.S. 1986. Clonal multiplication of tree species in the Thar desert through tissue culture. Forest *Ecology and Management,* 16(1–4), 201–208. https://doi.org/10.1016/0378-1127(86)90020-4

Barh, D. and Viswanathan, G. 2008. Syzygium cumini inhibits growth and induces apoptosis in cervical cancer cell lines: A primary study. *E-Cancer Medical Science*, 2, 83.

Batchelor, C.A., Yao, D., Kochler, M.J. and Harris, P.J.C. 1989. In vitro propagation of Prosopis species (P. chilensis, P. cineraria and P. juliflora). *Annals of Forest Science*, 46, 110–112.

Choudhri, N.A., Swamy, G.S.K., Jagdeesa, R.C., Mukesh, L.C., Mastiholi, A.B., Prabhuling, G. and Basavarajanna, H.R. 2013. Micropropagation studies in jamun (Syzygium cuminii L.). *International Journal of Applied Biotechnology and Biochemistry*, 3(1), 1–7.

Chun Hua, M., Xia, Y., Yan Hui, C., Jian Can, F., Xiao Li, S., Ji Dong, L., Ying Xia, W. and Jian Bin, H. 2012. Anatomical observations of adventitious bud regeneration from leaf explants of Ziziphus jujube Mill. 'Huizao'. *Horticulture, Environment and Biotechnology*, 53(4), 316–319.

Debnath, M. 2008. Clonal propagation and antimicrobial activity of an endemic medicinal plant Stevia rebaudiana. *Journal of Medicinal Plants Research*, 2, 45–51.

Dent, T.V. 1948. Seed storage with particular reference to the storage of seed of Indian forest plants. In: *Indian Forest Records: Silviculture* (New Series). New Delhi: Manager of Publications, p. 134.

Diengngan, S. and Hasan, M.A. 2015. Genetic diversity of underutilized fruits in India for environmental sustainability. *Advances in Plants and Agriculture Research*, 2(7), 1–6.

Feng, J.C., Yu, X.M., Shang, X.L., Li, J.D. and Wu, Y.X. 2010. Factors influencing efficiency of shoot regeneration in Ziziphus jujuba Mill. 'Huizao'. *Plant Cell Tissue and Organ Culture*, 101, 111–117.

Goyal, Y. and Arya, H.C. 1985. Tissue culture of desert trees II. Clonal multiplication of Ziziphus in vitro. *Journal of Physiology*, 119, 398–404.

Goyal, D., Bhadauria, S. and Kumar, A. 2006.A protocol for in vitro propagation of ber (Ziziphus jujuba). *Indian Journal of Plant Physiology*, 11(2), 178–181.

Gupta, R., Johri, S. and Saxena, A.M. 2009. Effect of ethanolic extract of Feronia elephantum Correa fruits on blood glucose levels in normal and streptozotocin-induced diabetic rats. *Natural Product Radiance*, 8(1), 32–36.

Handique, P.J. and Bhattacharjee, S. 2000. In vitro shoot proliferation of wood apple (Feronia elephantum Correa.). *Advances in Plant Sciences*, 13(1), 241–243.

Hansika, K.P., Ranawake, A.L. and Perer, R.N.I. 2017. Micropropagation of Ziziphus jujuba Mill. (Jujube) through shoot tip and nodal segment culture. *Tropical Agricultural Research & Extension*, 20(1 and 2), 52–58.

Haroon, A., Manzoor, N. and Iqbal, M. 2018. Micropropagation of Prosopis cineraria (L.) Druce, an endangered tree of Pakistan. International Journal of Biosciences *(Online),* 12(6), 443–449. www.innspub.net

Hiregoudar, L.V., Kumar Ashok, H.G. and Murthy, H.N. 2005. In vitro culture of Feronia limonia (L.) Swingle from hypocotyl and internodal explants. *Biological Plantarum*, 49, 41–45.

Hiregoudar, L.V., Murthy, H.N., Hema, B.P., Hahn, E.J. and Paek, K.Y. 2003. Multiple shoot induction and plant regeneration of Feronia limonia (L.) Swingle. *Scientia Horticulturae*, 98(4), 357–364. https://doi.org/10.1016/S0304-4238(03)00018-9

Hossain, M.A., Shamim, K.A.H.M., Jahan, T.A. and Hasan, M.N. 2008. Micropopagation of Stevia. *International Journal of Sustainable Crop Production*, 3(4), 1–9.

Hussain, S., Naseer, B., Qadri, T., Fatima, T. and Bhat, T. 2021. Ber/Jujube (Ziziphus mauritiana): Morphology, Taxonomy, Composition and Health Benefits. In: *Fruits grown in highland regions of Himalayas*. Husaain, S.Z. et.al. (Eds.). Switzerland AG: Springer Nature, pp. 157–158. doi: 10.1007/978-3-030-75502-7_12

Ibrahim, A., Ibrahim, M.I., Nasr, B.R. and Mohammed, M.E.Z. 2008. Plant growth regulator affecting in vitro cultivation of Stevia rebaudiana. *Sugar Tech: An International Journal of Sugar Crops and Related Industries*, 0(3), 254–259.

Ibrahim, M., Jasim, A. and Abbas, M. 2011. Somatic embryogenesis and plantlet regeneration in Indian jujube (Ziziphus mauritiana Lamk.) Cv. Zaytoni. *Genetics and Plant Physiology*, 1, 150–154.

Ibrahim, M.A., Jasim, A.M. and Fadhil, M. 2012. In vitro plant regeneration of Indian jujube (Ziziphus mauritiana Lamk.) cv. Zaytoni via indirect organogenesis. *Acta agriculturae Slovenica*, 99(1), 65–67.

Islam, M.B. and Simmons, M.P. 2006. A thorny dilemma: Testing alternative intrageneric classifications within Ziziphus (Rhamnaceae). *Systematic Botany*, 31(4), 826–842.

Islam, S., Zahan, M., Akter, S., Banu, A.T., Habib, A., Khan, S. and Jahan, A.A.M. 2010. Mass propagation of Feronia limonia L. through tissue culture. *Bangladesh Journal of Scientific and Industrial Research*, 45, 75–78.

Jagetia, G. and Baliga, M. 2002. Syzygium cumini (Jamun) reduces the radiation-induced DNA damage in the cul-tured human peripheral blood lymphocytes: A preliminary study. *Toxicology Letters*, 132(1), 19–25.

Jagetia, G., Shetty, P. and Vidyasagar, M.S. 2008. Treatment of mice with leaf extract of jamun (Syzygium cumini linn. Skeels) protects against the radiation-induced damage in the intestinal mucosa of mice exposed to different doses of γ-radiation. *PharmacologyOnline*, 1, 169–195.

Jain, M., Mishra, S.H., Albert, S. and Kapadia, R. 2011. Standardization of Feronia limonia L. leaves by HPLC, HPTLC, physico-chemical and histological parameters. *Boletín Latinoamericano y del Caribe de Plantas Medicinales y Aromáticas*, 10, 525–535.

Jain, N. and Babbar, S.B. 2003. Effect of carbon source on the shoot proliferation potential of epicotyl explants of Syzygium cuminii. *Biologia Plantarum*, 47(1), 133–136.

Jejurker, S.H., Tanpure, U.R., Pawar, S.D., Joshi, R.R. and Wadikar, M.S. 2016. Micropropogation of callus from Syzygium cumini explant. *Indian Journal of Pure and Applied Biosciences*, 4(1), 291–295. doi: http://dx.doi.org/10.18782/2320-7051.222

Jordan, M. and Balbao, O.1985. In vitro regeneration of Prosopis tamarugo Phil. and Prosopis chilensis (Mol.). *Stuntz from Nodal Sections Gartenbauwissenschaft,* 50, 138–141.

Kamboj, V., Tripathi, S. and Joshi, U. 2020. Tewari Vatsala underutilized fruits crops and technologies for meeting their market needs. *Journal of Postharvest Technology*, 8(4), 64–76. www.jpht.in

Khatak, S., Sharma, A. and Saini, R. 2018. Comparative analysis of plant growth regulators on bud break in prosopsis and tecomella for sustainable agriculture in arid and semi-arid regions of India. *Bioscience Biotechnology Research Communications*, 11(2), 208–215.

Kim, Y.W., Moon, H.K. and Son, S.C. 2006. Repetitive somatic embryogenesis and plant regeneration in Zizyphus jujuba Mill. *In Vitro Cellular & Developmental Biology-Plant*, 42(3), 247–251.

Kour, S., Bakshi, P., Sharma, A., Wali, V.K., Jasrotia, A. and Kumari, S. 2018. Strategies on conservation, improvement and utilization of underutilized fruit crops. *International Journal of Current Microbiology and Applied Science*, 7(3), 638–650. doi: https://doi.org/10.20546/ijcmas.2018.703.075

Kumar, S. and Singh, N. 2009. Micropropagation of Prosopis cineraria (L.) Druce – A multi-purpose desert tree. *Researcher*, 1(3), 28–32.

Kumar, V.R., Anshu, P. and Kumar, S. 2017. Effect of growth regulator on in vitro organogenesis of Prosopis cineraria (L.) Druce. *International Journal of Current Research*, 9(8), 55322–55325.

Lu, C.Y. 1993. The use of thidiazuron in tissue culture. *In Vitro Cellular and Developmental Biology*, 29, 92–96.

Lu, H., Yan, W., Yanhong, C., Xiaoming, P. and Yingyue, L. 2018. Optimisation of a highly efficient shoot regeneration system using leaf explants of Chinese jujube (*Ziziphus jujuba* Mill.) by response surface methodology. *Journal of Horticultural Science and Biotechnology,* 93(3), 289–295. doi: 10.1080/14620316.2017.1364980

Ma, C., Ye, X., Chen, Y., Feng, J., Shang, X., Li, J., Wu, Y. and Hu, J. 2012. Anatomical observations of adventitious bud regeneration from leaf explants of Ziziphus jujuba Mill. 'Huizao'. *Horticulture, Environment, and Biotechnology*, 53(4), 316–319.

Mahoney, D. 1990. *Trees of Somalia – A Field Guide for Development Workers*. Oxford: Oxfam/HDRA, pp. 133–136.

Mann, H.S. and Shankarnarayan, K.A. 1980. The role of Prosopis cineraria in an agropastora! system in Western Rajasthan. In *Browse in Africa* (Ed.) LeHouerou, H.N. International Livestock Centre for Africa, Addis Ababa, Ethiopia, pp. 437–442.

Mathur, N., Ramawat, K.G. and Nandwani, D. 1995. Rapid in vitro multiplication of jujube through mature stem explants. *Plant Cell Tissue and Organ Culture*, 43, 75–77. https://doi.org/10.1007/BF00042675

Mukherjee, P.K, and Wahile, A. (2006). Integrated approaches towards drug development from Ayurveda and other Indian system of medicines. *Journal of Ethnopharmacology*, 103(1), 25–35.

Naaz, A., Hussain, S.A., Naz, R., Anis, M. and Alatar, A.A. 2018. Successful plant regeneration system via de novo organogenesis in Syzygium cumini (L.) Skeels: An important medicinal tree. *Agroforestry Systems*, 93(4), 1–11.

Narang, A. and Shukla, A. 2022. Axenic propagation of a woody legume – Prosopis cineraria (l.) Duce through cotyledonary node explants. *International Journal of Multidisplinary Research*, 2(1), 16–21.

Navneet and Singh, A. 2018. Syzygium cumini Linn syn. Eugenia Jambolana (jamun) – A review, 6, 32–47. 10.21276/IJIPSR.2018.06.01.252.

Orwa, C., Mutua, A., Kindt, R, Jamnadass, R., and Simons, A. 2009. *Agroforestree Database: A Tree Reference and Selection Guide, version 4.0*. Nairobi: World Agroforestry Centre ICRAF.

Pareek, A., Sharma, D. and Pareek, L. 2012. In vitro micropropagation through cotyledonary nodal segments in *Prosopis cineraria. Research Journal of Pharmaceutical, Biological and Chemical Sciences,* 3(3), 309–313.

Pareek, A.K., Garg, D.S., Kumar, M. and Yadav, S.M. 2015. Prosopis cineraria: A gift of nature for pharmacy. *International Journal of Pharma Science and Research*, 6, 958–964.

Pasiecznik, N.M., Harris, P.J.C. and Smith, S.J. 2004. *Identifying Tropical Prosopis Species: A Field Guide*. Coventry, UK: Association HDR, HDRA Publishing; p. 26.

Prabhuling, G. and Carmelita, L. 2018. Aseptic culture establishment in Jamun (Syzygium cuminii L.). *International Journal of Tropical Agriculture,* 33(2), 339–342.

Prabhuling, G., Rashmi, H. and Babu, A.G. 2017. In vitro regeneration of jamun (Syzygium cumini L.). *International Journal of Science and Nature*, 8(4), 902–907.

Preece, J.E. and Read, P.E. 2003. Novel methods in micropropagation. Proceedings of the International Symposium on acclimatization and establishment of micropropagated plants. *Acta Horticulture*, 616, 71–76.

Purohit, S.D. and Tak, K. 1992. *In vitro* propagation of an adult tree *Feronia limonia* L. through axillary branching. *Indian Journal of Experimental Biology*, 30, 377–399.

Qing Rong, S., Hong Yan, S. and Guang Fang, Z. 2011. Induction and regeneration of adventitious shoot from in vitro leaves of Zizyphus jujuba 'Kongfusucui' *plantlet. Journal of Plant Resources and Environment*, 20(1), 83–85.

Rai, D.R., Chadha, S., Kaur, M.P., Jaiswal, P. and Patil, R.T. 2011. Biochemical, microbiological and physiological changes in Jamun (Syzyium cumini L.) kept for long term storage under modified atmosphere packaging. *Journal of Food Science and Technology,* 48(3), 357–365. doi: 10.1007/s13197-011-0254-y

Rama Krishna, D. and Shasthree, T. 2015. Adventitious rooting and proliferation from different explants of Citrullus colocynthis (L.) schrad an endangered medicinally important cucurbit. *Asian Journal of Biotechnology*, 7(2), 88–95. ISSN 1996-0700/DOI: 10.3923/ajbkr.2015.88.95

Rani, B., Singh, U., Sharma, R., Gupta, A., Dhawan, N.G., Sharma, A.K., Sharma, S. and Maheshwari, R.K. 2013. Prosopis cineraria (L) Druce: A desert to brace livelihood in Rajasthan. *Asian Journal of Pharmaceutical Research and Health Care*, 5, 58–64.

Rathore, T.S., Singh, R.P., Deora, N.S. and Shekhawat, N.S. 1992. Clonal propagation of Zizyphus species through tissue culture. Scientia *Horticulturae*, 51(1–2), 165–168.

Rathore, V., Shekhawat, N.S., Singh, R.P., Rathore, J.S. and Dagla, H.R. 2004. Cloning of adult trees of Jamun (Syzygium cuminii). *Indian Journal of Biotechnology*, 3, 241–245.

Remashree, A.B., Thomas, T.V., Raghu, A.V., Nabeesa, E. and Neelakandan, N. 2007. Micropropagation of Syzigium cumini (Linn.) skeels: A multipurpose tree. *Research Journal of Botany*, 2, 208–213.

Reynertson, K.A., Yang, H., Jiang, B., Basile, M.J. and Kennelly, M.E.J. 2008. Quantitative analysis of antiradical phenolic constituents from fourteen edible myrtaceae fruits. *Food Chemistry*, 109(4), 883–890. doi:10.1016/j.foodchem.2008.01.021

Rivera, C., Jean, C., Cabrera, P., Rosa, M. and Bulnes Soriano, F. 2020. Micropropagation of Prosopis pallida (Humb & Bonpl. Ex Willd.) Kunth from shoot tips. *Revista Colombiana de Biotechnologia*, 22(1), 18–26. https://doi.org/10.15446/rev.colomb.biote.v22n1.70949

Sagrawat, H., Mann, A. and Kharya, M. 2006. Pharmacological potential of Eugenia jambolana: A review. *Pharmacognosy Magazine*, 2, 96–104.

Saima, Y., Das, A.K., Sarkar, K.K., Sen, A.K.S. and Sur, P. 2000. An antitumor pectic polysaccharide from Feronia limonia L. *International Journal of Biological Macromolecules*, 27(5), 333–335.

Saini, C. and Tyagi, S. 2019. In vitro regeneration of under exploited ethno-medicinal plant Limonia acidissima L. *International Journal of Creative and Innovative Research in all Studies*, 2(6), 1–8.

Samadia, D., Haldhar, S., Verma, A., Gurjar, P., Berwal, M., Gora, J., Kumar, R. and Ram, H. 2021. Khejri (Prosopis cineraria) research for horticultural harnessing and environmental services: an appraisal. *Journal of Agriculture and Ecology*, 12(12), 1–26. Retrieved from https://journals.saaer.org.in/index.php/jae/article/view/399

Seeja, E., Edwin, E. and Smita, G. 2005. A comparative pharmacognostical and phytochemical studies on the leaves of Aegle marmelos and Feronia elephant. *Plant Archives,* 5(2), 549–552.

Sharma, S.B., Nasir, A., Prabhu, K.M., Murthy, P.S. and Dev, G. 2003. Hypoglycaemic and hypolipidemic effect of ethanolic extract of seeds of Eugenia jambolana in alloxan-induced diabetic rabbits. *Journal of Ethanopharmacology,* 85(2–3), 201–206. doi: 10.1016/S0378-8741(02)00366-5

Shekhawat, N.S., Rathore, T.S., Singh, R.P., Deora, N.S. and Rao, S.R. 1993. Factors affecting in vitro clonal propagation of Prosopis cineraria. *Plant Growth Regulators*, 12, 273–280.

Singh, S., Mishra, D.S., Singh, A., Pradhan, S., Tarai, R. and Panigrahi, I. 2019. Plant Growth Regulators in Jamun (Syzygium cumini Skeels). In: *Plant Growth Regulators in Tropical and subtropical fruits*. Delhi: Jaya Publishing House, pp. 263–284.

Soliman, H.I. and Hegazi, G.A.E. 2013. In Vitro clonal propagation and molecular characterization of jujube (Ziziphus jujuba Mill.). *Life Science Journal,* 10(2), 573–582.

Sudhersan, S., Aboel-Nil, M. and Hussian, J. 2001. In vitro propagation of Ziziphus mauritiana cultivar Umran by shoot tip and nodal multiplication. *Current Science*, 80, 290–292.

Sudhersan, S. and Hussian, J. 2003. In vitro clonal multiplication of multipurpose tree, Ziziphus spina-christi (L.). Desf. *Turkish Journal of Botany,* 27, 167–171.

Swami, S.N., Thakor, N., Patil, M. and Haldankar, P. 2012. Jamun (*Syzygium cumini* (L.)): A review of its food and medicinal uses. *Food and Nutrition Sciences*, 3(8), 1100–1117. doi: 10.4236/fns.2012.38146.

Teixeira da Silva, J.A., Gulyás, A., Magyar-Tábori, K., Rui Wang, M., Chun Wan, Q. and Dobranszki, J. 2019. In vitro tissue culture of apple and other *Malus* species: recent advances and applications. *Planta,* 249, 975–1006. https://doi.org/10.1007/s00425-019-03100-x

Tiwari, R.D. and Gupta, R.K. 1959. A note on the chemical examination of the constituents of the bark of Feronia elephantum. *Current Science*, 28, 213–214.

Vividha, M.S., Darasing, R.R., Santosh, J.G., Amrapali, A.A., Vijay, L.G., Neha, C.M., Balwant S.M., Dipika, A.P. and Aditya, V.R. 2019. In vitro shoot proliferation from nodal segments of Indian Blackberry (Syzygium cumini L.). International Journal of

Microbiology and Applied Science, 8(3), 2464–2469. https://doi.org/10.20546/ijcmas.2019.803.291

Vyas, S., Joshi, N., Tak, K. and Purohit, S. 2005. *In vitro* adventitious shoot bud differentiation and plantlet regeneration in *Feronia limonia* L. (Swingle). *In Vitro Cellular and Developmental Biology – Plant,* 41, 296–302. https://doi.org/10.1079/IVP2004607

Vyas, S. and Purohit, S.D. 2006.Effect of controlled carbon dioxide on *in vitro* shoot multiplication in *Feronia limonia* (L.) Swingle. *Acta Physiologiae Plantarum,* 28, 605–611. https://doi.org/10.1007/s11738-006-0056-4

Wang, G., Li, X., Chen, Q. and Tian, J. 2011. Somatic embryogenesis and plant regeneration from cotyledons of immature embryo of Chinese Jujube (Zizyphus jujuba Mill.) *Acta Horticulturae,* 923, 33–38.

Wang, J.R., Zhang, M. and Liu, M.J. 2009. Plant regeneration from immature embryo in Chinese Jujube (Ziziphus jujuba Mill.). *Acta Horticulturae,* 840, 262–272.

Yadav, R.S., Yadav, B.L. and Chhipa, B.R. 2008. Litter dynamics and soil properties under different tree species in a semi-arid region of Rajasthan, India. *Agroforestry System,* 73, 1–12.

Yadav, U., Bisht, S., Siddique, S.S. and Babu, R. 2014. Micropropagation of Syzygium cuminii L. using in vitro technique. *International Journal of Scientific & Engineering Research,* 5(10), 1517–1528.

Yadav, U., Lal, M. and Jaiswal, V.S. 1990. In vitro micropropagation of the tropical fruit tree Syzygium cuminii L. *Plant Cell, Tissue and Organ Culture*, 21(1), 87–92.

Ye, X., Chen, Y., Li, J., Yu, X., Feng, J. and Zheng, X. 2012. Callus induction and adventitious shoot regeneration in Zizyphus jujuba Mill. 'Huizao'. *African Journal of Biotechnology,* 11(16), 3888.

7 Nanotechnology in Horticulture
Milestones and Advancements

Shabnam Sircaik

7.1 INTRODUCTION

The word nanotechnology is derived from the Greek word "nano," which means dwarf. Precisely nanotechnology is an atomically precise technology at the nanoscale (1–100 nm). Nobel laureate Richard Feynman in his seminal talk at the annual meeting of the American Physical Society at Caltech on December 29, 1959, famously said that "there is plenty of room at the bottom" and this is believed to have laid the foundation of nanotechnology. Feynman suggested the huge potential in the field of nanotechnology and biology (Feynman 1960). The prefix "nano" means a billionth of a meter. There are various definitions of nanotechnology but the best definition comes from the U.S. Environmental Protection Agency (EPA, which defines nanotechnology as "the science, engineering, and application of materials and devices with structures and components that have at least one dimension in the nanometer range (approximately 1 to 100 nanometers) and exhibit unique properties because of their size." Nanotechnology is considered the fifth revolutionary technology that utilizes the difference in properties of nanosized particles from their bulk counterparts. These properties are differences in surface chemistry (higher surface-to-mass ratio, reduced lattice constants, and lower melting points), and electronic, optical, and magnetic features. Minimum Information on Nanoparticle Characterization (MINChar) is a set of guidelines that defines the physicochemical properties of nanomaterials such as core material, surface coating, size and size distribution, shape, etc. These guidelines help us standardize and compare nanomaterials across studies increasing transparency (Modena et al. 2019). Nanomaterials possess significant advantages over traditional materials, such as greater strength, a larger surface area-to-volume ratio, enhanced chemical reactivity, and superior conductivity. Consequently, nanotechnological inventions have begun to replace their conventional counterparts in every industry. Agriculture, like other major inductries, has greatly benefited from these nanotechnological advancements. This shift is largely driven by the belief that such innovations are contributing to an improved quality of life.

DOI: 10.1201/9781003343707-7

7.2 EARLY RESEARCH AND DEVELOPMENT

Nanotechnology has significantly impacted various scientific domains such as medicine, chemistry, physics, and agriculture, including horticulture. This chapter specifically discusses nanotechnology's role in horticulture, from its introduction to its applications and the challenges it presents in this field. Horticulture is a sector of agriculture that deals with the production of fruits, flowers, vegetables, spices, and medicinal and aromatic plants (Sonah et al. 2011). According to the American Society for Horticulture Science, "Horticultural science is the only plant science that incorporates both the science and aesthetics of plants." Horticulture is considered the most quality-conscious industry of all (Gamage 2019). Horticulture aims to improve the health, quality, production, and shelf life of the plants involved (Gaur et al. 2018). Horticultural produce in developing countries has faced severe losses (20–30%) due to pre- and post-harvest losses. Considering these losses, even a modest reduction of 5-10% could lead to significant improvements. The management of the losses will not only improve farmer's income but can also serves towards preserving the nutritional quality of the produce. Biotechnology and nanotechnological advances serve as promising tools to counter this problem. Nanotechnology has already been applied to various steps of the supply chain process to improve food quality by providing better fertilizers, pesticides, and post-harvest management with the use of packaging materials made of components developed through nanotechnological advances.

There is a considerable amount of data available on the effect of nanotechnology on the improvement of various agricultural crops (Islam 2019; Hossain et al. 2020); however, there is no comprehensive review of the production and post-harvest losses linked to horticultural crops. Thus this chapter focuses on the application of nanotechnology in enhancing productivity, quality, and protection, and reducing the post-harvest losses of horticultural crops. We highlight these contributions by dividing the chapter into subsections starting with crop growth enhancers and crop protection materials and continuing to the post-harvest use of nanotechnology products in improving the shelf-life and nutritional quality of horticultural crops. Further, we also discuss the potential and challenges associated with the use of nanotechnology in horticulture.

7.3 NANOTECHNOLOGY-BASED PRODUCTS FOR USE IN HORTICULTURE

The agricultural sector relies heavily on agrochemicals to meet the increasing food demand driven by global population growth and climate change. However, conventional agrochemicals pose environmental risks and pest resistance issues. Nanotechnology offers solutions to mitigate these risks in horticulture, with various applications aimed at minimizing environmental damage. Herein, we summarize nanotechnology's role in mitigating environmental harm from agrochemicals, categorizing its applications in horticulture based on their plant effects (Figure 7.1).

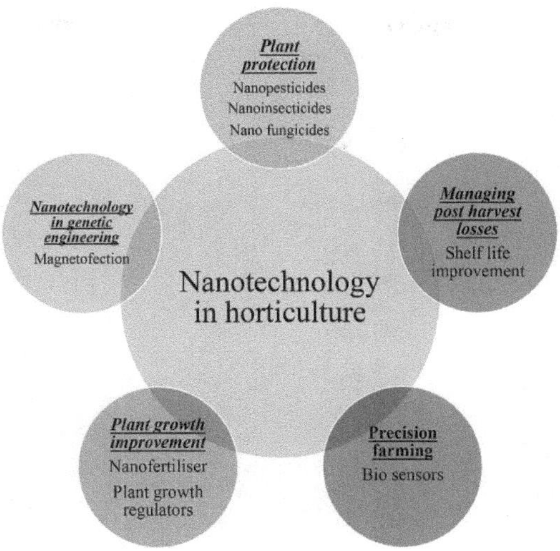

FIGURE 7.1 Overview of scope of the nanotechnology in horticulture.

7.3.1 Plant Growth Improvement

7.3.1.1 Nanofertilizers

Nanofertilizers serve as a cost-effective alternative to conventionally used fertilizers due to their increased efficiency and efficacy for entraping themselves in the soil. These can be classified into three main classes: nanoscale nanofertilizers, nanoadditives, and nanoscale coatings. The use of nanotechnology in the fertilizer industry is still in its early stages when compared to the use of nanotechnology in other areas of science like medicine and engineering. Nonetheless, some designs are adapted to increase the specificity of fertilizers and have characteristics like slow, quick, and specific fertilizer release. In addition, developments are underway to add additional characteristics like release induced by environmental parameters like moisture, heat, or pH (Manjunatha etal. 2016). Based on the type of their use, different types of nanofertilizers are summarized in Figure 7.2.

There are different ways in which a nanofertilizer can benefit a crop. Some of the routes by which nanoparticles can supply nutrients to the host plants are getting to the plant in different forms: 1. in the form of nanoporous material or via the route of nanotubes; 2. entrapping the nutrient into a polymer film made up of a nanoparticle, which is easy to absorb by the plant; and 3. in the form of an emulsion/paste made of the components with nanoscale measurements.

The use of nanomaterials such as carbon nanotubes, copper, silver, manganese, molybdenum, zinc, molybdate, iron, silicon, and titanium (Cu, Ag, Mn, Mo, Zn, Fe, Si, and Ti) as nanofertilizers are revolutionizing conventional farming into precision farming (Taha and Alsharef 2018). Some well-known examples of the nanofertilizers used in the past include use of the zinc oxide nanoparticles to improve the production of *Arachis hypogaea* (Peanut) (Prasad et al. 2012). The

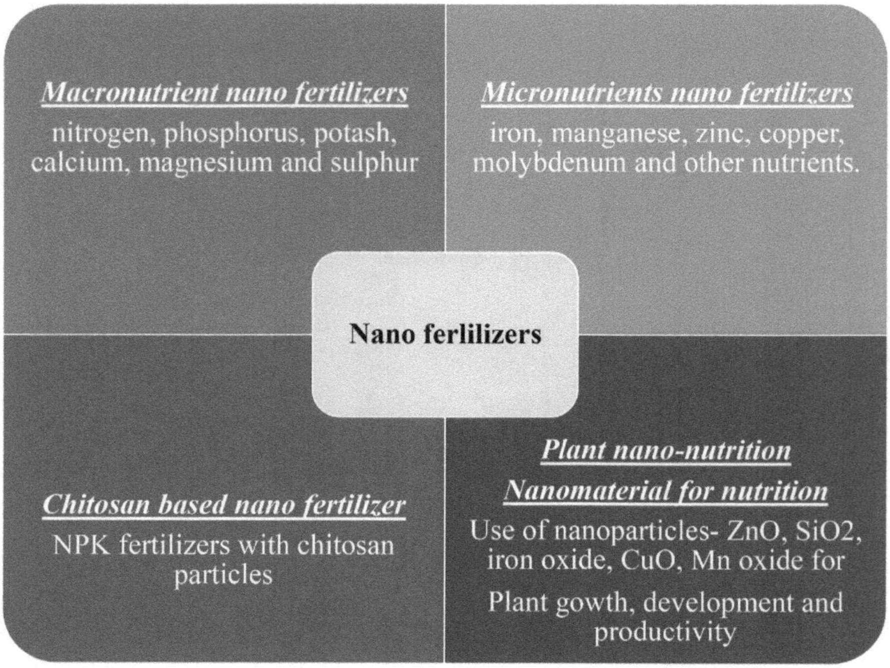

FIGURE 7.2 Classification of nanofertilisers based on their components and nutrient deliver.

use of silicon dioxide nanoparticles for increasing the plant biomass in maize to improve crop mass in terms of increased biomolecules is another such example (Suriyaprabha et al. 2012).

The use of carbon nanotubes is another way of imparting carbon to the plant. Briefly, carbon naotubes are cylindrical molecules that have hexagonal arrangement of hybridized carbon atoms, that are made by rolling up the sheets of graphene (building block and an allotrope of carbon) made up of the layers of carbon atoms arranged in a specific hexagonal lattice nanostructure. Carbon nanotubes are known to impact the growth regulation by their efficient delivery of the nutrient to the plant pertaining to their small structure and ability to cross plant cell wall. Carbon nanotubes have demonstrated a promising impact on seed growth, with notable effects observed in the seedling growth of mustard plants (Ghodake et al. 2010). The list also includes other horticultural crops like rice and black gram (Nair et al. 2010). The potential use of nanofertilizers in agriculture has been extensively reviewed (Naderi and Danesh-Shahraki 2013; Raliya et al. 2017). However, this chapter provides detailed information on the application of nanofertilizers in various horticultural crops, emphasizing their significant role in enhancing crop cultivation and boosting the production of horticultural crops. The use of nanofertilizers in horticulture has resulted in better plant growth, and flower fertility resulting in better yields and better-quality fruits (Zahedi et al. 2020; Zagzog and Gad 2017). The effect of nanomaterials on the enhancement of chlorophyll and essential nutrients such as nitrogen (N), phosphorus (P), potassium (K), manganese (Mn), magnesium (Mg), boron (B), zinc (Zn), and iron (Fe) in the plant leaves has also been observed (Abdelaziz et al. 2019). A similar quality improvement effect has also been seen in the fruit quality in terms of better quality, n umber, and total soluble sugars (TSS) and phenol quantity in the fruits (Davarpanah et al. 2017). Table 7.1 summarizes the beneficial effect of the application of nanofertilizer on these horticultural plants.

7.3.1.2 Nano Plant Growth Stimulator

Plant growth regulators are the chemicals that are used to benefit plant growth in various ways (i.e., increasing biomass, increasing flowering, controlling fruit growth, and helping in fruit maturity). The overall application of nanoplant growth regulators is similar to that of nanofertilizers, with both aiming to improve plant yield and health. This area of application has advanced extraordinarily in recent decades. The report on the positive results on the plants includes the effect on the garden pea where the use of zinc oxide (ZnO) showed a positive effect on nodulation and root length (De La Torre-Roche et al. 2013). The use of nanotitanium dioxide (TiO_2) to promote growth and photosynthesis has been also reported on spinach (Yang et al. 2006). Nonetheless, the effects are not always beneficial. Studies have observed that some side effects can affect the plant quality. This is because understanding of the full mechanism of action is still limited. It seems like higher doses of the nanoparticles may be harmful to plant health. Some failed examples include the use of CeO_2 on cucumber and TiO_2 on spinach, which resulted in an overall effect on the plant growth or quality (López-Moreno et al. 2016; Moghadam et al. 2012; Cañas et al. 2008). Carbon nanotubes have also been used to improve the plant quality in horticultural crops like onion and cucumber. However, there are instances where their use has had negative effects, such as on corn and soybeans, resulting in low biomass (Cañas et al. 2008). (Cañas et al. 2008). Similarly, the use of nano-Ag on faba beans and radish has shown no effect or even negative impacts, leading to a decline in germination capacity or no effect on germination (Raliya et al. 2017; Abd-Alla et al. 2016). Some of the studies of nanomaterials as plant growth regulators are summarized in Table 7.2.

7.3.2 Plant Protection

7.3.2.1 Nanopesticide

Pesticides, as defined by the U.S. EPA, are substances used to prevent, destroy, or mitigate pests, regulate plant growth, or stabilize nitrogen. They contain an active ingredient crucial for pest management and product effectiveness, categorized as chemical or non-chemical based on composition and toxicity levels, ranging from minimal to highly hazardous. The extensive classification can be found on the WHO homepage (Zobir et al. 2021; Raliya et al. 2017). Pesticides encapsulated in nanoparticles are termed nanopesticides. The use of nanopesticides has shown higher performance and reduction in pest-induced diseases in various studies (Chhipa and Joshi 2016). Nanopesticides are considered eco-friendly

TABLE 7.1
Examples of beneficial effects of the application of nano fertilizer on horticultural plants

	Plant	Improved quality	Nanofertilizer used	Reference
Vegetables				
1.	Potato *cv.* Arizona	yielded highest number of fruits per plant, fruit weight, fruit diameter	Nano NPK fertilizers	Hayyawi W A Al-Juthery et al. (2021)
	Tomato	highest plant height and stem diameter	300 kg/ha K Nanofertilizer, 400 kg/ha K Nanofertilizer	Ajirloo et al. (n.d.)
2.	Cucumber	fruit diameter	Ferbanat nanotechnology liquid fertilizers @ 3 L /ha	Ekinci et al. (2014)
3.	Cucumber	highest TSS	Nanonat @ 3.0 L/ ha	Ekinci et al. (2014)
4.	Sugar beet	increased biomass	Bio-organic nanofertilizers 0.5 or 1 litre/ha	Jakienė et al. (2015)
5.	Drumsticks	production of α-tocopherols, stigmasterols and campesterol	Nano iron 0, 1, 2, 3 and 4 mg/Litre	Kadim (2018)
6.	Faba bean	increasing nano iron concentration	Nano iron with 6 mg/L spray	Muktadir (2019)
7.	Spinach variety	Increased wet weight	Varamin 88 and Viroflay	Moghadam et al. (2012)
8.	French bean	absorption of nutrient	Carbon nanotubes NPK and chitosan nanoparticles NPK	Hasaneen et al. (2016)
9.	Bhindi	0.4 % recorded higher nutrient	Foliar organic nano NPK fertilization	Pm and Ushakumari (2019)
10.	Tomato	increase in antioxidant capacity and lycopene content increased yield by 30 % and fruit lycopene content increased by 12 %	Chitosal polymer Copper chitosan PVA Cu-chitosan nanomaterial	Jubeir and Ahmed (2019); Hernández-Hernández et al. (2019)

Fruits				
11.	Date palm	improved vegetative growth	Nano NPK fertilizers	Jubeir and Ahmed (2019)
12.	Pomegranate	increase in pomegranate fruit yield, fruit quality	Nano-Zn Nano-B	Davarpanah et al. (2016)
13.	Pomegranate	increase in pomegranate fruit yield	foliar Ca nano fertilization	Davarpanah et al. (2017)
14.	Mango	improving yield and fruit quality	Nano-zinc	Zakzouk (2017)
15.	Olive seedlings *cv.* Kalamat	growth	Nano NPK	Hagagg et al. (2018)
16.	Bitter almond seeds	improves seed germination		Badran and Savin (2018)
Spices, medicinal & aromatic plants				
17.	Sweet basil	enhanced plant growth, iron and oil content	Fe_3O4 nanoparticles	Elfeky et al. (2013)
18.	Saffron	saffron flowering and improves yield	Nanofertilizers Fe, P, K	Amirnia et al. (2014)
19.	Peppermint	highest plant height, number and branches of leaves	Nanofertilizers Fe, Zn, K	Hassani et al. (2015)
20.	German chamomile	increased yield	Synthetic nanozeolite/ nanohydroxyapatite	Mikhak et al. (2017)
21.	Dill	seed yield, oil yield percentage	Fe nanofertilizer	Gholinezhad (2017)
22.	Black cumin	increase the yield and yield components		Safaei et al. (2017)
Flowers				
23.	Gerbera	vase life and flower quality	Calcium nano fertilizer	Mohammadbagheri and Naderi (2017)
24.	Poinsetta	more number of leaves, shoots, root length and volume	Iron chelate fertilizers	Kaviani et al. (2016)

TABLE 7.2
Examples of nanomaterials as PGRs

Plant	Growth stimulator	Effect on plant	References
Tomato	Carbon nanotubes (MWCNT)	Better germination and growth rate	Cañas et al. (2008); De La Torre-Roche et al. (2013)
Onion and cucumber		Negative effect on root elongation	
		Root elongation	
Zucchini, tomato,		No effect	
Corn, soybean		Reduces biomass	
Cucumber	CeO2	No overall effect on plant growth	López-Moreno et al. (2016)
Garden pea	ZnO	Better nodulation and root length	Raliya et al. (2017)

with target-specific effects due to controlled release to the soil. For instance, nanomicrocapsule formations of these nanopesticides have enabled better adhesion to the plant surfaces that imparts better dispersion and bioactivity of the pesticide. This feature is due to their small size, improved droplet ductility, and better target absorption. Nanoencapsulation is one such method to target nanopesticides in the plant. This method involves coating the nanosized active ingredient with a thin protective coating increasing the effectiveness with a reduction in the amount of pesticide going to the plant.

7.3.2.2 Nanoinsecticides

Like nanopesticides, nanoinsecticides are also delivered to plants via nanoencapsulation. Nanoencapsulation allows slow discharge of these chemicals to the plants. There are various ways these insecticides can be derived to the plants. These include dissolution, biodegradation, and delivery through osmotic pressure or diffusion due to a pH difference (Vidhyalakshmi et al. 2009). Nanoinsectides have been extensively used to control insects in agricultural crops; however, there are limited studies of the use of nanopesticides on horticultural crops. Some examples include the use of commercial nanoformulations to control steom fly (*Melanagromyza sojae*) and whitefly (*Bemisia tabaci*) on soybean and promising results were observed over the commonly used commercial formulations. Another such example is the use of imidacloprid and carbofuran CR formulations for the control of leafhopper *Amrasca biguttula* Ishida, and aphid on potato plants (Adak et al. 2012). Management of fruit flies by use of pheromones is an eco-friendly alternative to the use of chemical insecticides that can trap insects quite effectively. Nanopheromones formed of nanogel containing methyl eugenol and pheromone with a low-molecular-weight organic gelator (e.g., all-trans tri (p-phenylene vinylene) bis-aldoxime) are a promising alternative to the commercial pheromones to trap insect *Bactrocera dorsalis* causing a huge crop loss, for example, in guava and mango crop (Bhagat et al. 2013).

7.3.2.3 NanoFugicides

Conventional fungicides use formulations including cupric salt, dinitropheol, trialoes, etc. These are used quite commonly due to their target specificity; however, the incorporation of nanotechnology can significantly increase its target specificity. Studies are still limited on the application of this area of nanotechnology in horticulture. One such study included the development of the nanoform of the fungicide that is commercially available. The formulation contains 25% and 50% mixture of trifloxystrobin and tebuconazole to control the fungus *Macrophomina phaseolina* (Kumar et al. 2016). The formulation is observed to have an inhibitory effect on the hyphal growth of the fungus. Another such example is the use of Cu-Zn bimetallic nanoparticles to control fungus *Saccharomyces cerevisiae* in tomato plants (Antonoglou et al. 2018).

7.4 NANOTECHNOLOGY IN POST-HARVEST MANAGEMENT

The shelf-life of horticultural crops has been a major concern in the post-harvest management of these crops. The conventional post-harvest management of fresh produce includes cold storage, preservation with chemicals, and techniques including irradiation and bio-preservation (Ma et al. 2017). The basic principles of improving shelf-life include the inactivation of microbial growth, controlling respiration, or minimizing evaporation. Though these are routinely used for food protection, all these methods have limitations. One of the most common problems contributing to post-harvest losses is a chilling injury to the plant (Zhang et al. 2021). Chemical preservation is the most conventionally used technique to manage this issue, because of its high impact and low cost. Nonetheless, in recent years this method has raised some issues due to safety reasons and potential side effects of the chemical used in the process (Meireles et al. 2016). The use of ionizing and non-ionizing radiations is another conventional method for crop preservation, but the treatment needs to be controlled as it has more operation time and requires high cost of management (Jeong and Jeong 2018). Bearing in mind the limitation of the available preservation methods, it is vital to look for more cost-effective and sustainable methods to satisfy consumers' needs without compromising on the quality of the product and with health-benefiting value to the produce. Recent decades have seen logarithmic growth in the use of nanotechnology in post-harvest management pipelines including processing, packaging, storage, and transport. Using nano-based packaging serves as the first phase of this pipeline with a reduction in food waste. Various forms of nanomaterials used in post-harvest food management are: 1) nanoparticles, 2) nanoemulsions, and 3) nanoclays. Silver-based nanoparticles are the most commonly used and extensively studied nanoparticles used in post-harvest treatments. Various studies have reported the beneficial effect of treatment of these particles on the shelf-life of horticultural crops (Tavakoli et al. 2017).

Nanostructured materials (NSMs) are safe packaging materials produced from inorganic metal, metal oxides, and composites made of the bioactive agent like chitosan or starch. Studies shows that silver, zinc, titanium, and gold (and their oxides) are the most commonly used nanoparticles for the food treatment (Chellaram et al. 2014; Souto et al. 2013). These materials have been reported to have better antimicrobial properties with broad spectrum effect. Nanoemulsions are routinely used to make packaging biofilms (Otoni et al. 2014). These are used as transporters for nanoparticles that serve as a protective barrier from the enviornment. This helps in minimizing the respiration rate slowing down the decomposition and extending the crop shelf-life. Studies have shown that chitosan-based and alginate-based biofilms have antioxidant and antimicrobial properties (Alexandre et al. 2016; Dammak et al. 2017). Nanoclays are bio-based polymers that are shown to improve the barrier properties by improving the mechanical strength of biopolymer without compromising the biodegradability (Bhuyan et al. 2010; Mallakpour and Dinari 2012). Selected successful examples of the frequently used nanomaterials for improving the shelf-life of horticultural crops by different methods are summarized in Table 7.3.

7.5 PRECISION FARMING VIS USE OF NANOSENSORS

Plant diseases and pest attacks have increased and are major threats to food security and the economy. It is believed that if the plant health is monitored in real-time and identified at early stages, crop yield could be significantly improved (Rana et al. 2021). Recent developments in nanotechnology include portable nanosensors that can detect harmful chemical and toxins in real-time, allowing *in situ* detection of the problems associated with the plants. These nanosensors also allow the early detection of other contaminants like pests, nutrient scarcity, moisture levels, and pathogens. This capability empowers farmers to take timely and targeted actions, such as applying chemicals, nutrients, or water, to protect their crops when needed. Nanotechnological biosensors have reduced the time and money consumption required for getting results when compared to their conventional counterparts of sample testing that involves sample collection and sending samples to traditional laboratories for testing and results. Nanosensors like DNA biochips can be used to detect the presence of pathogens like fungi and bacteria in the soil. Similar nanosenors have also been developed and used to analyze food safety at different levels of the supply chain starting from the farm itself. Pathogen detection is one such application of nanosensors. Various researchers have developed immunosorbent liposomal nanostructures that can detect pathogens and pathogen-related toxins like *Salmonella* and *Cronobacter* in tomatoes (Shukla et al. 2014). One such example of use of a nanosensor is lanthanide chelate-loaded silica nanoparticles for the detection of corn-wilt-causing pathogen *Pantoea stewartia*. The biosensors can detect the *P. stewartia* in corn seeds under the detection levels (Zhu et al. 2015).

Nanosensors also play an important role in managing the post-harvest losses of the horticultural crops. Ethylene production during food ripening is one such concern in post-harvest losses, and management of this problem is one of the applications of nanosensors in horticulture. Ethylene efflux has been traditionally checked in the laboratory by chromatography for measuring fruit metabolism (Janssen et al. 2014). However, the use of ethylene sensors in the detection of fruit metabolism can facilitate quick detection of ethylene levels in the ripening fruits (Scariot et al. 2022). Use of nanosensors is a promising approach for an affordable and cost-effective alternative for the detection of ethylene production in various fruits like apple, pear, kiwi,

TABLE 7.3
Examples of nanomaterials used for improving the shelf life of horticultural crops

Nanomaterials	Commodity	Imparted characterstic		References
Nano-ZnO/ Nano-ZnOs	Apple	Enhanced shelf life of fresh-cut apples up to 6 days	Nanopakaging made of ZnO	Li et al. (2011)
	Kiwifruit	Maintained the texture of fruit	Coating with ZnO	Meng et al. (2014)
	Carrot	Increased the shelf life up to 60 days	Product sterilization with ZnO	Xu et al. (2017)
	Pomegranate	Reduced weight loss	Using ZnO and carboxymethyl cellulose suspension coating	Koushesh Saba and Amini (2017)
	Soybean (green)	Nanoparticles inhibit the load of yeast, mold, bacteria and coliform A	ZnO treatment and UV irradiation	Yu et al. (2015)
	Strawberry	Inhibit microbial load	ZnO treatment	Sogvar et al. (2016)
Nano-CaCO3	Chinese yam	Delayed the ripening	Packed into Nano-CaCO3 packaging	Luo et al. (2015)
Chitosan	Cucumber	Increased antioxidant activity	Coating with chitosan	Mohammadi et al. (2016)
Nano-silica	Longan fruit	Reduced fruit degradation due to high oxidase activity	Chitosan/ nanosilica biofim	Shi et al. (2013)
Nano-Ag	Mushroom (Shiitake)	Decreased the rate of electrolyte leakage	Coating with Alginate/nanoAg	Jiang et al. (2012)

and avocado. Researchers have already developed a carbon nanotube sensor to detect emitted ethylene by the horticulture produce during the storage period (Esser et al. 2012). Similarly, E-nose sensors can detect the presence of ethylene from contamination or spoilage in food products (Valdés et al. 2009). The physiological states of the fruits are dependent on the ripening index, ethylene production, releasing procedure of volatile compounds, and storage conditions (Wilson and Baietto 2009). Consistent with the detection of food ripening by nanosensors, nanotubes have also been designed to detect pesticides. These nanotubes have been used to detect diazinon-based organophosphate pesticides. The review by a group of researchers summarizes the state-of-the-art use of nanosensors to detect pesticides in medicinal and crop-based plants (Saini er al. 2017) .While commercial use of these nanobiosensors is not yet widespread, future advancements in nanotechnology include the development of DNA nanochips for detecting agrochemical contaminants in horticultural crops.

7.6 NANOTECHNOLOGY AND GENETIC ENGINEERING

Recent years have seen significant advancements in the incorporation of nanotechnological tools and techniques in improving plant genetic engineering. The sole purpose of these technological advances is to overcome the limitations related to the available genetic engineering tools. Herein, we describe some of the advances of nanotechnology that have been used in the field of horticulture.

7.6.1 Magnetofection

The use of magnetic iron oxide nanoparticles coupled with DNA (deoxyribonucleic acid) for DNA delivery is one invention of nanotechnology that has been routinely used for gene transformation in mammalian cell culture techniques; However, there are challenges in adapting this gene delivery technique to plant cells, primarily due to the cell wall's resistance to endocytosis (Wang et al. 2016). A successful example of this DNA delivery has been reported for the DNA delivery into cotton plant pollen to integrate for insect resistance (Cui et al. 2011). The use of these techniques for gene transformation in horticulture crops is still limited. Some successful examples include the use of pollen magnetofection in pumpkin plants (Zhao et al. 2017), and nano iron oxide in magnetofection on okra (Farooq et al. 2022).

7.6.2 "DNA Free" Genome Editing

The use of genome editing tools are well known to increase the overall throughput of plant molecular biology. However, there are concerns related to this technology because of the integration of the transgene into the host plant in this process. Due to this issue, DNA-free genome editing systems are attracting a lot of attention, especially from a regulatory perspective, as they eliminate the use of any transgene for genome editing. The recent use of nanoparticles to facilitate genome editing and target the delivery of RNA (Ribonucleic acids), nucleases, and plasmids has received great interest (Liu et al. 2017). Nanoparticles have been extensively used for the delivery of these biomolecules to the mammalian cell lines; however the developments are still underway to design a universal system that can be used in plant systems. The system faces more challenges in plant cells due to the diverse nature and presence of the cell wall in the plant cells. Some studies that have shown the use of nanotechnological formulations to facilitate the intake of DNA, RNA, or protein to agricultural crops involve the use of polymeric nanoparticles and carbon nanotubes (Raliya et al. 2016; L. Jiang et al. 2014; Demirer et al. 2017).

7.7 REGULATORY CHALLENGES AND ETHICAL CONCERNS

Nanotechnology raises ethical concerns regarding human health and environmental safety, but current evidence suggests no known harmful effects due to the non-toxic and non-reactive nature of nanoparticles. Nanotoxicology has emerged as a specialized field to address these concerns, aiming to identify and mitigate potential risks associated with nanotechnology. However, there is a significant number of reports that suggest a potential risk associated with the use of nanomaterials in the food sector (Roberto and Christofoletti 2019; Sahu and Hayes 2017). As nanomaterials have a very small size, they have the potential of passing through the cell membrane and can accumulate into cytosol, thus leading to some side effects on the viability of the cell (Ou et al. 2016). More concerns are related to the use of nanomaterials in food packaging and food fortification as they have the potential of entering these nanomaterials into the body by ingestion of skin exposure (Maisanaba et al. 2015). If they penetrate the human body, they can encounter biomolecules like proteins, sugars, and lipids and enter the blood or lymph (Farhoodi 2016). Some studies have shown the toxic accumulation of titanium and silver nanoparticles in body organs (Carrero-Sanchez et al. 2006; Kim et al. 2008; Rhim et al. 2013). There are not many studies showing the possible toxic effects of these nanomaterials. Reports have shown that the particle migration are within limits by the European Commission (EC).

Overall, studies on the potential toxic effects of nanomaterials remain limited, largely due to their small size and surface chemistry. It is crucial to investigate these effects individually, as nanomaterials operate through diverse mechanisms (Daloiso and Spagnolo 2014). One ongoing concern in nanotechnology is the accumulation and mobility of nanoparticles within biological systems. However, we still do not have enough evidence to prove the negative impacts of nanomaterials on the environment and still need to develop proper guidelines for the disposal of the nanomaterials to avoid the potentially harmful effects.

7.8 CONCLUSION

Nanotechnology represents a cutting-edge tool across various scientific disciplines, including agriculture, where it has demonstrated substantial contributions. In horticulture, nanotechnology offers innovative solutions to enhance crop productivity and mitigate post-harvest losses of fruits and vegetables. The cost-effective development of nanomaterials allows for rapid creation with minimal resource expenditure, providing sustainability advantages over traditional alternatives. Utilizing nanomaterial-based packaging biofilms and nanosensors aids in early detection of food-spoiling pathogens, thereby reducing spoilage losses. Additionally, nanotechnology facilitates advancements in genetic engineering, although further research is warranted for dependable techniques. Despite initial economic challenges and the need for quality assessment, nanotechnology holds promise for addressing food security concerns amidst rising global population and environmental changes. Future research should prioritize sustainable alternatives to agrochemicals, ecotoxicological assessments, and regulatory frameworks. Overall, leveraging nanotechnology in horticulture promises a future of enhanced food availability and quality, akin to its success in other scientific domains.

REFERENCES

Abd-Alla, Mohamed Hemida, Nivien Allam Nafady, and David Mamdouh Khalaf. 2016. "Assessment of Silver Nanoparticles Contamination on Faba Bean-Rhizobium Leguminosarum Bv. Viciae-Glomus Aggregatum Symbiosis: Implications for Induction of Autophagy Process in Root Nodule." *Agriculture, Ecosystems & Environment* 218 (February): 163–77. doi:10.1016/j.agee.2015.11.022

Abdelaziz, F. H., AMMA Akl, A. Y. Mohamed, and M. A. Zakier. 2019. "Response of Keitte Mango Trees to Spray Boron Prepared by Nanotechnology Technique." *New York Science Journal* 12: 48–55.

Adak, Totan, Jitendra Kumar, Debjani Dey, N. A. Shakil, and S. Walia. 2012. "Residue and Bio-Efficacy Evaluation of Controlled Release Formulations of Imidacloprid against Pests in Soybean (G *Lycine Max*)." *Journal of Environmental Science and Health, Part B* 47 (3): 226–31. doi:10.1080/03601234.2012.634368

Alexandre, Elisabete Maria Cruz, Rodrigo Vinícius Lourenço, Ana Mônica Quinta Barbosa Bittante, Izabel Cristina Freitas Moraes, and Paulo José do Amaral Sobral. 2016. "Gelatin-Based Films Reinforced with Montmorillonite and Activated with Nanoemulsion of Ginger Essential Oil for Food Packaging Applications." *Food Packaging and Shelf Life* 10 (December): 87–96. doi:10.1016/j.fpsl.2016.10.004

Amirnia, Reza, Mahdi Bayat, and Mahdi Tajbakhsh. 2014. "Effects of Nano Fertilizer Application and Maternal Corm Weight on Flowering at Some Saffron (Crocus Sativus L.) Ecotypes." *Turkish Journal Of Field Crops* 19 (2): 158. doi:10.17557/tjfc.46269

Antonoglou, Orestis, Julietta Moustaka, Ioannis-Dimosthenis S. Adamakis, Ilektra Sperdouli, Anastasia A. Pantazaki, Michael Moustakas, and Catherine Dendrinou-Samara. 2018. "Nanobrass CuZn Nanoparticles as Foliar Spray Nonphytotoxic Fungicides." *ACS Applied Materials & Interfaces* 10 (5): 4450–61. doi:10.1021/acsami.7b17017

Badran, A., and I. Savin. 2018. Effect of Nano-Fertilizer on Germination of Bitter Almond Seeds. *BioNanoScience* 8: 742–51. https://doi.org/10.1007/s12668-018-0531-6

Bhagat, Deepa, Suman K. Samanta, and Santanu Bhattacharya. 2013. "Efficient Management of Fruit Pests by Pheromone Nanogels." *Scientific Reports* 3 (1). Nature Publishing Group: 1294. doi:10.1038/srep01294

Bhuyan, Satyam, S. Sundararajan, Yongshang Lu, and Richard C. Larock. 2010. "A Study of the Physical and Tribological Properties of Biobased Polymer–Clay Nanocomposites at Different Clay Concentrations." *Wear* 268 (5): 797–802. doi:10.1016/j.wear.2009.12.021

Cañas, Jaclyn E., Monique Long, Shawna Nations, Rodica Vadan, Lenore Dai, Mingxiang Luo, Ramya Ambikapathi, E. Henry Lee, and David Olszyk. 2008. "Effects of Functionalized and Nonfunctionalized Single-Walled Carbon Nanotubes on Root Elongation of Select Crop Species." *Environmental Toxicology and Chemistry* 27 (9): 1922. doi:10.1897/08-117.1

Carrero-Sanchez, J. C., A. L. Elias, R. Mancilla, G. Arrellin, H. Terrones, J. P. Laclette, and M. J. N. L. Terrones. 2006. "Biocompatibility and Toxicological Studies of Carbon Nanotubes Doped with Nitrogen." *Nano Letters* 6 (8). ACS Publications: 1609–16.

Chellaram, C., G. Murugaboopathi, A. A. John, R. Sivakumar, S. Ganesan, S. Krithika, and G. Priya. 2014. "Significance of Nanotechnology in Food Industry." *APCBEE Procedia* 8. Elsevier: 109–13.

Chhipa, Hemraj, and Piyush Joshi. 2016. "Nanofertilisers, Nanopesticides and Nanosensors in Agriculture." In *Nanoscience in Food and Agriculture 1*, edited by Shivendu Ranjan, Nandita Dasgupta, and Eric Lichtfouse, 247–82. Sustainable Agriculture Reviews. Cham: Springer International Publishing. doi:10.1007/978-3-319-39303-2_9

Cui, Jinjie, Junyu Luo, Wopke Van Der Werf, Yan Ma, and Jingyuan Xia. 2011. "Effect of Pyramiding Bt and CpTI Genes on Resistance of Cotton to Helicoverpa Armigera (Lepidoptera: Noctuidae) Under Laboratory and Field Conditions." *Journal of Economic Entomology* 104 (2): 673–84. doi:10.1603/EC09228

Daloiso, Viviana, and Antonio G. Spagnolo. 2014. "Ethics Research Committees in Reviewing Nanotechnology Clinical Trials Protocols." In *Responsibility in Nanotechnology Development*, edited by Simone Arnaldi, Arianna Ferrari, Paolo Magaudda, and Francesca Marin, 13: 97–109. The International Library of Ethics, Law and Technology. Dordrecht: Springer Netherlands. doi:10.1007/978-94-017-9103-8_6

Dammak, Ilyes, Rosemary Aparecida de Carvalho, Carmen Sílvia Fávaro Trindade, Rodrigo Vinicius Lourenço, and Paulo José do Amaral Sobral. 2017. "Properties of Active Gelatin Films Incorporated with Rutin-Loaded Nanoemulsions." *International Journal of Biological Macromolecules* 98 (May): 39–49. doi:10.1016/j.ijbiomac.2017.01.094

Davarpanah, Sohrab, Ali Tehranifar, Gholamhossein Davarynejad, Javier Abadía, and Reza Khorasani. 2016. "Effects of Foliar Applications of Zinc and Boron Nano-Fertilizers on Pomegranate (Punica Granatum Cv. Ardestani) Fruit Yield and Quality." *Scientia Horticulturae* 210 (October): 57–64. doi:10.1016/j.scienta.2016.07.003

Davarpanah, Sohrab, Ali Tehranifar, Gholamhossein Davarynejad, Mehdi Aran, Javier Abadía, and Reza Khorassani. 2017. "Effects of Foliar Nano-Nitrogen and Urea Fertilizers on the Physical and Chemical Properties of Pomegranate (Punica Granatum Cv. Ardestani) Fruits." *HortScience* 52 (2): 288–94. doi:10.21273/HORTSCI11248-16

De La Torre-Roche, Roberto, Joseph Hawthorne, Yingqing Deng, Baoshan Xing, Wenjun Cai, Lee A. Newman, Qiang Wang, Xingmao Ma, Helmi Hamdi, and Jason C. White. 2013. "Multiwalled Carbon Nanotubes and C60 Fullerenes Differentially Impact the Accumulation of Weathered Pesticides in Four Agricultural Plants." *Environmental Science & Technology* 47 (21). American Chemical Society: 12539–47. doi:10.1021/es4034809

Demirer, Gozde S., Huan Zhang, Juliana L. Matos, Natalie Goh, Francis Cunningham, Younghun Sung, Roger Chang, et al. 2017. "High Aspect Ratio Nanomaterials Enable Delivery of Functional Genetic Material Without DNA Integration in Mature Plants." Preprint." *Nature Nanotechnology* 14 (5): 456–64. doi:10.1038/s41565-019-0382-5

Ekinci, Melek, Atilla Dursun, Ertan Yildirim, and Fazilet Parlakova. 2014. "Effects of Nanotechnology Liquid Fertilizers on the Plant Growth and Yield of Cucumber (Cucumis Sativus L.)." *Acta Scientiarum Polonorum Hortorum Cultus* 13 (3): 135–41.

Elfeky, Souad, Mohammed Mohammed, Mohammed Khater, Yasser Osman, and Alsayed Elsherbini. 2013. "Effect of Magnetite Nano-Fertilizer on Growth and Yield of Ocimum Basilicum L." *Plants*, 46 (3): 1286–93. ISSN: 2051-4263

Esser, Birgit, Jan M. Schnorr, and Timothy M. Swager. 2012. "Selective Detection of Ethylene Gas Using Carbon Nanotube-Based Devices: Utility in Determination of Fruit Ripeness." *Angewandte Chemie International Edition* 51 (23). Wiley Online Library: 5752–56.

Farhoodi, Mehdi. 2016. "Nanocomposite Materials for Food Packaging Applications: Characterization and Safety Evaluation." *Food Engineering Reviews* 8 (1): 35–51. doi:10.1007/s12393-015-9114-2

Farooq, Naila, Laraib Ather, Muhammad Shafiq, Muhammad Shah Nawaz-ul-Rehman, Muhammad Haseeb, Tehmina Anjum, Qamar Abbas, et al. 2022. "Magnetofection Approach for the Transformation of Okra Using Green Iron Nanoparticles." *Scientific Reports* 12 (1): 16568. doi:10.1038/s41598-022-20569-x

Feynman, Richard P. 1960. "There's Plenty of Room at the Bottom." *Engineering and Science* 23 (5). Pasadena, CA: California Institute of Technology: 22–36.

Gamage, T. V. 2019. "Disinfestation of Fruits and Vegetables: An Overview." In *Reference Module in Food Science*, B9780081005965212000. Elsevier. doi:10.1016/B978-0-08-100596-5.21194-3

Gaur, Rajarshi K., Rakesh K. Verma, and Satyendra M. P. Khurana. 2018. "Genetic Engineering of Horticultural Crops." In *Genetic Engineering of Horticultural Crops*, edited by Gyana Zanjan Rout and K. V Peter, 23–46. Elsevier. doi:10.1016/B978-0-12-810439-2.00002-7

Ghodake, Gajanan, Yeong Deuk Seo, Donghee Park, and Dae Sung Lee. 2010. "Phytotoxicity of Carbon Nanotubes Assessed by *Brassica Juncea* and *Phaseolus Mungo*." *Journal of Nanoelectronics and Optoelectronics* 5 (2): 157–60. doi:10.1166/jno.2010.1084

Gholinezhad, Esmaeil. 2017. "Effect of Drought Stress and Fe Nano-Fertilizer on Seed Yield, Morphological Traits, Essential Oil Percentage and Yield of Dill (Anethum Graveolens L.)." *Journal of Essential Oil Bearing Plants* 20 (4). Taylor & Francis: 1006–17.

Hagagg, F. Laila, Nabil Mustafa, M. F. M. Shahin, and Eman El-Hady. 2018. "Impact of Nanotechnology Application on Decreasing Used Rate of Mineral Fertilizers and Improving Vegetative Growth of Aggizi Olive Seedlings." *Bioscience Research* 15 (January): 1304–11.

Hasaneen, Mohammed Nagib Abdel-Ghany, Heba Mahmoud Mohammad Abdel-aziz, and Moheb Omer. 2016. "Effect of Foliar Application of Engineered Nanomaterials: Carbon Nanotubes NPK and Chitosan Nanoparticles NPK Fertilizer on the Growth of French Bean Plant." *Biochemistry and Biotechnology Research* 4(4): 68–76.

Hassani, Ali, Ali A. Tajali, and Seyed M. Hosseni Mazinani. 2015. "Studying the Conventional Chemical Fertilizers and Nano-Fertilizer of Iron, Zinc and Potassium on Quantitative Yield of the Medicinal Plant of Peppermint (Mentha Piperita L.) in Khuzestan." *International Journal of Agriculture Innovations and Research* 3 (4). International Journal of Agriculture Innovations and Research: 1078–82.

Hayyawi W. A. Al-Juthery, E. A. H. M. Ali, Rafid N Al-Ubori, Q. M. Nal-Shami, and Duraid K. A. Al-Taey. 2021. "Role of Foliar Application of Nano Npk, Micro Fertilizers and Yeast Extract on Growth and Yield of Wheat." *International Journal of Agricultural and Statistical Sciences*. DocID: https://connectjournals.com/03899.2020.16.1295

Hernández-Hernández, Hipólito, Tomasa Quiterio-Gutiérrez, Gregorio Cadenas-Pliego, Hortensia Ortega-Ortiz, Alma Delia Hernández-Fuentes, Marcelino Cabrera de la Fuente, Jesús Valdés-Reyna, and Antonio Juárez-Maldonado. 2019. "Impact of Selenium and Copper Nanoparticles on Yield, Antioxidant System, and Fruit Quality of Tomato Plants." *Plants* 8 (10). Multidisciplinary Digital Publishing Institute: 355. doi:10.3390/plants8100355

Hossain, Akbar, Rout George Kerry, Muhammad Farooq, Nawfel Abdullah, and M. Tofazzal Islam. 2020. "Application of Nanotechnology for Sustainable Crop Production Systems." In *Nanotechnology for Food, Agriculture, and Environment*, edited by Devarajan Thangadurai, Jeyabalan Sangeetha, and Ram Prasad, 135–59. Nanotechnology in the Life Sciences. Cham: Springer International Publishing. doi:10.1007/978-3-030-31938-0_7

Islam, Md Tofazzal. 2019. "Applications of Nanomaterials for Future Food Security: Challenges and Prospects." *Malaysian Journal of Halal Research* 2 (1): 6–9. doi:10.2478/mjhr-2019-0002

Jakienė, Elena, Vidmantas Spruogis, Kęstutis Romaneckas, Anželika Dautartė, and Dovilė Avižienytė. 2015. "The Bio-Organic Nano Fertilizer Improves Sugar Beet Photosynthesis Process and Productivity." *Zemdirbyste-Agriculture* 102 (2): 141–46.

Janssen, S., K. Schmitt, M. Blanke, M. L. Bauersfeld, J. Wöllenstein, and W. Lang. 2014. "Ethylene Detection in Fruit Supply Chains." *Philosophical Transactions. Series A, Mathematical, Physical, and Engineering Sciences* 372 (2017): 20130311. doi:10.1098/rsta.2013.0311

Jeong, M.-A., and R.-D. Jeong. 2018. "Applications of Ionizing Radiation for the Control of Postharvest Diseases in Fresh Produce: Recent Advances." *Plant Pathology* 67 (1): 18–29. doi:10.1111/ppa.12739

Jiang, Li, Lian Ding, Bicheng He, Jie Shen, Zejun Xu, Meizhen Yin, and Xiaolan Zhang. 2014. "Systemic Gene Silencing in Plants Triggered by Fluorescent Nanoparticle-Delivered Double-Stranded RNA." *Nanoscale* 6 (17): 9965–69. doi:10.1039/c4nr03481c

Jiang, Tianjia, Lifang Feng, and Jianrong Li. 2012. "Changes in Microbial and Postharvest Quality of Shiitake Mushroom (Lentinus Edodes) Treated with Chitosan–Glucose Complex Coating under Cold Storage." *Food Chemistry* 131 (3): 780–86. doi:10.1016/j.foodchem.2011.08.087

Jubeir, Sh M., and W. A. Ahmed. 2019. "Effect of Nanofertilizers and Application Methods on Vegetative Growth and Yield of Date Palm." *The Iraqi Journal of Agricultural Science* 50 (1). Jami'at Baghdad: 267–74.

Kadim, Abdulameer Ali Yaseen Akhlass Mery. 2018. "Impact of Nano Chelated Iron, GA3 and Organic Fertilizer (Acadian) in Moringa Leaves Content of α-Tocopherol and Phytosterols." *Research Journal of Pharmacy and Technology* 11 (5). A & V Publications: 1840–46.

Kaviani, Behzad, Mohammad Vali Fakouri Ghaziani, and Naser Negahdar. 2016. "The Effect Fo Iron Nano-Chelate Fertilizer and Cycocel (CCC) on Some Quantity and Quality Characters of Euphorbia Pulcherrima Willd." *Journal of Medical and Bioengineering* 5 (1): 41–44. doi:10.12720/jomb.5.1.41-44

Kim, Yong Soon, Jin Sik Kim, Hyun Sun Cho, Dae Sik Rha, Jae Min Kim, Jung Duck Park, Byung Sun Choi, et al. 2008. "Twenty-Eight-Day Oral Toxicity, Genotoxicity, and Gender-Related Tissue Distribution of Silver Nanoparticles in Sprague-Dawley Rats." *Inhalation Toxicology* 20 (6): 575–83. doi:10.1080/08958370701874663

Koushesh Saba, Mahmoud, and Rasoul Amini. 2017. "Nano-ZnO/Carboxymethyl Cellulose-Based Active Coating Impact on Ready-to-Use Pomegranate during Cold Storage." *Food Chemistry* 232 (October): 721–26. doi:10.1016/j.foodchem.2017.04.076

Kumar, G. Dileep, N. Natarajan, and S. Nakkeeran. 2016. "Antifungal Activity of Nanofungicide Trifloxystrobin 25% + Tebuconazole 50% against Macrophomina Phaseolina." *African Journal of Microbiology Research* 10 (4). Academic Journals: 100–105. doi:10.5897/AJMR2015.7692

Li, Xihong, Weili Li, Yunhong Jiang, Yulong Ding, Juan Yun, Yao Tang, and Peipei Zhang. 2011. "Effect of Nano-ZnO-Coated Active Packaging on Quality of Fresh-Cut 'Fuji' Apple: Effect of Nano-ZnO-Coated Active Packaging on Quality." *International Journal of Food Science & Technology* 46 (9): 1947–55. doi:10.1111/j.1365-2621.2011.02706.x

Liu, Chang, Li Zhang, Hao Liu, and Kun Cheng. 2017. "Delivery Strategies of the CRISPR-Cas9 Gene-Editing System for Therapeutic Applications." *Journal of Controlled Release: Official Journal of the Controlled Release Society* 266 (November): 17–26. doi:10.1016/j.jconrel.2017.09.012

López-Moreno, Martha L., Leany Lugo Avilés, Nitza Guzmán Pérez, Bianca Álamo Irizarry, Oscar Perales, Yarilyn Cedeno-Mattei, and Félix Román. 2016. "Effect of Cobalt Ferrite (CoFe2O4) Nanoparticles on the Growth and Development of Lycopersicon

Lycopersicum (Tomato Plants)." *Science of The Total Environment* 550 (April): 45–52. doi:10.1016/j.scitotenv.2016.01.063

Luo, Zisheng, Yansheng Wang, Lei Jiang, and Xiaoling Xu. 2015. "Effect of Nano-CaCO3-LDPE Packaging on Quality and Browning of Fresh-Cut Yam." *LWT – Food Science and Technology*, Food Innovation in China, 60 (2, Part 2): 1155–61. doi:10.1016/j.lwt.2014.09.021

Ma, Liang, Min Zhang, Bhesh Bhandari, and Zhongxue Gao. 2017. "Recent Developments in Novel Shelf Life Extension Technologies of Fresh-Cut Fruits and Vegetables." *Trends in Food Science & Technology* 64 (June): 23–38. doi:10.1016/j.tifs.2017.03.005

Maisanaba, Sara, Silvia Pichardo, María Puerto, Daniel Gutiérrez-Praena, Ana M. Cameán, and Angeles Jos. 2015. "Toxicological Evaluation of Clay Minerals and Derived Nanocomposites: A Review." *Environmental Research* 138 (April): 233–54. doi:10.1016/j.envres.2014.12.024

Mallakpour, Shadpour, and Mohammad Dinari. 2012. "Fabrication of Polyimide/Titania Nanocomposites Containing Benzimidazole Side Groups via Sol–Gel Process." *Progress in Organic Coatings* 75 (4): 373–78. doi:10.1016/j.porgcoat.2012.07.012

Manjunatha, S. B., D. P. Biradar, and Yallappa R. Aladakatti. 2016. "Nanotechnology and Its Applications in Agriculture: A Review." *Journal of Farm Science* 29 (1): 1–13.

Meireles, Ana, Efstathios Giaouris, and Manuel Simões. 2016. "Alternative Disinfection Methods to Chlorine for Use in the Fresh-Cut Industry." *Food Research International* 82 (April): 71–85. doi:10.1016/j.foodres.2016.01.021

Meng, Xiangyong, Min Zhang, and Benu Adhikari. 2014. "The Effects of Ultrasound Treatment and Nano-Zinc Oxide Coating on the Physiological Activities of Fresh-Cut Kiwifruit." *Food and Bioprocess Technology* 7 (1): 126–32. doi:10.1007/s11947-013-1081-0

Mikhak, Azadeh, Akbar Sohrabi, Mohammad Zaman Kassaee, and Mohammad Feizian. 2017. "Synthetic Nanozeolite/Nanohydroxyapatite as a Phosphorus Fertilizer for German Chamomile (Matricariachamomilla L.)." *Industrial Crops and Products* 95 (January): 444–52. doi:10.1016/j.indcrop.2016.10.054

Modena, Mario M., Bastian Rühle, Thomas P. Burg, and Stefan Wuttke. 2019. "Nanoparticle Characterization: What to Measure?" *Advanced Materials*, May, 31 (32): 1901556. doi:10.1002/adma.201901556

Moghadam, Alirezaladan, Hesam Vattani, Nasim Baghaei, and Narges Keshavarz. 2012. "Effect of Different Levels of Fertilizer Nano_Iron Chelates on Growth and Yield Characteristics of Two Varieties of Spinach ('Spinacia Oleracea' L.): Varamin 88 and Viroflay." *Research Journal of Applied Sciences, Engineering and Technology* 4 (22). Maxwell Science Publishing: 4813–18.

Mohammadbagheri, Leila, and Davood Naderi. 2017. "Effect of Growth Medium and Calcium Nano-Fertilizer on Quality and Some Characteristics of Gerbera Cut Flower." *Journal of Ornamental Plants* 7 (3). Islamic Azad University: 205–13.

Mohammadi, Ali, Maryam Hashemi, and Seyed Masoud Hosseini. 2016. "Integration between Chitosan and Zataria Multiflora or Cinnamomum Zeylanicum Essential Oil for Controlling Phytophthora Drechsleri, the Causal Agent of Cucumber Fruit Rot." *LWT – Food Science and Technology* 65 (January): 349–56. doi:10.1016/j.lwt.2015.08.015

Muktadir, Md Abdul. 2019. "Understanding Morphological, Physiological and Biochemical Characteristics of Faba Bean (Vicia Faba L.) under Drought Condition." PhD Thesis.

Naderi, M. R., and A. Danesh-Shahraki. 2013. "Nanofertilizers and Their Roles in Sustainable Agriculture." *International Journal of Agriculture and Crop Sciences (IJACS)* 5 (19). Science Explorer Publications: 2229–32.

Nair, Remya, Saino Hanna Varghese, Baiju G. Nair, T. Maekawa, Y. Yoshida, and D. Sakthi Kumar. 2010. "Nanoparticulate Material Delivery to Plants." *Plant Science* 179 (3): 154–63. doi:10.1016/j.plantsci.2010.04.012

Otoni, Caio G., Márcia R. de Moura, Fauze A. Aouada, Geany P. Camilloto, Renato S. Cruz, Marcos V. Lorevice, Nilda de F. F. Soares, and Luiz H. C. Mattoso. 2014. "Antimicrobial and Physical-Mechanical Properties of Pectin/Papaya Puree/Cinnamaldehyde Nanoemulsion Edible Composite Films." *Food Hydrocolloids* 41 (December): 188–94. doi:10.1016/j.foodhyd.2014.04.013

Ou, Lingling, Bin Song, Huimin Liang, Jia Liu, Xiaoli Feng, Bin Deng, Ting Sun, and Longquan Shao. 2016. "Toxicity of Graphene-Family Nanoparticles: A General Review of the Origins and Mechanisms." *Particle and Fibre Toxicology* 13 (1): 57. doi:10.1186/s12989-016-0168-y

Pm, Nibin, and Ushakumari. 2019. "Organic Nano NPK Formulations for Enhancing Soil Post-Harvest Nutrient Status of Bhindi." *Journal of Pharmacognosy and Phytochemistry* 8 (2S). AkiNik Publications: 22–25.

Prasad, T. N. V. K. V., P. Sudhakar, Y. Sreenivasulu, P. Latha, V. Munaswamy, K. Raja Reddy, T. S. Sreeprasad, P. R. Sajanlal, and T. Pradeep. 2012. "Effect of Nanoscale Zinc Oxide Particles on the Germination, Growth and Yield of Peanut." *Journal of Plant Nutrition* 35 (6): 905–27. doi:10.1080/01904167.2012.663443

Raliya, Ramesh, Christina Franke, Sanmathi Chavalmane, Remya Nair, Nathan Reed, and Pratim Biswas. 2016. "Quantitative Understanding of Nanoparticle Uptake in Watermelon Plants." *Frontiers in Plant Science* 7: 1288. www.frontiersin.org/articles/10.3389/fpls.2016.01288

Raliya, Ramesh, Vinod Saharan, Christian Dimkpa, and Pratim Biswas. 2017. "Nanofertilizer for Precision and Sustainable Agriculture: Current State and Future Perspectives." *Journal of Agricultural and Food Chemistry* 66 (26). ACS Publications: 6487–6503.

Rana, Ruhul, Md. Siddiqui, Milan Skalicky, Marian Brestic, Akbar Hossain, Emrul Kayesh, Marek Popov, et al. 2021. "Prospects of Nanotechnology in Improving the Productivity and Quality of Horticultural Crops." *Horticulturae* 7 (10): 332. doi:10.3390/horticulturae7100332

Rhim, Jong-Whan, Hwan-Man Park, and Chang-Sik Ha. 2013. "Bio-Nanocomposites for Food Packaging Applications." *Progress in Polymer Science*, Progress in Bionanocomposites: from green plastics to biomedical applications, 38 (10): 1629–52. doi:10.1016/j.progpolymsci.2013.05.008

Roberto, Matheus M., and Cintya A. Christofoletti. 2019. *How to Assess Nanomaterial Toxicity? An Environmental and Human Health Approach*. London: IntechOpen.

Safaei, Zeinab, Majid Azizi, Gholamhossein Davarynejad, and Hossein Aroiee. 2017. "The Effect of Planting Seasons on Quantitative and Qualitative Characteristics of Black Cumin (Nigella Sativa L.)." *Journal of Medicinal Plants and By-Product* 6 (1). Iranian Medicinal Plants Society: 27–33.

Sahu, Saura C., and A. Wallace Hayes. 2017. "Toxicity of Nanomaterials Found in Human Environment: A Literature Review." *Toxicology Research and Application* 1. SAGE Publications Sage UK: London, England: 2397847317726352.

Saini, Rajesh Kumar, Laxmi P. Bagri, and Anil Kumar Bajpai. 2017. "14 - Smart Nanosensors for Pesticide Detection." In *New Pesticides and Soil Sensors*, edited by Alexandru Mihai Grumezescu, 519–59. Academic Press. doi:10.1016/B978-0-12-804299-1.00015-1

Scariot, Valentina, Antonio Ferrante, and Daniela Romano. 2022. "Editorial: Edible Flowers: Understanding the Effect of Genotype, Preharvest, and Postharvest on Quality, Safety, and Consumption." *Frontiers in Plant Science* 13: 1025196. www.frontiersin.org/articles/10.3389/fpls.2022.1025196

Shi, Shengyou, Wei Wang, Liqin Liu, Shijia Wu, Yongzan Wei, and Weicai Li. 2013. "Effect of Chitosan/Nano-Silica Coating on the Physicochemical Characteristics of Longan Fruit under Ambient Temperature." *Journal of Food Engineering* 118 (1): 125–31. doi:10.1016/j.jfoodeng.2013.03.029

Shukla, Shruti, Hyerim Leem, Jong-Suk Lee, and Myunghee Kim. 2014. "Immunochromatographic Strip Assay for the Rapid and Sensitive Detection of *Salmonella* Typhimurium in Artificially Contaminated Tomato Samples." *Canadian Journal of Microbiology* 60 (6): 399–406. doi:10.1139/cjm-2014-0223

Sogvar, Ommol Banin, Mahmoud Koushesh Saba, Aryou Emamifar, and Rahman Hallaj. 2016. "Influence of Nano-ZnO on Microbial Growth, Bioactive Content and Postharvest Quality of Strawberries during Storage." *Innovative Food Science & Emerging Technologies* 35 (June): 168–76. doi:10.1016/j.ifset.2016.05.005

Sonah, Humira, Rupesh K. Deshmukh, Vinay P. Singh, Deepak K. Gupta, Nagendra K. Singh, and Tilak R. Sharma. 2011. "Genomic Resources in Horticultural Crops: Status, Utility and Challenges." *Biotechnology Advances* 29 (2): 199–209. doi:10.1016/j.biotechadv.2010.11.002

Souto, Eliana B., Patrícia Severino, Rafael Basso, and Maria Helena A. Santana. 2013. "Encapsulation of Antioxidants in Gastrointestinal-Resistant Nanoparticulate Carriers." In *Oxidative Stress and Nanotechnology: Methods and Protocols*, edited by Donald Armstrong and Drub J. Bharali, 37–46. Springer.

Suriyaprabha, R., G. Karunakaran, R. Yuvakkumar, P. Prabu, V. Rajendran, and N. Kannan. 2012. "Growth and Physiological Responses of Maize (Zea Mays L.) to Porous Silica Nanoparticles in Soil." *Journal of Nanoparticle Research* 14 (12): 1294. doi:10.1007/s11051-012-1294-6

Taha, Mohd, and Jamal Alsharef. 2018. "Performance of Soil Stabilized with Carbon Nanomaterials." *Chemical Engineering Transactions* 63 (January): 757–62. doi:10.3303/CET1863127

Tavakoli, Hamidreza, Hossein Rastegar, Mahdi Taherian, Mohammad Samadi, and Hossein Rostami. 2017. "The Effect of Nano-Silver Packaging in Increasing the Shelf Life of Nuts: An in Vitro Model." *Italian Journal of Food Safety* 6 (4): 6847. doi:10.4081/ijfs.2017.6874

Valdés, Mayra Granda, Arístides Camilo Valdés González, Josefa Angela García Calzón, and Marta Elena Díaz-García. 2009. "Analytical Nanotechnology for Food Analysis." *Microchimica Acta* 166 (1): 1–19. doi:10.1007/s00604-009-0165-z

Vidhyalakshmi, R., R. Bhakyaraj, and R. S. Subhasree. 2009. "Encapsulation 'the Future of Probiotics' - A Review." *Advances in Biological Research* 3 (3–4): 96–103.

Wang, Peng, Enzo Lombi, Fang-Jie Zhao, and Peter M. Kopittke. 2016. "Nanotechnology: A New Opportunity in Plant Sciences." *Trends in Plant Science* 21 (8): 699–712. doi:10.1016/j.tplants.2016.04.005

Wilson, Alphus, and Manuela Baietto. 2009. "Applications and Advances in Electronic-Nose Technologies." *Sensors* 9 (7): 5099–148. doi:10.3390/s90705099

Xu, Jicheng, Min Zhang, Bhesh Bhandari, and Robert Kachele. 2017. "ZnO Nanoparticles Combined Radio Frequency Heating: A Novel Method to Control Microorganism and Improve Product Quality of Prepared Carrots." *Innovative Food Science & Emerging Technologies*, Emerging Technologies with reference to IUFoST, 44 (December): 46–53. doi:10.1016/j.ifset.2017.07.025

Yang, Fan, Fashui Hong, Wenjuan You, Chao Liu, Fengqing Gao, Cheng Wu, and Ping Yang. 2006. "Influences of Nano-Anatase TiO_2 on the Nitrogen Metabolism of Growing Spinach." *Biological Trace Element Research* 110 (2): 179–90. doi:10.1385/BTER:110:2:179

Yu, Ning, Min Zhang, Md Nahidul Islam, Liqun Lu, Qian Liu, and Xinfeng Cheng. 2015. "Combined Sterilizing Effects of Nano-ZnO and Ultraviolet on Convenient Vegetable Dishes." *LWT – Food Science and Technology* 61 (2): 638–43. doi:10.1016/j.lwt.2014.12.042

Zagzog, Osama, and Mohamed Gad. 2017. "Improving Growth, Flowering, Fruiting and Resistance of Malformation of Mango Trees Using Nano-Zinc." *Middle East Journal of Agriculture Research* 6 (July): 673–81.

Zahedi, Seyed Morteza, Mahdieh Karimi, and Jaime A Teixeira da Silva. 2020. "The Use of Nanotechnology to Increase Quality and Yield of Fruit Crops." *Journal of the Science of Food and Agriculture* 100 (1): 25–31. doi:10.1002/jsfa.10004

Zakzouk, Usama Ahmed Ibrahim. 2017. "Improving Growth, Flowering, Fruiting and Resistance of Malformation of Mango Trees Using Nano-Zinc." *Middle East Journal of Agriculture Research* 6(3): 673–81. www.publications.zu.edu.eg/Pages/PubShow.aspx?ID=26850&pubID=18

Zhang, Wanli, Haitao Jiang, Jiankang Cao, and Weibo Jiang. 2021. "Advances in Biochemical Mechanisms and Control Technologies to Treat Chilling Injury in Postharvest Fruits and Vegetables." *Trends in Food Science & Technology* 113 (July): 355–65. doi:10.1016/j.tifs.2021.05.009

Zhao, Xiang, Zhigang Meng, Yan Wang, Wenjie Chen, Changjiao Sun, Bo Cui, Jinhui Cui, et al. 2017. "Pollen Magnetofection for Genetic Modification with Magnetic Nanoparticles as Gene Carriers." *Nature Plants* 3 (12): 956–64. doi:10.1038/s41477-017-0063-z

Zhu, Chengzhou, Guohai Yang, He Li, Dan Du, and Yuehe Lin. 2015. "Electrochemical Sensors and Biosensors Based on Nanomaterials and Nanostructures." *Analytical Chemistry* 87 (1): 230–49. doi:10.1021/ac5039863

Zobir, Syazwan Afif Mohd, Asgar Ali, Fariz Adzmi, Mohd Roslan Sulaiman, and Khairulmazmi Ahmad. 2021. "A Review on Nanopesticides for Plant Protection Synthesized Using the Supramolecular Chemistry of Layered Hydroxide Hosts." *Biology* 10 (11): 1077. doi:10.3390/biology10111077

8 Prospects of Microbial Diversity in Horticulture

Amit Sharma and Sapna Tyagi

8.1 INTRODUCTION

Horticultural crops including fruits, vegetables, ornamentals, and aromatic medicinal plants provide nutrients, biologically active substances, and aesthetic values. They play a vital role in the economy and prosperity of a nation (Jun Xu et al., 2022). Horticultural crops have carved an indelible mark on the daily dietary requirement as an alternative source of nutritional security to the extent that the per capita consumption of cereal crops has reduced remarkably. Horticulture is recognized to have the potential to raise farm income, provide livelihood security and earn foreign exchange, considering as much as 37% contribution of horticulture out of the total export of agricultural commodities (NHB, 2020). The importance of horticulture in day-to-day life is like a panacea, and has made unexpected strides in the recent past, thereby surpassing food grain production by nearly 30 million tons, which in the coming years is expected to make even bigger strides. India's horticultural production registered a 1% increase during 2018–19 to record 314.67 million tons from 25.87 million ha area (NHB, 2020; Bora and Bora,2020). Despite such success, quality production coupled with low productivity and profitability in several horticultural crops are still a cause of serious concern. Of them, outbreaks of diseases account for maximum damage to crops. Most of the growers basically rely on chemicals for disease management as a quick remedy. However, the development of resistance against chemicals by pathogenic microbes coupled with environmental and public concern over hazardous chemicals have posed newer challenges to disease management of horticultural crops. The world trend in recent times is moving towards chemical residue-free fruits, vegetables, and tea plantations (Bora and Bora, 2016).

India ranks second globally in horticultural production, necessitating eco-friendly technologies to enhance productivity and ensure nutritional food security (Pathak et al., 2017). The imperative of the 21st century is sustainable agricultural practices, optimizing resource efficiency and biological processes through integrated management and green technologies (Weekley et al., 2012). To meet escalating demand and foster economic growth, the adoption of sustainable green technologies is essential, ensuring increased production without environmental degradation or compromising product quality (Saharan et al., 2021). Microorganisms play a pivotal role in plant health and

DOI: 10.1201/9781003343707-8

ecosystem balance, facilitating nutrient recycling, soil improvement, and environmental detoxification (Sengupta and Gunri, 2015). While beneficial, certain pathogens and pests pose threats to agricultural productivity (Umokoro, 2020). Microbial communities influence soil structure and nutrient cycles, affecting plant growth and health (Bhargava et al., 2019; Basu et al., 2021). Their regulation of soil mineral elements is crucial for maintaining plant physiological health (Dar et al., 2021).

8.2 MICROBIAL DIVERSITY IN HORTICULTURAL CROPS

The diverse groups of microbes are associated with crops and is considered important for maintaining production systems for environmental sustainability. The associated microbial communities help the plants in growth, yield improvement, and adaptation to harsh environmental conditions (Verma et al., 2017; Kour et al., 2022). In horticultural production, plant–microbe interactions may be considered the main factor for plant-growth promotion and protection against abiotic and biotic stresses and resistance against adverse conditions (Finkle et al., 2017; Meena et al., 2017). The positive plant–microbe interactions develop microbial inoculants for application in various horticultural crops (Srivastava et al., 2015). Depending on their mode of action and effects, these products can be employed as biofertilizers, plant strengtheners, phytosimulators, and biopesticides. Rhizosphere alteration by soil microorganism exudation through roots is an essential feature that influences the availability of nutrients in the soil and their acquisition by plants (Sharma, 2021). Microbial diversity in horticultural crops can be categorized into different types based on the microbial groups or niches they belong to.

8.2.1 Rhizosphere Microbial Diversity

Rhizosphere is a site of higher microbial activity in and around the root of the soil; it harbors a great diversity of microorganisms affecting plant growth and health (Sharma and Pethe, 2017). The rhizosphere hosts a diverse community of bacteria including plant growth-promoting rhizobacteria (PGPR), phosphate solubilizers and nitrogen fixers such as *Pseudomonas, Bacillus, Azospirillum,* and *Rhizobium,* which promote plant growth and protect against pathogens. Mycorrhizal associations involve arbuscular mycorrhizal fungi (AMF) such as *Glomus, Rhizophagus,* and *Funneliformis* as well as ectomycorrhizal fungi (EMF) such as Amanita, Boletus, and Laccaria. Each species contributes to nutrient uptake and plant symbiosis differently. The rhizosphere is also inhabited by other microorganisms like oomycetes (*Phytophthora, Pythium*), protozoa (ciliates, amoebae), and nematodes (free-living, plant-parasitic). These organisms contribute to the overall microbial diversity and interact with plants in various ways.

8.2.2 Phyllosphere Microbial Diversity

Phyllosphere or aerial surface of plants represents the globally largest and peculiar microbial habitat that inhabits diverse and rich communities of bacteria, fungi, viruses, cyanobacteria, actinobacteria, nematodes, and protozoans. These hyperdiverse microbial communities are related to the host's specific functional traits and influence the host's physiology and the ecosystem's functioning (Bashir et al., 2022). It harbors diverse bacterial communities such as *Pseudomonas, Pantoea, Erwinia, Xanthomonas,* etc., and can have both beneficial and pathogenic interactions with plants. Filamentous fungi and yeasts colonize the phyllosphere and exhibit diversity. Genera like *Alternaria, Cladosporium, Colletotrichum,* and *Penicillium* are common phyllosphere fungi, with some having pathogenic potential. Phyllosphere viruses are diverse, and different viruses may cause specific diseases in horticultural crops. Generally, microbial diversity associated with horticultural crops has been reported belonging to different species of *Burkholderia, Enterobacter, Pseudomonas,* and *Rahnella* from apple (Passos et al., 2014; Soni et al., 2016), *Azotobacter, Bacillus, Micrococcus,* and *Pseudomonas* from pea (Nazir et al., 2020), *Aeromonas, Bacillus, Chryseobacterium, Delftia, Enterobacter, Klebsiella, Kosakonia, Pseudomonas, Sphingobacterium,* and *Stenotrophomonas* from tomato (Guerrieri et al., 2020; Hariprasad et al., 2014). Understanding the types of microbial diversity in horticultural crops is essential for studying their roles in plant–microbe interactions, nutrient cycling, disease suppression and overall ecosystem functioning. In this chapter, the diverse microbial consortium and their beneficial role associated with horticultural plants has been discussed, to promote their production and quality in some horticultural crops.

8.3 PLANT GROWTH PROMOTING RHIZOBIA

Plant growth-promoting rhizobacteria (PGPR) outperform other soil microflora, offering significant benefits to horticultural and agricultural plants. They synthesize essential phytohormones, including abscisic acid, cytokinins, gibberellins, and indole acetic acid (IAA), enhancing plant growth and immunity. Moreover, PGPR increase plant stress tolerance during drought and release volatile compounds like butanol and pentanol, rendering plants resistant to pathogens. Additionally, they produce siderophores that inhibit pathogenic microbes (Ruzzi & Aroca, 2015; Kloepper et al., 1980). The organisms may be symbiotic, free-living, and/or associative diazotrophs (i.e. the strains of *Variovorax, Serratia, Rhizobium, Pseudomonas, Pantoea, Comamonas, Burkholderia, Bacillus, Azotobacter, Azospirillum, Arthobacter, Alcaligens,* and *Agrobacterium)* (Jimenez-Gomez et al., 2017). Saharan and Nehra (2011) reported a cocktail of biostimulator containing *Rhizobium leguminosarum, Pseudomonas sp., P. putida, Bacillus licheniformis, B. lentus, Azotobacter chroococcum, Azospirillum lipoferum* and *A. brasilens* in tomato, basil, strawberry and capsicum with a tremendous increase in the antioxidant activity and nitrogen content. Singh et al. (2017) reported PGPR strains *Bacillus, Azotobacter, Pseudomonas* and *Acinetobacter* isolated from the rhizosphere of *Momordica charantia.*

A number of PGPRs including *Pseudomonas* (Loper et al., 2007), *Azospirillum* (Cassan and García Salamone, 2008), *Bacillus* (Jacobson et al., 2004), *Stenotrophomonas* (Ryan et al., 2009), Rhizobium (Long, 2001), *Serratia* (De Vleeschauwer and Hofte, 2007), *Azotobacter, Klebsiella, Enterobacter, Alcaligens, Arthrobacter, Burkholderia*

(Joseph et al., 2007) and *Streptomyces* (Schrey and Tarkka, 2008) with various crops and vegetables have been commercialized as biofertilizer to promote plant growth, development and protection. Glick (1995) considered PGPRs as a bioprotectant, since they suppress plant disease through the induction of systemic resistance and the production of antibiotics and siderophores. PGPR strains have been found as an effective tool to increase the quality of different ornamental plants in the horticulture industry. *Bacillus subtilis* and *Glomus* sp. increase the total flower number of *Tagetes erecta* (Flores et al., 2007) and *Pseudomonas putida* and *P. fluorescens* strains increase plant size and flower number of *Pelargonium peltatum, Chrysanthemum* sp. and *Dahlia variabilis* plants cultivated under greenhouse conditions (Gore and Altin, 2006). Lamoureux et al. (2020) investigated the population dynamics of *B. subtilis* and *B. velezensis* on spores of different ornamental plants *Echinacea purpurea* 'Salsa red', *Echinacea purpurea* 'Fatal attraction', and *Lavandula angustifolia* and found rapid germination of spores. Nathan and Jones (2020) found *Arthrobacter sp.* (Coleus), *Pseudarthrobacter sp.* (Geranium), *Pseudomonas brassicacearum* (Petunia) and *Pseudomonas corrugate* (Zinnia) bacteria in water-stressed greenhouse ornamentals to have biostimulant activity. Common genera *Bacillus, Paenibacillus, Pseudomonas, Acinetobacter, Agrobacterium,* and *Stenotrophomonas* having PGPR characteristics reported from the rhizosphere of wild orchids plants (Hilal and Ibrahim, 2020). Medicinal aromatic plants hold a special significance in sustainable agriculture systems (Bhattacharjee et al., 2020). Therefore, the demand for cultivation of medicinal plants is steadily increasing worldwide. Hence, the focus has currently been shifted to the cultivation of different therapeutically useful medicinal plants such as ashwagandha, opium, aloe, isubgol, aonla, medicinal solanum, stevia, etc., using sustainable approaches (Strzemsk et al., 2021). Plant growth-promoting rhizobacteria associated with medicinal plants belong to the genera *Azotobacter, Acinetobacter, Bacillus, Brevibacterium, Burkholderia, Exiguobacterium, Pseudomonas, Pantoea, Mycobacterium, Methylobacterium,* and *Serratia*. These microbes enhance plant-growth parameters by producing secondary metabolites, including enzymes and antibiotics, which help in nutrient uptake, enhance soil fertility, improve plant growth, and protect against plant pathogens (Vaghela and Gohel, 2023).

8.4 NITROGEN-FIXING MICROORGANISMS

Nitrogen is required for the biosynthesis of proteins, phytohormones, nucleic acids, enzymes, and many other biomolecules. Nitrogen (N2) fixation is an important phenomenon that involves many microbes, including blue-green algae (*Nostoc, Anabaena*), *Rhizobium* (pulses), *Frankia,* symbiotic relationship between *Anabaena–Azolla*, cyanobacteria and *Anthoceros*. Nitrogen is assimilated (conversion of nitrogen to nitrates and nitrites) by microbes associated with the plants. The process is carried out by Actinomycetes and *Bacillus spp.* (*B. ramosus, B. valgarius*), which act upon the dead and decaying matter and convert complex nitrogenous compounds into ammonia. Bacteria such as *Nitrosomonas, Nitrococcus, Nitrosogoloea, Nitobacter,* and *Nitrocysts* convert ammonia into nitrites and nitrates using a process referred to as nitrification. In this way, the nitrogen pool is maintained, and nutrient recycling takes place. It is reported that *Azotobacter* is a very important organism that alone contributes up to 50% of the nitrogen requirement to plants. It increases biometric parameters such as runners count, heavier fruit, fruit length, and fruit breadth in strawberries (Saharan and Nehra, 2011). Various free-living nitrogen-fixing bacteria can contribute to nitrogen availability in horticultural systems. These bacteria fix nitrogen in the soil independently and can indirectly benefit horticultural crops including *Azotobacter, Azospirillum,* and *Klebsiella.* Reports were found showing the presence of *Azospirillum sp.* in the rhizosphere of various fruit crops including banana and pineapple plants (Ghai and Thomas 1989; Weber et al., 1999; Pedraza et al., 2007). Manynitrogen-fixing bacteria including *Allorhizobium, Neorhizobium, Pararhizobium, Rhizobium, Brevundimonas,* and *Microvirga* have been found to be associated with papaya crop (Pang et al., 2021; Pang et al., 2022).The diazotrophs *Klebsiella* spp. and *Rhizobium* sp. are most frequently found in banana (Martínez et al., 2003). *Azospirillum amazonense, Azospirillum lipoferum, Burkholderia sp.,* and *Herbaspirillum* were detected in banana and pineapple fruit crops (Weber et al., 1999). *Rhizobium* and *Bradyrhizobium* are well-known nitrogen fixers and are relatively abundant in apple roots soil (Macik 2020). Diverse groups of diazotrophic bacteria were observed in the rhizospheric soils of the citrus-poplar cropping system identified as *Stenotrophomonas maltophilia, Stenotrophomonas rhizophila, Rhizobium spp., Pseudomonas fulva, Xanthomonas spp., Agrobacterium spp., Achromobacter spp., Ensifer adhaerens, Bacillus amyloliquefaciens, Xanthomonas theicola,* and *Azotobacter spp.* (Vijaya Gopal et al., 2021).

Pseudomonas, Klebsiella, Rhodococcus, Paenarthrobacter, Bacillus, Variovorax, Kosakonia and *Paracoccus* as nitrogen-fixing strains were found to be associated with potato cropland (Wang et al., 2019). A high-level diversity of diazotrophic *Burkholderia* species including *B. unamae*, *B. xenovorans*, and *B. tropica* was found to be involved in N_2-fixing, bioremediation and plant growth-promoting activities in the rhizospheres and rhizoplanes of tomato plants (Jesus caballero et al., 2007). Martin Paul et al. (2019) identified and reported *Pseudomonas putida* TS18 from the water calyx fluid of African tulip, *Spathodea campanulata* playing a unique role in the flower development by showing nitrogen fixing potential under *in vitro* conditions. Two isolates of *Rhizobium* were found associated with Babchi (*Psorolea corylifolia*) root nodules, a medicinal plant of immense biological importance against several skin diseases such as psoriasis, leukoderma, and leprosy (Shengepallu et al., 2018).

8.5 PHOSPHATE SOLUBILIZING MICROORGANISMS

Phosphate solubilizing microorganisms (PSMs) are a group of beneficial microorganisms capable of hydrolyzing organic and inorganic phosphorus compounds from insoluble compounds. Among these PSMs, strains from bacterial genera (*Bacillus*, *Pseudomonas*, and *Rhizobium*), fungal genera (*Penicillium* and *Aspergillus*), actinomycetes, and arbuscular mycorrhizal (AM) are notable (Kalayu, 2019). A wide range of bacteria and fungi, known as phosphate solubilizing microbes, convert elemental phosphate into organic phosphate, which can be easily utilized by plants (Mahanty et al., 2017). A number of bacterial strains belonging to the genera of *Pseudomonas, Rhizobium, Agrobacterium, Flavobacterium, Bacillus, Burkholderia, Aerobacter, Achromobacter, Erwinia,* and *Micrococcus* have been found to have the ability to solubilize insoluble inorganic phosphate compounds such as rock phosphate, dicalcium phosphate, tricalcium phosphate, and hydroxyl apatite in various horticultural crops (Rodriguez and Fraga, 1999; Rodriguez et al., 2006). Tricalcium phosphate and hydroxyl apatite have been reported to be more degradable substrates than rock phosphate, while the most powerful phosphate solibilizers are *Pseudomonas, Bacillus,* and *Rhizobium* (Banerjee et al., 2006; Umokoro, 2020). Several Trichoderma, *T. harzianum,* and *T. virens* were found helpful in promoting crop development and yield as a phosphate solubilizer and microbial antagonist in cucumber, cotton seedlings, sweet corn, bell pepper, and strawberry (Bal and Altintas, 2006; Elad et al., 2006; Sharma et al., 2021). PSM strains of *Scutellospora sp., Sclerocystis sp., Pseudomonas striata, Pisolithus sp., Penicillium sp., Laccaria sp., Glomus sp., Gigaspora sp., Boletus sp., Bacillus megaterium, Bacillus circulans, Aspergillus awamori, Amanita sp.,* and *Acaulospora sp.* are reported to increase crop yield of tomato, tobacco, soybean, potato, pea, oats, and groundnut by about 10–15% (Pathak et al., 2017). *Pseudomonas putida* and *P. fluorescens* are associated with tomatoes, peppers, cucurbits, and leafy greens. *Phyllobacterium sp., Paenibacillus polymyxa, Bacillus megaterium,*and *Bacillus amyloliquefaciens* were reported to enhance phosphate availability in horticultural crops like tomatoes, strawberry, peppers, and cucurbits (Flores-Felix et al., 2015; Erturk et al., 2012; Shen et al., 2016). *Enterobacter sp.* is also known for phosphate solubilization in some fruits and vegetable crops such as tomatoes, peppers, and strawberries (Kidoglu et al., 2007). Endophytic bacterial strains *Bacillus licheniformis, Agrobacterium tumefaciens, Bacillus subtilis* and *Lysinibacillus fusiformi* were isolated from the roots of *Momordica charantia* L. showed phosphate solubilization activity in *M. charantia L.(*Singh et al., 2022). The diversity and efficiency of phosphate-solubilizing bacteria *Klebsiella pneumoniae* strain M-AI-2 and *Gluconacetobacter sp.* was found to have strong potential from eggplant fields (Sungthongwises et al., 2015). The phosphate-solubilizing activity of *Burkholderia ambifaria* and *B. tropica* reported in organic and conventionally grown chili (*Capsicum frutescens*) and found useful as a bio-inoculant for chili production (Surapat et al., 2013). Zaidi et al. (2016) reported adequate information on the impact of PSM on the production and quality of ornamental plants. The rhizosphere of sunflower was evidenced by the presence of phosphate-solubilizing bacterial species of *Enterobacter* and *Burkholderia* (Ambrosini et al., 2012). Gore and Altin (2006) evaluated the ability of Pseudomonas flourescens to promote plant growth of ornamental plants like Pelargonium, Chrysanthamum and dahelia. *Bacillus subtilis* and *Bacillus megaterium* promote growth and nutrient uptake and enhance phosphate availability in various ornamental plants like roses, gerberas, chrysanthemums, etc., *Pseudomonas fluorescens* and *P. putida*, in roses, geraniums, and orchids. Actinomycetes have been found in association with chrysanthemums, marigolds and in aromatic plants such as lavender (*Lavandula spp.*), rosemary (*Rosmarinus officinalis*) and mint (*Mentha spp.*).Tariq et al. (2022) reported PSMs in many medicinal plants including *Aloe vera, Bauhinia variegata, Cannabis sativa, Lantana camara*, and *Mentha viridis.*

8.6 POTASSIUM-SOLUBILIZING MICROORGANISMS

Potassium is a fundamental macronutrient for plant growth and plays an important role in making several metabolic processes including protein synthesis, photosynthesis, and enzyme activation (Zahedi, 2016). A wide range of bacteria, namely *Pseudomonas, Burkholderia, Acidothiobacillus ferrooxidans, Bacillus mucilaginosus, Bacillus edaphicus, B. circulans,* and *Paenibacillus sp.,* have been reported to release potassium in accessible form from potassium bearing minerals in soils (Liu et al., 2012). *Aspergillus niger* has shown the highest potassium solubilization and acid production by utilizing feldspar and potassium aluminum silicate as an insoluble source of potassium (Prajapati et al., 2012). Potassium-solubilizing bacteria (KSB) are an important group in the soil ecosystem and have multiple growth-promoting activities for many horticultural plants. Several KSBs were evaluated with high K soluble activity in banana, dragon fruit, and mango (Xuyen et al., 2021). Nadaf and Navale (2016) isolated and characterized potassium solubilizing bacteria from the root rhizosphere soil of onion and found *Pseudomonas spp.* as the most efficient KSB under *in vitro* conditions. To detect the population properties and potassium-dissolving characteristics of potassium-dissolving bacteria existing in apple rhizosphere soil, the highly efficient potassium-dissolving bacterial isolates *Acinetobacter sp., Pseudomonas sp.,* and *Bacillus sp.* were gained from the rhizosphere of apple trees (Jiang et al., 2017). *Bacillus* and *Pseudomonas* strains were also reported to solubilize potassium from rhizosphere of apple var. delicious (Ahmad and Zargar, 2017). Saha et al. (2016) reported *Bacillus licheniformis* and *Pseudomonas azotoformans* from *Oryza sativa* and *Musa paradisiaca*. Raji and Thangavelu (2021) found *B. licheniformis* and *B. cenocepacia* as an excellent potassium solubilizing prospect in tomato plants.

8.7 DROUGHT-TOLERANT MICROORGANISMS

Drought adaptive microbes play a potential role in mitigating the adverse effects of the drought stress in horticultural plants. These microbes with different mechanisms ensure the survival of the plants under water deficit conditions. Diverse drought adaptive microbes belonging to genera *Achromobacter, Arthrobacter, Aspergillus, Bacillus, Pseudomonas, Penicillium* and *Streptomyces* have been reported worldwide. The applications of drought adaptive plant growth-promoting (PGP) microbes as biofertilizers and biopesticides for horticulture is a potential strategy to improve the productivity and protection of horticultural crops from abiotic and biotic stresses for agricultural sustainability (Kour et al., 2022). In horticultural crops, the increased level of ripening hormone, ethylene is lowered by the various drought-tolerant microbes. *Bacillus subtilis* strain of bacterial species was reported to enhance the growth of tomato seedlings (Xu et al., 2014). The osmotolerant Actinobacterium *Citricoccus zhacaiensis* was found in the onion seeds to improve the germination rate (Selvakumar et al., 2015). Drought-tolerant microbes play a key role in maintaining the cell turgor through osmatic adjustment. Asrar et al. (2012) noted *Glomus deserticola* fungi in snapdragon flowers grown under water-stressed conditions. *Rhizophagus sp.* and *Burkholderia seminalis* were proclaimed to mitigate the drought stress in tomato and bell pepper plants (Tallapragada et al., 2016). *Bacillus subtilis* in drought-stressed okra was found to increase compatible solutes including total sugar and proline, which helps to resist stress (Puthiyottil and Akkara, 2021). Drought-tolerant microbiomes help in producing phytohormones that cope up with water-deficit conditions (Vurukonda et al., 2016). Different microbes have been found to produce various phytohormones that help in alleviating drought stress in horticultural crops. *Bacillus pumilus* and *Achromobacter xylosoxidans,* the endophytic bacteria isolated from sunflower was disclosed for producing phytohormones jasmonic acid and salicylic acid that helps in enhancing growth of water-stressed sunflower seedlings (Forchetti et al., 2010). Cherni et al. (2019) isolated rhizospheric microbiota *Ensifer adhaerens* and *Pseudomonas resinovorans* associated with *Citrus sinensis*. These microbes were reported to alleviate stress caused by water shortage in tomatoes as well as capsicum. Chiappero et al. (2019) reported the potential of alleviating the adverse effects of drought stress and plant-growth promotion in *Mentha piperita* with *Bacillus amyloliquefaciens* and *Pseudomonas fluorescens* through improved antioxidant status. *Klebsiella sp.* strain San01 isolated from sweet potato roots exhibited diverse plant-growth-promoting activities and drought tolerance (Quing et al., 2022).

8.8 PLANT GROWTH-PROMOTING BACTERIA

Plant growth-promoting bacteria are the beneficial bacteria that stimulate plant growth by activating various physiological characteristics of the metabolism of host plants without causing a negative impact on them, the environment, and consumer health (Numan et al., 2018, Seifikalhor et al., 2019). PGPB live freely in the soil or inhabit the surface (rhizosphere and phyllosphere) and/or the interior of the tissues (endophytes) of host plants. They are capable of enhancing the growth and inducing systemic resistance of plants to a wide range of pathogens (Yang et al., 2006; Droby, 2006; Alfonzo et al., 2009; Beneduzi et al., 2012; Maksimov et al., 2015; Waewthongrak et al., 2015; Lastochkina, 2019). Among PGPBs is *Bacillus subtilis*, which is the most attractive and safe (GRAS – Generally Recognized As Safe) agents for use in the food industry and crop production, as natural fertilizers and plant protection products (Pandey et al., 2017; Lastochkina et al., 2017; 2019; Sarma et al., 2018). The beneficial association of *B. subtilis* has been shown for various horticultural plant species, including sweet basil (Sayed et al., 2011), cucumber (Egamberdieva et al., 2011), strawberry (Zhao et al., 2007), melon (Wang et al., 2010; García-Gutiérrez et al., 2013), mango (Govender et al., 2005c), apple (Kim et al., 2015; 2016), kiwi (Kim et al., 2015), cabbage (Turan et al., 2014), muskmelons (Yang et al., 2006), artichoke (Saleh et al., 2005), potato (Lastochkina et al., 2020; 2020), lettuce (Seifikalhor et al., 2018), sugar beet (Pusenkova et al., 2016; Lastochkina et al., 2018) etc. Joao Frerico et al. (2014) identified the diversity of plant growth-promoting (PGP) bacterial strains belonging to *Enterobacter, Pseudomonas, Burkholderia, Cedecia, Leclercia, Stenotrophomonas, Rhanella* and *Ewingella* genera associated with apple trees. Ajeethan (2023) showed abundance of PGPBs such as *Bradyrhizobium*, *Rhizobium* group, *Sphingomonas, Streptomyces* and *Nonomuraea* in apple roots. Somtrakoon et al. (2019) performed the ability of plant growth-promoting bacteria *Paenibacillus sp., Streptomyces sp.,* in ornamental plants namely *Impatiens balsamina* L., *Tagetes erecta* L. and *Crotalaria juncea* L. Samayoa et al. (2020) isolated *Bacillus sp.* and *Pseudomonas sp.* from the roots, bulbs and rhizosphere of onion.

8.9 MICROORGANISMS AS BIOAGENTS

Beneficial plant microbes like *Trichoderma spp., fluorescent Pseudomonas*, and *Bacillus spp.* promote crop growth by suppressing diseases and enhancing plant development (Bora et al., 2013). Endophytic microbes found in horticultural crops exhibit antimicrobial properties, with genera like *Pseudomonas* and *Bacillus* showing antagonistic effects against pathogens (Compant et al., 2011). Microbial biological control agents protect plants through direct antagonism and by priming host defense mechanisms (Conrath et al., 2015). *Pseudomonas fluorescens* and *P. putida* have been reported to exhibit biocontrol efficacy against a wide range of phytopathogens. Their antagonistic activity is attributed to antibiotics, siderophore production and ability to compete for nutrition or space besides producing an array of antifungal metabolites and lytic enzymes (chitinase and glucanase) that inhibit the mycelial growth of pathogenic fungi (Bora and Deka, 2007; Bora et al.,2016]. *Bacillus* spp. also produces a wide range of antibiotics effective against many plant pathogens. These bacterial bioagents are effective against many bacterial and fungal pathogens (Bora and Bhagabati, 1999; Bora et al., 2019). Some of the destructive fruit diseases causing huge losses to the fruit industry worldwide include apple scab, mango malformation, guava wilt, fire blights, *Fusarium* wilt of banana, brown rot of stone fruits, crown gall, downy and powdery mildews, etc. (Singh, 1996; Bora and Deka, 2014). Significant progress has been made in the biological management of these diseases through BCAs like *Trihcoderma* spp., *Gliocladium* spp., *Verticillum* spp., *Bacillus subtilis*, *Agrobacterium radiobacter, Pseudomonas* spp. (Rishbeth, 1988). Antagonists such as *B. subtilis* and *T. koningii* have been also used to control apple canker (Lima et al., 1994). Ecofriendly management of vegetable diseases through microbe-based strategy has been quite successful. Bacterial wilt (*R. solanacearum*), fungal wilt (*Fusarium* spp.), damping off and root rot (*Pythium* spp, *Phytopthora* spp., *Rhizoctonia solani*), anthracnose (*Colletotrichum* spp.), etc. are few examples of diseases with successful management through different bioagents under field conditions (Bora and Deka, 2007). *Trichoderma* sp. is widely used against many diseases of spice crops including ginger, turmeric, pepper, cardamom, cumin, turmeric, etc. (Chattopadhyay et al., 2018).

8.10 MYCORRHIZAL SYMBIOSIS

Mycorrhizal symbiosis is a mutualistic relationship between a fungus and a plant, the symbiont fungus receives trophic support from the host plant and in turn, benefits the plant in the form of improved uptake of mineral elements, ensuring better growth and excellent resistance to insects and telluric pathogens (Gupta et al., 2018; He et al., 2020). Arbuscular mycorrhizal fungi establish symbioses with plant roots, which helps to improve nutrient uptake by the host plant and alter its physiology to withstand external abiotic factors and pathogens (Nadeem et al., 2017). Arbuscular mycorrhizal fungi interactions with bacteria have been reported to enhance plant growth through phosphate solubilization, nitrogen fixation, increased AMF spore germination, and colonization of plant roots (Selvakumar et al., 2016). The potential of these two organisms in promoting the growth of horticultural crops like tomato (Bona et al., 2018), strawberry (Todeschini et al., 2018), lettuce (Kohler et al., 2010), and spinach (Khalid et al., 2017) have been reported. The AMF are estimated to colonize the roots of up to 80% of terrestrial plant species (Tarkka et al., 2018) and are beneficial for many horticultural plants such as citrus, apple, peach, strawberry, lettuce, pineapple, onions, melon, and pepper (Thamsurakul et al., 2000; Wu et al., 2013).

8.11 CONCLUSION

Microbial diversity is integral to horticultural crop management, offering benefits across various fronts. Understanding crop-specific microbial interactions facilitates targeted interventions for improved performance. Beneficial microbes aid in disease management, reducing reliance on chemical pesticides and fostering sustainability. They also enhance soil health by improving nutrient availability, structure, and moisture retention. Moreover, microbes bolster stress tolerance, augmenting plant defense mechanisms against environmental pressures. Additionally, they enhance crop quality, prolong shelf life, and align with sustainable farming practices. Overall, harnessing microbial interactions optimizes productivity, minimizes environmental impact, and ensures a resilient horticulture industry.

REFERENCES

Ahmad, M.S. and Zargar, M.Y. 2017. Characterization of potassium solubilizing bacteria (KSB) in rhizospheric soils of apple (*Malus domestica* Borkh.) in temperate Kashmir. *Journal of Applied Life Sciences International* 15(1): 1–7.

Ajeethan, N., Ali, S., Fuller, K.D., Abbey, A. and Yurgel, S.N. 2023. Apple root microbiome as indicator of plant adaptation to Apple replant diseased soil. Microorganisms 11(6): 1–16. doi.org/10.3390/microorganisms11061372

Alfonzo, A., Conigliaro, G., Torta, L., Burruano, S. and Moschetti, G. 2009. Antagonism of *Bacillus subtilis* strain AG1 against vine wood fungal pathogens. *Phytopathologia Mediterranea* 48: 155–158.

Ambrosini, A., Beneduzi, A., Stefanski, T., Pinheiro, F.G., Vargas, L.K. and Passaglia, L.M.P.2012. Screening of plant growth promoting rhizobacteria isolated from sunflower (*Helianthus annuus* L.). *Plant and Soil* 356: 245–264.

Asrar, A.A., Abdel-Fattah, G.M. and Elhindi, K.M. 2012. Improving growth, flower yield and water relations of snapdragon (*Antirhinum majus* L.) plants grown under well-watered and water-stress conditions using arbuscular mycorrhizal fungi. *Photosynthetica* 50(2): 305–316.

Bal, U. and Altintas, S. 2006. Application of the antagonistic fungus *Trichoderma harzianum* (TrichoFlow WPTM) to root zone increases yield of bell peppers grown in soil. *Biological Agriculture and Horticulture* 24(2): 149–163.

Banerjee, M.R., Yesmin, L. and Vessey, J.K. 2006. Plant growth promoting rhizobacteria as biofertilizers and biopesticides. In: *Handbook of Microbial Biofertilizers*. Rai, M.K. (Ed.). New York: Haworth Press, pp.137–181.

Bashir, I., Aadil, F.W., Iflah, R., Zafar, A.R., Irfan, R. and Yogesh, S.S. 2022. Phyllosphere microbiome: Diversity and functions. *Microbiological Research* 254:126888. https://doi.org/10.1016/j.micres.2021.126888

Basu, S., Kumar, G., Chhabra, S. and Prasad, R. 2021. Role of soil microbes in biogeochemical cycle for enhancing soil fertility. In: *New and Future Developments in Microbial Biotechnology and Bioengineering*. Verma, J.P. (Ed.). The Netherland: Elsevier, pp. 149–157.

Bharagava, R.N., Purchase, D., Saxena, G. and Mulla, S.I. 2019. Applications of metagenomics in microbial bioremediation of pollutants: From genomics to environmental cleanup. In: *Microbial Diversity in the Genomic Era*. Academic Press, pp. 459–477.

Bhattacharjee, T., Sen, S., Chakraborty, R., Maurya, P.K. and Chattopadhyay, A. 2020. Cultivation of medicinal plants: Special reference to important medicinal plants of India. In: *Herbal Medicine in India*. Sen, S. and Chakraborty, R. (Eds.). Berlin/Heidelberg: Springer, pp. 101–115.

Bona, E., Todeschini, V., Cantamessa, S., Cesaro, P., Copetta, A., Lingua, G., Gamalero, E., Berta, G. and Massa, N. 2018. Combined bacterial and mycorrhizal inocula improve tomato quality at reduced fertilization. *Scientia Horticulturae* 234: 160–165.

Bora, L.C. and Bhagabati, K.N. 1999. Quantitative aspects of assessing virulence of Xanthomonas campestris pv. citri (Hasse) Dye, the incitant of citrus canker in Assam. In: *Hi Tech Citrus Management: Proceedings of International Symposium on Citriculture*. Singh, S. (Ed.). Nagpur: National Research Centre for Citrus, Indian Society of Citriculture, p. 82.

Bora, L.C and Deka, S.N. 2007. Wilt disease suppression and yield enhancement in tomato (Lycopersicon esculentum) by application of Pseudomonas fluorescens based bio pesticide (Biofor-Pf) in Assam. *Indian Journal of Agricultural Sciences* 77: 490–494.

Bora, P. and Bora, L.C. 2016. *Hand Book of Organic Farming*. Guwahati: Sabda Publishers.

Bora, P. and Bora, L.C. 2020. Disease management in horticulture crops through microbial interventions: An overview. *Indian Journal of Agricultural Sciences* 90(8): 1389–1396.

Bora, P., Bora, L.C. and Begum, M. 2013. Eco-friendly management of soil borne diseases in brinjal through application of antagonistic microbial population. *Journal of Biological Control* 27: 29–34.

Bora, P., Bora, L.C. and Begum, M. 2013. Eco-friendly management of soil borne diseases in brinjal through application of antagonistic microbial population. *Journal of Biological Control* 27: 29–34.

Bora, P. and Deka, P.C. 2014. Biointensive approaches for disease management of rabi vegetables. In: *Agriculture for Sustaining Livelihood*. Bhattacharya, H., Neog, M. and Saud, B. (Eds.). Jorhat: Assam Agricultural University Press, pp. 123–126.

Bora, P., Saikia, K., Hazarkia, H. and Gavas, R. 2019. Exploring potential of bacterial endophydes in disease management of horticultural crops. *Current Horticulture* 7: 32–37.

Cassán, F. and García Salamone, I. 2008. *Azospirillum sp. Cell Physiology, Plant Response, Agronomic and Environmental Research in Argentina*. Buenos Aires: Asociacion Argentina de Microbiologia.

Chattopadhyay, A., Purohit, J., Patel, D.S, and Tetarwal, M. L. 2018. Management of diseases of spice crops through microbes. In: *Biological Control of Crop Diseases: Recent Advances & Perspective*. Singh, D., Chakraborty, B.N., Pandey, R.N. and Sharma, P. (Eds.). New Delhi: Today & Tomorrow's Printers and Publishers, pp. 249–270.

Cherni, M., Ferjani, R., Mapelli, F., Boudabous, A., Borin, S. and Ouzari, H.-I., 2019. Soil parameters drive the diversity of Citrus sinensis rhizosphere microbiota which exhibits a potential in plant drought stress alleviation. *Applied Soil Ecology* 135: 182–193.

Chiappero, J., Cappellari, L.d.R., Sosa Alderete, L.G., Palermo, T.B. and Banchio, E. 2019. Plant growth promoting rhizobacteria improve the antioxidant status in Mentha piperita grown under drought stress leading to an enhancement of plant growth and total phenolic content. *Industrial Crops and Products* 139: 111553.

Compant, S., Mitter, B., Colli-Mull, J.G., Gangl, H. and Sessitsch, A. 2011. Endophytes of grapevine flowers, berries, and seeds: identification of cultivable bacteria, comparison with other plant parts, and visualization of niches of colonization. *Microbial Ecology* 62: 188–197.

Conrath, W., Beckers, G.J.M., Langenbach, C.J.G., and Jaskiewicz, M.R. 2015. Priming for enhanced defense. *Annual Review of Phytopathology* 53: 97–119.

Dar, S.A, Bhat, R.A., Dervash, M.A. and Dar, Z.A. 2021. Azotobacter as biofertilizer for sustainable soil and plant health under saline environmental conditions. In: *Microbiota and Biofertilizers*. Hakeem, K.R., Dar, G.H., Mahmood, M.A. and Bhat, R.A. (Eds.). Switzerland AG: Springer, pp. 231–254.

De Vleeschauwer, D. and Höfte, M. 2007. Using Serratia plymuthica to control fungal pathogens of plants. *CAB Reviews* 2: 46.

Droby, S. 2006. Improving quality and safety of fresh fruit and vegetables after harvest by the use of biocontrol agents and natural materials. *Acta Horticulturae* 709: 45–51.

Egamberdieva, D., Kucharova, Z., Davranov, K., Berg, G., Makarova, N., Azarova, T., Chebotar, V., Tikhonovich, I., Kamilova, F., Validov, S.Z. and Lugtenberg, B. 2011. Bacteria able to control foot and root rot and to promote growth of cucumber in salinated soils. *Biology and Fertility of Soils* 47: 197–205.

Elad, Y., Chet, I., and Henis, Y. 2006. Biological control of Rhizoctonia solani in strawberry fields by Trichoderma harzianum. *Plant and Soil* 60: 245–254.

Erturk, Y., Ercisli, S. and Cakmakci, R. 2012. Yield and growth response of strawberry to plant growth-promoting rhizobacteria inoculation. *Journal of Plant Nutrition* 35: 817–826.

Finkel, O.M., Castrillo, G., Herrera Paredes, S., Salas González, I. and Dangl, J.L. 2017. Understanding and exploiting plant beneficial microbes. *Current Opinion in Plant Biology* 38: 155–163.

Flores, A.C., Luna, A.A.E. and Portugal, V.O. 2007. Yield and quality enhancement of marigold flowers by inoculation with Bacillus subtilis and Glomus fasciculatum. *Journal of Sustainable Agriculture* 31: 21–31. 10.1300/J064v31n01

Flores-Felix, J.D., Silva, L.R. and Rivera, L.P. 2015. Plants probiotics as a tool to produce highly functional fruits: The case of Phyllobacterium and vitamin C in strawberries. *PLoS One* 10: e0122281.

Forchetti, G., Masciarelli, O., Izaguirre, M.J., Alemano, S., Alvarez, D. and Abdala, G. 2010. Endophytic bacteria improve seedling growth of sunflower under water stress, produce salicylic acid, and inhibit growth of pathogenic fungi. *Current Microbiology* 61(6): 485–493.

García-Gutierrez, L., Zeriouh, H., Romero, D., Cubero, J., Vicente, A. and Pérez-García, A. 2013. The antagonistic strain Bacillus subtilis UMAF6639 also confers protection to melon plants against cucurbit powdery mildew by activation of jasmonate – and salicylic acid-dependent defense responses. *Microbial Biotechnology* 6(3): 264–274.

Ghai, S.K. and Thomas, G.V. 1989. Occurrence of Azospirillum spp. in coconut-based farming systems. *Plant Soil* 114: 235–241.

Glick, B.R. 1995. The enhancement of plant growth by free-living bacteria. *Canadian Journal of Microbiology* 41: 109–117.

Gore, M.E. and Altin, N. 2006. Growth promoting of some ornamental plants by root treatment with specific fluorescent pseudomonads. *Journal of Biological Sciences* 6: 610–615.

Govender, V., Korsten, L. and Sivakumar, D. 2005. Semi-commercial evaluation of Bacillus licheniformis to control mango postharvest diseases in South Africa. *Postharvest Biology and Technology* 38(1): 57–65.

Guerrieri, M.C., Fanfoni, E., Fiorini, A., Trevisan, M. and Puglisi, E. 2020. Isolation and screening of extracellular PGPR from the rhizosphere of tomato plants after long-term reduced tillage and cover crops. *Plants* 9(5): 668.

Gupta, N., Vats, S. and Bhargava, P. 2018. *Sustainable Agriculture: Role of Metagenomics and Metabolomics in Exploring the Soil Microbiota Silico Approach for Sustainable Agriculture*. Singapore: Springer, pp. 183–199.

Hariprasad, P., Venkateswaran, G. and Niranjana, S. 2014. Diversity of cultivable rhizobacteria across tomato growing regions of Karnataka. *Biological Control* 72: 9–16.

He, X., Xu, M., Wei, Q., Tang, M., Guan, L., Lou, L., Xu, X., Hu, Z., Chen, Y., Shen, Z., and Xia, Y. 2020. Promotion of growth and phytoextraction of cadmium and lead in Solanum nigrum L. mediated by plant-growth-promoting rhizobacteria. *Ecotoxicology and Environmental Safety* 205: 111333.

Hilal, A. and Ibrahim, O. 2020. Isolation and molecular characterization of plant growth promoting bacteria from the rhizosphere of orchids in turkey. *Rhizosphere* 16: 100280. doi:10.1016/j.rhisph.2020.100280

Jacobsen, B.J., Zidack, N.K. and Larson, B.J. 2004. The role of Bacillus-based biological control agents in integrated pest management systems: plant diseases. *Phytopathology* 94: 1272–1275.

Jesus, C.-M., Janette, O.-L., Paulina, E.-d and Lourdes, M.-A. 2007. The tomato rhizosphere, an environment rich in nitrogen-fixing Burkholderia species with capabilities of interest for agriculture and bioremediation. *Applied and Environmental Microbiology – Plant Microbiology* 73(16): 5308–5319. https://doi.org/10.1128/AEM.00324-07

Jiang, J.H., Peng, X.W., Yan, Z.X., He, B.W., Zhu, C.X., Guo, H. and Geng, B. 2017. Isolation and identification of potassium-solubilizing bacteria from rhizosphere soil of apple tree. *Chinese Journal of Agrometerology* 38(11): 738–748. doi:10.3969/j.issn.1000-6362.2017.11.006

Jiménez-Gómez, A., Celador-Lera, L., Fradejas-Bayón, M. and Rivas, R. 2017. Plant probiotic bacteria enhance the quality of fruit and horticultural crops. *AIMS Microbiology* 3(3): 483.

Joseph, B., Patra, R.R. and Lawrence, R. 2007. Characterization of plant growth promoting Rhizobacteria associated with chickpea (Cicer arietinum L). *International Journal of Plant Production* 1: 141–152.

Jun, X., Ningyuan, Z., Ke, W., Qianqian, X., Jingping, D. and Xuehao, C. 2022. Exploring new strategies in diseases resistance of horticultural crops. *Frontiers in Sustainable Food Systems* 6: 1–10. doi:10.3389/fsufs.2022.1021350

Kalayu, G. 2019. Phosphate solubilizing microorganisms: Promising approach as biofertilizers. *International Journal of Agronomy* 2019(1): 1–7. https://doi.org/10.1155/2019/4917256

Khalid, M., Hassani, D., Bilal, M., Asad, F. and Huang, D. 2017. Influence of bio-fertilizer containing beneficial fungi and rhizospheric bacteria on health promoting compounds and antioxidant activity of Spinacia oleracea L. *Botanical Studies* 58(1): 35.

Kidoglu, F., Gül, A., Ozaktan, H. and Tüzel, Y. 2007. Effect of rhizobacteria on plant growth of different vegetables. *Acta Horticultarae* 801: 1471–1478. doi:10.17660/ActaHortic.2008.801.181

Kim, G.H., Koh, Y.J., Jung, J.S. and Hur, J.S. 2015. Control of postharvest fruit rot diseases of kiwifruit by antagonistic bacterium Bacillus subtilis. *Acta Horticulturae* 1096: 377–382.

Kim, H.M., Lee, K.J. and Chae, J.C. 2015. Postharvest biological control of Colletotrichum acutatum on apple by Bacillus subtilis HM1 and the structural identification of antagonists. *Journal of Microbiology and Biotechnology* 25(11): 1954–1959.

Kim, Y.S., Balaraju, K. and Jeon, Y. 2016. Effects of rhizobacteria Paenibacillus polymyxa APEC136 and Bacillus subtilis APEC170 on biocontrol of postharvest pathogens of apple fruits. *Journal of Zhejiang University Science B.* 17(12): 931–940.

Kloepper, J. W., Leong, J., Teintze, M. and Schroth, M.N. 1980. Enhanced plant growth by siderophores produced by plant growth-promoting rhizobacteria. *Nature* 286: 885–886.

Kohler, J., Caravaca, F. and Roldán, A. 2010. An AM fungus and a PGPR intensify the adverse effects of salinity on the stability of rhizosphere soil aggregates of Lactuca sativa. *Soil Biology and Biochemistry* 42 (3): 429–434.

Kour, D., Sofia, S.K., Tanvir, K., Harpreet, K., Gagandeep, S., Ashok, Y. and Ajar, N.Y. 2022. Drought adaptive microbes as bioinoculants for the horticultural crops. *Heliyon* 8(5): 1–14. doi:https://doi.org/10.1016/j.heliyon.2022.e09493

Lamoureux, V.C., Maude, T., Assena, K. and Pascale, B.B. 2020. Bacillus subtilis and Bacillus velezensis population dynamics and quantification of spores after inoculation on ornamental plants. *Canadian Journal of Microbiology* 66(11): 664–669. https://doi.org/10.1139/cjm-2020-0174

Lastochkina, O. 2019. Bacillus subtilis-mediated abiotic stress tolerance in plants. In: *Bacilli and Agrobiotechnology: Phytostimulation and Biocontrol*. Islam, M.T., Rahman, M.M., Pandey, P., Boehme, M.H. and Haesaert, G. (Eds.). Switzerland AG: Springer Nature, pp. 97–133.

Lastochkina, O., Baymiev, A., Shayahmetova, A., Garshina, D., Koryakov, I., Shpirnaya, I., Pusenkova, L., Mardanshin, I., Kasnak, C. and Palamutoglu, R. 2020. Effects of endophytic Bacillus subtilis and salicylic acid on postharvest diseases (Phytophthora infestans, Fusarium oxysporum) development in stored potato tubers. *Plants* 9: 76.

Lastochkina, O., Pusenkova, L., Garshina, D., Yuldashev, R., Shpirnaya, I., Kasnak, C., Palamutoglu, R., Mardanshin, I., Garipova, S., Sobhani, M. and Aliniaeifard, S. 2020. The effect of endophytic bacteria Bacillus subtilis and salicylic acid on some resistance and quality traits of stored Solanum tuberosum L. tubers infected with fusarium dry rot. *Plants* 9(6): 738.

Lastochkina, O., Pusenkova, L., Yuldashev, R., Babaev, M., Garipova, S., Blagova, D., Khairullin, R. and Aliniaeifard, S. 2017. Effects of Bacillus subtilis on some physiological and biochemical parameters of Triticum aestivum L. (wheat) under salinity. *Plant Physiology and Biochemistry* 121: 80–88.

Lastochkina, O.V., Pusenkova, L.I., Il'yasova, E.Y. and Aliniaeifard, S. 2018. Effect of Bacillus subtilis based biologicals on physiological and biochemical parameters of sugar beet (Beta vulgaris L.) plants infected with Alternaria alternata. *Agrobiology* 53(5): 958–968.

Lima, G., Ippolito, A. and Nigro, F. 1994. Attempting at biological control of citrus mal secco (Phoma tracheiphila) with endophytic bacteria. *Difesadelle–piante* 17: 43–49.

Liu, D., Lian, B. and Dong, H. 2012. Isolation of Paenibacillus spp. and assessment of its potential for enhancing mineral weathering. *Geomicrobiology Journal* 29(5): 413–421.

Long, S.R. 2001. Gene and signals in the Rhizobium-legume symbiosis. *Plant Physiology* 125: 69–72.

Loper, J.E., Kobayashi, D.Y. and Paulsen, I.T. 2007. The genomic sequence of Pseudomonas fluorescens Pf-5: Insights into biological control. *Phytopathology* 97: 233–238.

Mącik, M., Gryta, A. and Frąc, M. 2020. Biofertilizers in agriculture: An overview on concepts, strategies and effects on soil microorganisms. *Advances in Agronomy* 162: 31–87.

Mahanty, T., Bhattacharjee, S., Goswami, M., Bhattacharyya, P., Das, B., Ghosh, A. and Tribedi, P. 2017. Biofertilizers: A potential approach for sustainable agriculture development. *Environmental Science and Pollution Research* 24(4): 3315–3335.

Maksimov, I.V., Veselova, S.V., Nuzhnaya, T.V., Sarvarova, E.R. and Khairullin, R.M. 2015. Plant growth promoting bacteria in regulation of plant resistance to stress factors. *Russian Journal of Plant Physiology* 62(6): 715–726.

Martin Paul, A., Jayanthi, D. and Thamizhseran, N. 2019. Isolation and proto-cooperation of *Pseudomonas putida* TS 18 from Water Calyx Fluid of *Spathodea campanulata* P. Beauv. *Journal of Applied Microbiology* 13(4): 2027–2033. https://doi.org/10.22207/JPAM.13.4.14

Martínez, L., Jesús, C-M, Josè, O. and Esperanza, M.-R. 2003. Diazotrophic bacteria associated with banana (Musa spp.). *Plant and Soil* 257: 35–47.

Meena, K.K, Sorty, A.M., Bitla, U.M., Choudhary, K., Gupta, P., Pareek, A.,Singh, D., Prabha, R., Sahu, P.K. and Gupta, V.K. 2017. Abiotic stress responses and microbe-mediated mitigation in plants: The omics strategies. *Frontiers in Plant Science* 8: 172.

Nadaf Jasmin, M. and Navale, A. 2016. Isolation and characterization of potassium solubilizing bacteria from rhizospheric soil of onion. *Indian Phytopathology* 69(4s): 744–745.

Nadeem, S.M., Khan, M.Y., Waqas, M.R., Binyamin, R., Akhtar, S. and Zahir, Z.A. 2017. Arbuscular mycorrhizas: an overview. In: *Arbuscular Mycorrhizas and Stress Tolerance of Plants*. Springer, pp. 1–24.

Nathan, P.N. and Michelle, L.J. 2020. Isolation of rhizosphere bacteria that improve quality and water stress tolerance in greenhouse ornamentals. *Frontiers in Plant Science* 11: 1–16. doi:10.3389/fpls.2020.00826

National Horticulture Board. 2020. *Area and Production of Horticultural Crops for 2018–19* (1st Advance Estimate). www.nhb.gov.in

Nazir, U., Zargar, M., Baba, Z., Mir, S., Mohiddin, F. and Bhat, N. 2020. Isolation and characterization of plant growth promoting rhizobacteria associated with pea rhizosphere in North Himalayan region. *International Journal of Chemical Studies* 8: 1131–1135.

Numan, M., Bashir, S., Khan, Y., Mumtaz, R., Shinwari, K., Khan, A.L., Khan, A. and AL-Harrasi, A. 2018. Plant growth promoting bacteria as an alternative strategy for salt tolerance in plants: A review. *Microbiological Research* 209: 21–32.

Pandey, P.K., Singh, M.C., Singh, S.S., Kumar, M., Pathak, M., Shakywar, R.C. and Pandey, A.K. 2017. Inside the plants: Endophytic bacteria and their functional attributes for plant growth promotion. *International Journal of Current Microbiology and Applied Sciences* 6(2): 11–21.

Pang, Z., Dong, F., Liu, Q., Lin, W., Hu, C. and Yuan, Z. 2021. Soil metagenomics reveals effects of continuous sugarcane cropping on the structure and functional pathway of rhizospheric microbial community. *Frontiers in Microbiology* 12, 1–21. doi:10.3389/fmicb.2021.627569

Pang, Z.Q., Huang, J.W., Fallah, N., Lin, W.X., Yuan, Z.N. and Hu, C.H. 2022. Combining n fertilization with biochar affects root-shoot growth, rhizosphere soil properties and bacterial communities under sugarcane monocropping. *Industrial Crops and Products* 182: 114899. doi:10.1016/j.indcrop.2022.114899

Passos, J.F.M.d., Costa, P.B.d., Costa, M.D., Zaffari, G.R., Nava, G., Boneti, J.I., Oliveira, A.M.R.d. and Passaglia, L.M. 2014. Cultivable bacteria isolated from apple trees cultivated under different crop systems: Diversity and antagonistic activity against Colletotrichum gloeosporioides. *Genetics and Molecular Biology* 37(3): 560–572. doi:10.1590/s1415-47572014000400013

Pathak, D.V., Mukesh, K. and Kusum, R. 2017. Biofertilizer application in horticultural crops. In *Microorganisms for Green Revolution*, Microorganisms for Sustainability

6. Panpatte, D.G. et al. (Eds.). Springer Nature Singapore, pp. 215–227. doi:10.1007/978-981-10-6241-4_11

Pedraza, R.O., Motok, J., Tortora, M.L., Salazar, S.M. and Diaz-Ricci, J.C. 2007. Natural occurrence of Azospirillum brasilense in strawberry plants. *Plant Soil* 295: 169–178. doi:10.1007/s11104-007-9273-x

Prajapati, K., Sharma, M.C. and Modi, J.A. 2012. Isolation of two potassium solubilizing fungi from ceramic industry soils. *Lifescience Leaflet* 5: 71–75.

Pusenkova, L.I., Il'yasova, E., Lastochkina, O.V., Maksimov, I.V. and Leonova, S.A. 2016. Changes in the species composition of the rhizosphere and phyllosphere of sugar beet under the impact of biological preparations based on endophytic bacteria and their metabolites. *Eurasian Soil Sciences* 49(10): 1136–1144.

Puthiyottil, P. and Akkara, Y. 2021. Pre-treatment with Bacillus subtilis mitigates drought induced photo-oxidative damages in okra by modulating antioxidant system and photochemical activity. *Physiology and Molecular biology of Plants* 27(5): 945–957.

Qing, L., Huang, Z., Deng, C., Lin, K.H., Hua, S. and Chen, S.P. 2022. Endophytic Klebsiella sp. San01 promotes growth performance and induces salinity and drought tolerance in sweet potato (Ipomoea batatas L.). *Journal of Plant Interactions* 17(1): 608–619. doi:10.1080/17429145.2022.2077464

Raji, M. and Thangavelu, M. 2021. Isolation and screening of potassium solubilizing bacteria from saxicolous habitat and their impact on tomato growth in different soil types. *Archives in Microbiology* 203: 3147–3161. https://doi.org/10.1007/s00203-021-02284-9

Rishbeth, J. 1988. Biological control of air borne pathogens. In Biological Control of Pests. Pathogens and Weeds: Development and Prospects. Proceedings of Royal Society Discussion Meeting, Londan (Eds. Wood, RKS and MJ Way). London Royal Society.

Rodríguez, H., and Fraga, R. 1999. Phosphate solubilizing bacteria and their role in plant growth promotion. *Biotechnology Advances* 17: 319–339.

Rodríguez, H., Fraga, R., Gonzalez, T. and Bashan, T. 2006. Genetics of phosphate solubilization and its potential applications for improving plant growth-promoting bacteria. *Plant and Soil* 287: 15–21.

Ruzzi, M. and Aroca, R. 2015. Plant growth-promoting rhizobacteria act as biostimulants in horticulture. *Scientia Horticulturae* 196: 124–134.

Ryan, R.P., Monchy, S., Cardinale, M. and Taghavi, S. 2009. The versatility and adaptation of bacteria from the genus Stenotrophomonas. *Nature Reviews Microbiology* 7: 514–525.

Saha, M., Maurya, B.R., Meena, V.S., Bahadur, I. and Ashok, K. 2016. Identification and characterization of potassium solubilizing bacteria from Indo-Gangetic Plains of India. Biocatalysis and Agricultural Biotechnology 7: 202–209.

Saharan, B.S. and Nehra, V. 2011 Plant growth promoting rhizobacteria: A critical review. *Life Sciences and Medicine Research*, 21: 1–30.

Saharan, B.S., Parshad, J., Kumar, D., Kanika, N., and Nidhi, S. 2021. Plant-microbial interactions in natural/organic cultivation of horticultural plants. In *Plant-Microbial Interactions and Smart Agricultural Biotechnology*. Tyagi, S. et al. (Eds.). Abingdon (UK): Taylor & Francis, pp. 115–128. doi:10.1201/9781003213864-6

Saleh, S.A., Heuberger, H. and Schnitzler, W.H. 2005. Alleviation of salinity effect on artichoke productivity by Bacillus subtilis FZB24, supplemental Ca and micronutrients. *Journal of Applied Botany and Food Quality* 79: 24–32.

Samayoa, B.E., Shen, F.T., Lai, W.A. and Chen, W.C. 2020. Screening and assessment of potential plant growth-promoting bacteria associated with Allium cepa Linn. *Microbes and Environment* 35(2): 1v10. doi:10.1264/jsme2.ME19147e

Sarma, B.K., Yadav, K.S., Singh, D.P. and Singh, H.B. 2018. Rhizobacteria mediated induced systemic tolerance in plants: Prospects for abiotic stress management. In: *Bacteria in Agrobiology: Stress Management*. Maheshwari, D. (Ed.). Berlin: Springer, pp. 225–238.

Sayed, S.A., Atef, A.S. and Soha, E. 2011. Response of three sweet basil cultivars to inoculation with Bacillus subtilis and arbuscular mycorrhizal fungi under salt stress conditions. *Natural Sciences* 9(6): 31–36.

Schrey, S.D. and Tarkka, M.T. 2008. Friends and foes: Streptomycetes as modulators of plant disease and symbiosis. *Antonie Van Leeuwenhoek* 94: 11–19.

Seifikalhor, M., Aliniaeifard, S., Shomali, A., Azad, N., Hassani, B., Lastochkina, O. and Li, T. 2019. Calcium signaling and salt tolerance are diversely entwined in plants. *Plant Signaling and Behavior* 14: 1665455.

Seifikalhor, M., Aliniaeifard, S., Seif, M., Asayesh, J.E., Li, T. and Lastochkina, O.V. 2018. Rhizobacteria Bacillus subtilis reduces toxic effects of high electrical conductivity in soilless culture of lettuce. *Acta Horticulturae* 1227: 471–478.

Selvakumar, G., Bhatt, R.M., Upreti, K.K., Bindu, G.H. and Shweta, K. 2015. Citricoccus zhacaiensis B-4 (MTCC 12119) a novel osmotolerant plant growth promoting actinobacterium enhances onion (Allium cepa L.) seed germination under osmotic stress conditions. *World Journal of Microbiology and Biotechnology* 31(5): 833–839.

Selvakumar, G., Krishnamoorthy, R., Kim, K. and Sa, T.M. 2016. Genetic diversity and association characters of bacteria isolated from arbuscular mycorrhizal fungal spore walls. *PLoS One* 11 (8):e0160356.

Sengupta, A. and Gunri, S.K. 2015. Microbial intervention in agriculture: An overview. *African Journal of Microbiology Research* 9(18): 1215–1226.

Sharma, L.D., Indira, S., Rahul, S., Herojit Singh, Y., Sawant, C.G. and Lalhming, S. 2021. Microbes in fruit crops – An analysis. *International Journal of Plant Biotechnology* 7(2). doi:10.37628/IJPB

Sharma, N. and Pethe, A. 2017. Isolation and identification of rhizospheric microflora of some medicinal plants. *International Journal of Advance Research and Innovative Ideas in Education* 3(2): 5595–5601.

Shen, F., Zhu, T.B. and Teng, M.J. 2016. Effects of interaction between vermicompost and probiotics on soil nronerty, yield and quality of tomato. *Yingyong Sheng Tai Xue Bao.* 27(2): 484–490.

Shengepallu, M.D., Ravindra, T.G., Visvas, A.C. and Yumlembam, R.A. 2018. Isolation and characterization of nitrogen fixing bacteria from Babchi (Psorolea corylifolia L.) and testing them for plant growth promotion traits in vitro. *International Journal of Current Microbiology and Applied Sciences* 7(3): 441–447.

Singh, R., Kapil, D.P., Ajay, K. and Monika, S. 2017. PGPR isolates from the rhizosphere of vegetable crop Momordica charantia: Characterization and application as biofertilizer. *International Journal of Current Microbiology and Applied Sciences* 6(3): 1789–1802. https://doi.org/10.20546/ijcmas.2017.603.205

Singh, R., Pandey, K.D., Singh, M., Singh, S.K., Hashem, A., Al-Arjani, A.B.F., Abd_Allah, E.F., Singh, P.K. and Kumar, A. 2022. Isolation and characterization of endophytes bacterial strains of Momordica charantia L. and their possible approach in stress management. *Microorganisms* 10: 290.https://doi.org/10.3390/ microorganisms10020290

Singh, S.J. 1996. *Advances in Disease of Fruit Crops in India*. Ludhiana: Kalyani Publisher.

Somtrakoon, K., Sangdee, A. and Chouychai, W. 2019. Roles of plant growth promoting bacteria on growth of ornamental plants grown in anthracene-spiked soil. *Journal of Agricultural Research and Extension* 36 (2): 11–21.

Soni, R., Kapoor, R. and Kaur, M. 2016. Evaluation of siderophore production and antimicrobial activity by fluorescent pseudomonas diversity associated with rhizosphere of apple and pear. *International Journal of Agriculture Environment and Biotechnology* 9(6): 1109–1115.

Srivastava, A.K., Singh, S. and Huchche, A.D. 2015. Evaluation of INM in citrus on Vertic Ustochrept: biometric response and soil health. *Journal of Plant Nutrition* 38(6): 854–867.

Strzemski, M., Dzida, K., Dresler, S., Sowa, I., Kurzepa, J., Szymczak, G. and Wójciak, M. 2021. Nitrogen fertilisation decreases the yield of bioactive compounds in Carlina acaulis L. grown in the field. *Industrial Crops and Products* 170: 113698.

Sungthongwises, K., Matsuoka, M., Ohnishi, K., Tanaka, S. and Iwai, C.B. 2015. The potential of phosphate-solubilizing bacteria from the eggplant fields. *Acta Hortacultarae* 1099: 333–337. doi:10.17660/ActaHortic.2015.1099.38

Surapat, W., Pukahuta, C., Rattanachaikunsopon, P., Aimi, T. and Sophon, B. 2013. Characteristics of phosphate solubilization by phosphate-solubilizing bacteria isolated from agricultural chili soil and their efficiency on the growth of chili (Capsicum frutescens L. cv. Hua Rua). *Chiang Mai Journal of Science* 40(1): 11–25. http://it.science.cmu.ac.th/ejournal/

Tallapragada, P., Dikshit, R. and Seshagiri, S. 2016. Influence of Rhizophagus spp. and Burkholderia seminalis on the growth of tomato (Lycopersicon esculatum) and bell pepper (Capsicum annuum) under drought stress. *Communications in Soil Science and Plant Analysis* 47(17): 1975–1984.

Tariq, M.R. Shaheen, F., Mustafa, S., Ali, S., Fatima, A., Shafiq, M., Safdar, W., Sheas, M.N., Hameed, A. and Nasir, M.A. 2022. Phosphate solubilizing microorganisms isolated from medicinal plants improve growth of mint. *Peer J* 10(1): e13782. doi:10.7717/peerj.13782. PMID: 35996668; PMCID: PMC9392447.

Tarkka, M.T., Drigo, B. and Deveau, A. 2018. Mycorrhizal microbiomes. *Mycorrhiza* 28(5): 403–409.

Thamsurakul, S., Nopamonbodi, O., Charoensook, S. and Roenrungroeng, S. 2000. *Increasing Pineapple Yield Using VA Mycorrhizal Fungi*. Leuven: International Society for Horticultural Science (ISHS), pp. 199–202.

Todeschini, V., AitLahmidi, N., Mazzucco, E., Marsano, F., Gosetti, F., Robotti, E., Bona, E., Massa, N., Bonneau, L. and Marengo, E. 2018. Impact of beneficial microorganisms on strawberry growth, fruit production, nutritional quality and volatilome. *Frontiers in Plant Science* 9: 1611.

Turan, M., Ekinci, M., Yıldırım, E., Güneş, K., Karagöz, K., Kotan, R. and Dursun, A. 2014. Plant growth-promoting rhizobacteria improved growth, nutrient, and hormone content in cabbage (Brassica oleracea) seedlings. *Turkish Journal of Agricultural Forestry* 38: 327–333.

Umukoro, B.O-J. 2020. Tropical crops and microbes. In *Microorganisms*. Eds. Blumenberg et al. United Kingdom: Intechopen, pp. 1–22. doi:10.5772/intechopen.89531

Vaghela, N. and Sangeeta, G. 2023. Medicinal plant-associated rhizobacteria enhance the production of pharmaceutically important bioactive compounds under abiotic stress conditions. *Journal of Basic Microbiology* 63(3–4): 308–325. doi:10.1002/jobm.202200361

Verma, P., Yadav, A.N., Kumar, V., Singh, D.P. and Saxena, A.K. 2017. Beneficial plant–microbes interactions: Biodiversity of microbes from diverse extreme environments and its impact for crop improvement. In: *Plant-Microbe Interactions in Agro-Ecological Perspectives: Volume 2: Microbial Interactions and Agro-Ecological Impacts*. Singh, D.P., Singh, H.B. andPrabha, R. (Eds.). Singapore: Springer Singapore, pp. 543–580.

Vijaya Gopal, A., Gosal, S.K. and Jupinder, K. 2021. Diversity of nitrogen fixing bacteria in rhizospheric soils of citrus–poplar cropping system of Punjab. *Indian Journal of Agroforestry* 23(1): 5–12.

Vurukonda, S.S.K.P., Vardharajula, S., Shrivastava, M. and SkZ, A. 2016. Enhancement of drought stress tolerance in crops by plant growth promoting rhizobacteria. *Microbiology Research* 184: 13–24.

Waewthongrak, W., Pisuchpen, S. and Leelasuphakul, W. 2015. Effect of Bacillus subtilis and chitosan applications on green mold (Penicilium digitatum Sacc.) decay in citrus fruit. *Postharvest Biology and Technology* 99: 44–49.

Wang, W.S., Yang, S.Q. and Yang, Z.L. 2019. Isolation and characterization of nitrogen fixing and phosphate solubilizing bacteria from potato rhizosphere of reclaimed cropland. *Journal of Northwest A & F University – Natural Science* 47 (8): 127–133.

Wang, Y., Xu, Z., Zhu, P., Liu, P., Zhang, Z., Mastuda, Y., Toyoda, H. and Xu, L. 2010. Postharvest biological control of melon pathogens using Bacillus subtilis EXWB1. *Journal of Plant Pathology* 92: 645–652.

Weber, O.B., Baldani, V.L.D, Teixeira, K.R.S, Kirchhof, G., Baldani, J.I. and Dobereiner, J. 1999. Isolation and characterization of diazotrophic bacteria from banana and pineapple plants. *Plant Soil* 210: 103–113.

Weekley, V., Gabbard, J. and Nowak, J. 2012. Micro-level management of agricultural inputs: Emerging approaches. *Agronomy* 2: 321–357.

Wu, Q.S., Srivastava, A.K. and Zou, Y.N. 2013. AMF-induced tolerance to drought stress in citrus: A review. *Scientia Horticulturae* 164: 77–87.

Xu, M., Sheng, J., Chen, L., Men, Y., Gan, L., Guo, S. and Shen, L. 2014. Bacterial community compositions of tomato (*Lycopersicum esculentum* Mill.) seeds and plant growth promoting activity of ACC deaminase producing Bacillus subtilis (HYT-12-1) on tomato seedlings. *World Journal of Microbiology and Biotechnology*, 30 (3): 835–845.

Xuyen, T.V., Nguyen, Q.D., Danh, C.V. and Nhut, N.N. 2021. Presence and potential for isolation of potassium solubilizing bacteria in mango, pitaya, and banana rhizospheric soils. In 1st Van Lang International Conference on Heritage and Technology Conference Proceeding, AIP Conference Proceedings. https://doi.org/10.1063/5.0067228 2406, 040007

Yang, D.M., Bi, Y., Chen, X.R., Ge, Y.H. and Zhao, J. 2006. Biological control of postharvest diseases with *Bacillus subtilis* (B1 strain) on muskmelons (Cucumis melo L. cv. Yindi). *Acta Horticulturae*, 712 (2): 735–739.

Zahedi, H. 2016. Growth-promoting effect of potassium-solubilizing microorganisms on some crop species. In: *Potassium Solubilizing Microorganisms for Sustainable Agriculture*. Meena, V., Maurya, B., Verma, J. and Meena, R. (Eds.). New Delhi: Springer, pp. 31–42. https://doi.org/10.1007/978-81-322-2776-2_3

Zaidi, A., Khan, M.S., Ahmad, E., Saima, S., Asfa, R. and Shahid, M. 2016. Growth stimulation and management of diseases of ornamental plants using phosphate solubilizing microorganisms: Current perspective. *Acta Physiologiae Plantarum* 38: 117. https://doi.org/10.1007/s11738-016-2133-7

Zhao, Y., Shao, X.F., Tu, K. and Chen, J.K. 2007. Inhibitory effect of Bacillus subtilis B10 on the diseases of postharvest strawberry. *Journal of Fruit Science* 24(3): 339–343.

9 Microbial Consortium as Biofertilizer and Biocontrol Agent for Sustainable Agriculture

Monica Durairaj, Joseph Sahayarayan, Vidhyavathi RM, Sivasakthivelan Panneerselvam, and Kavitha Thangavel

9.1 INTRODUCTION

The current population is constantly growing, and it will reach 9.7 billion within the next 30 years. India would contribute to the majority of the population. Population growth is directly and indirectly related to the demand for contaminant-free, healthy food. The World Health Organization has pledged to increase global food production by 50% by 2029. Green Revolution practices are also increasing agricultural productivity while lowering the risk of chemical-based fertilizers to human health and the environment (Chittora *et al.,* 2020). However, the overuse of chemical fertilizers has resulted in negative environmental impacts and biodiversity challenges. Water body pollution/eutrophication and algal bloom, land degradation, food and food crop-product contamination, adverse human health effects, greenhouse effect/ozone layer depletion, and general ecosystem imbalance are some of the negative impacts. Because of these negative effects, researchers have used green technology to create eco-friendly environments through the use of microbes (Onyia *et al.*, 2020). Microbial inoculants can be used as biofertilizers, biopesticides, bioherbicides, and biocontrol agents. These microbes are benign and effective in small quantities, have more targeted activity and a faster breakdown process, and induce plant protection mechanisms. Plant growth-promoting microbes, nitrogen-fixing microbes and phosphate solubilizers play an important role in plant growth and development. Nitrogen is a crucial element that can limit plant growth. It is most abundant and ubiquitous in the air, yet becomes a limiting nutrient due to the difficulty of its fixation and uptake by the plants. Specific microorganisms, some of which can form various connections with plants, can however, fix a significant amount of nitrogen. This property enables more efficient availability to plants of fixed nitrogen while reducing losses due to denitrification, leaching, and volatilization.

DOI: 10.1201/9781003343707-9

Cyanobacteria are one of the most efficient nitrogen (N_2)-fixing microorganisms in a variety of agricultural soils. In addition to the organic matter added to the soil rhizosphere, free-living cyanobacteria can fix 10-30 kg of N/ha/year. As a result, they are regarded as an important ingredient of naturally available biofertilizers (Hamed *et al.,* 2022). Phosphorous is a second very important factor that may limit plant growth. Phosphorus is degraded by forming highly robust complex with the salts in the soil. Phosphorus in acid soils is converted to aluminium phosphates and iron phosphates, whereas in alkaline soils, it is transformed to calcium phosphates or sodium phosphates and becomes unavailable to plants. In order to make these forms of phosphorus available to plants, some bio-inoculants produce organic acids that convert them to soluble forms such as hypophosphites (Elavarasi *et al.,* 2020).

Plant growth-promoting microorganisms (PGPMs) have been demonstrated to be essential for the success of long-term agriculture. Plant growth is aided by PGPMs via a variety of mechanisms (Debasis *et al.,* 2019). Plant growth-promoting rhizobacteria (PGPR) are found in nature as soil bacteria that improve plant productivity and immunity. These bacteria are linked to the rhizosphere, the area of soil influenced by plant roots and exudates. PGPRs are classified into two types based on their relationships with plants: symbiotic bacteria that live inside plants and exchange metabolites with them directly, and free-living rhizobacteria that live outside plant cells. Most symbiotic bacteria live in the intercellular spaces of host plant, but some bacteria can form truly mutualistic interactions with their hosts and permeate plant cells (Garcia-Fraile *et al.*, 2015). The essential and useful mechanisms showed by PGPR that promote plant growth are atmospheric nitrogen fixation, phosphorus solubilization, nutrient uptake enhancement, and plant growth hormone production.

Biofertilizers are a combination of live or latent microorganisms that can fix atmospheric nitrogen, decompose organic matter, and increase soil mineral solubility. Biofertilizers augment the abundance of beneficial microorganisms, which improve plant growth. Recently described microbial biofertilizers contain a high concentration of essential micro and macronutrients, growth stimulants, vitamins, and growth-promoting hormones, resulting in remarkable results. Bio-fertilizers not only provide an alternative to chemical fertilizers, but they also are known to improve soil and plant productivity, as we will discuss further in this book chapter. They play an important role in soil development and nutrient mobilisation by expediting microbial processes that help the whole plant growth system (Kumar, 2016). The term 'biopesticide' refers to the use of beneficial microorganisms to control insects. The main constraint, however, is the scarcity of biopesticides in comparison to cropped areas. Pesticides derived from naturally occurring materials such as animals, plants, bacteria, and minerals are known as biopesticides. Biological control of plant diseases refers to the inhibition of plant pathogen communities by living organisms. Microbial biological control agents (MBCA) protect crops from disease damage in a variety of ways. They may stimulate or prime augmented resistance to pathogen infections in plant tissues without direct antagonistic interaction with the pathogen. Competition for nutrients and space is another indirect interaction with pathogens. Thus, antagonistic microorganisms can limit pathogen populations in a variety of ways. The nature of the method of action does not only evaluate how an antagonist impacts a pathogen community. The properties of MBCA are also affected by the mode of action used. Risks to humans or the environment, risks of resistance development against the biocontrol agent, pathogen specificity, and reliance on environmental conditions and crop physiology may differ between modes of action. Preferences for specific modes of action for the intended application of a biocontrol agent will influence the screening methods used to select new antagonists. (Lahlali *et al.*, 2022). Biopesticides are divided into three categories: biochemical pesticides, plant pesticides, and microbial pesticides. 90% of all biopesticides are used all over the world. The most successful and popular biopesticide in the industry is *Bacillus thuringiensis* (Bt). Because of the demand for residue-free crop production, modern agriculture necessitates the use of biopesticides and biofertilizers. To meet the demand, the government and non-governmental organisations must encourage entrepreneurs to produce biofertilizers and biopesticides (Seenivasagan & Babalola, 2021). The purpose of this book chapter is to discuss the methods to enhance the development and yield of plants by using microbes as biofertilizers (phosphate solubilizers, nitrogen fixers, zinc solubilizers, sulphur solubilizers, siderophore solubilizers, cyanobacteria and phytohormones, PGPR, AM) and biopesticides (microbial, plant incorporated protection).

9.2 BIOFERTILIZER

Due to the commendable progress made in the studies of the association between microorganisms and plants over the last two decades, the term biofertilizer or bioinoculant has been deduced in various ways. When applied to seed, plant surfaces, or soil, a biofertilizer colonizes the rhizosphere or the interior of the plant and stimulates growth by enhancing the supply or accessibility of primary nutrients to the host plant (Basu *et al.,* 2021). As a biofertilizer, many different types of microorganisms are used.

9.2.1 Types of Biofertilizer

Biofertilizers can be grouped in different ways based on their nature and function. N_2-fixing microbes, phosphate solubilizing microbes, sulfur solubilizing microbes, zinc solubilizing microbes, plant growth-promoting rhizobacteria (PGPR), arbuscular mycorrhiza (AM), siderophore-producing microbes and phytohormones producing microbes, etc.

9.2.1.1 N_2-Fixing Microbes as Biofertilizers

The technique of converting elemental nitrogen into the plant-usable form of nitrogen is biological nitrogen fixation. Nitrogen-fixing bacteria (NFB) convert insoluble atmospheric N_2 to organic compounds. Diazotrophs are the bacteria that fix nitrogen; they encode nitrogenase, the enzyme complex that catalyzes the conversion of N_2 gas to ammonia. In both free-living and symbiotic diazotrophs, the nitrogenase complex is functionally important. Nitrogen fixers, also known as N-fixing organisms, are used in biofertilizers as a living fertilizer made up of microbial inoculants or groups of microorganisms that can fix atmospheric nitrogen. They are classified as

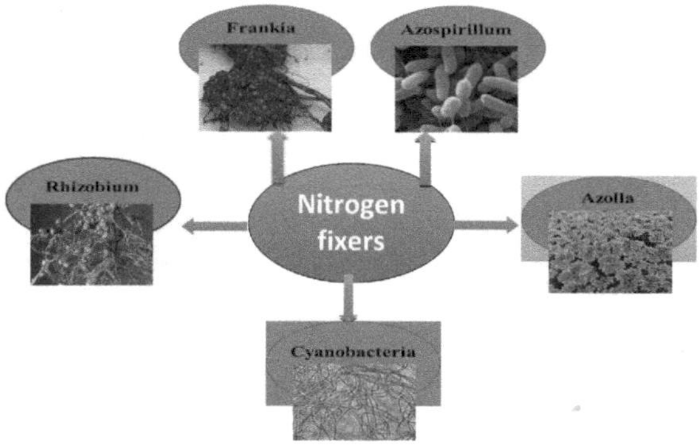

FIGURE 9.1 N_2-fixing microorganisms.

free-living bacteria (*Azospirillium*), blue-green algae (cyanobacteria), and symbionts (*Rhizobium, Frankia*, and *Azolla*) (Bhat *et al.,* 2015). Figure 9.1 represents N_2-fixing microorganisms.

9.2.1.1.1 Azospirillum *as Nitrogen Fixers*

Azospirillum is the most studied and used PGPB, comprising bacteria with an impressive ability to benefit a variety of plant species. In 1925, Beijerinck described the genus *Spirillum*, which was later reclassified as *Azospirillum* due to its ability to fix atmospheric nitrogen (N_2), which was encountered and reported by group of Dr. Johanna Dobereiner in Brazil in the 1970s. Following the discovery that *Azospirillum* was diazotrophic, several studies were conducted to assess its ability to fix N_2 and replace N-fertilizers when linked with grasses such as sugarcane (*Saccharum* spp.), grain crops such as maize (*Zea mays L.*), wheat (*Triticum aestivum* L.), and rice (*Oryza sativa* L.). So far, twenty Azospirillum species have been reported, but *A. brasilense* and *A. lipoferum* have received the most physiological and genetic attention (Fukami *et al.,* 2018). Azospirillum species such as *A. lipoferum, A. brasilense, A. oryzae*, and *A. halopraeferens* can promote plant growth in a variety of ways. Mechanisms such as the release and biosynthesis of indol-3-acetic acid, amino acids, gibberellins, polyamines, and cytokinins promote root growth, nutrient and water uptake in the plant. Apart from these characteristics, nitrogen-fixing ability of *Azospirillum* spp. directly assists in making atmospheric N available to plants. Azospirillum, like other PGPR, enhances N compound assimilation in plants and facilitates a reduction in mineral fertilizer application and N_2O release while increasing crop production significantly (Naqqash *et al.,* 2022).

9.2.1.1.2 Cyanobacteria as Nitrogen Fixers

Cyanobacteria are a group of gram-negative photoautotrophic bacteria that were responsible for producing our oxygenic atmosphere through photosynthetic activities during the Archaean and Proterozoic Eras (2.7 billion years ago). Cyanobacteria were the first organisms to split H_2O instead of H_2S using two photosystems (photosystems

I and II). Cyanobacteria, like algae and plants, use chlorophyll and other accessory pigments, particularly C-phycocyanin, to capture sunlight and perform photosynthesis (Ananya & Ahmad, 2014). Many cyanobacteria perform photosynthesis and can only grow in the presence of light (obligate photoautotrophs), but some species can grow in the dark on certain substances (facultative heterotrophs) (Kumar, 2016). Green technology discusses several aspects of the use of cyanobacteria to improve crop productivity and soil fertility. Notably, cyanobacteria are found to grow on various saline habitats based on saline tolerance such as freshwater, backwater, estuarine and marine environments. Cyanobacterial species can be easily adapted to the various salt environments as they can tolerate and grow in wide range of salinity. Cyanobacteria can degrade a wide range of pollutants and perform different roles in the soil ecosystem to sustain soil fertility (Singh *et al.*, 2017). Cyanobacteria, in addition to fixing atmospheric nitrogen, play an important role in reducing soil erosion due to their ability to secrete polysaccharides that bind soil. They also control soil runoff, increase soil organic matter content, and produce certain substances that promote plant growth. Because of this critical nitrogen fixation property, the utility of cyanobacteria in agriculture to boost production is undeniable. Cyanobacteria evolved very early in the history of life, and they share some characteristics with both gliding bacteria and higher plants. Cyanobacteria have the potential to contribute to productivity in a wide range of agricultural and environmental situations (Mohan & Kumar, 2019). Cyanobacteria are emerging microorganisms that can help with sustainable agricultural development. Diazotrophs are cyanobacteria that can be used to produce inexpensive and environmentally friendly biofertilizers. They can regulate nitrogen deficiency in plants, enhance plant aeration, and water retention capacity, and add vitamin B_{12}. In the rice crop cultivated areas, the most efficient nitrogen-fixing cyanobacteria are *Nostoc linkia, Anabaena variabilis, Aulosira fertilisima, Calothrix* sp., *Tolypothrix* sp., and *Scytonema* sp. Researchers indicate that the biofertilization with cyanobacteria and green algae is able to increase the rooting of grapes cuttings and germination of sunflower seeds, and improve plant growth, as observed in rice, barley, oats, tomato, radish, cotton, sugarcane, maize, chili, lettuce, wheat, gillyflower, grapevine, and corn (Grzesik *et al.*, 2017). It has been proposed that cyanobacteria and algae can synthesize polysaccharides, as well as microalgae polysaccharides varies between 0.5 and 20 g/L. The method by which these polysaccharides are produced and microalgae and cyanobacteria were used to make this extract. There are various types of polysaccharides derived from cyanobacteria and algae (Singh *et al.,* 2017). Cyanobacteria-produced exopolysaccharides (EPSs) have been shown to aid in the reclamation of desert soils. Cyanobacteria-produced EPSs aid in the binding of soil particles and the enhancement of soil moisture. There have been reports of extracellular polysaccharide production leading to desiccation tolerance. Several mechanisms for desiccation tolerance in cyanobacteria have been proposed, including the production of extracellular polysaccharides. Cyanobacteria can improve plant growth by improving the soil structure as they have the potential to secrete extracellular polysaccharides that help in soil aggregation and water retention.

Cyanobacteria can boost plant growth both directly and indirectly. The production of various plant growth-promoting compounds that are biologically active, such as auxin, gibberellins, and cytokinins, is one of the direct methods. Cyanobacteria indirectly promote plant growth by preventing or counteracting the negative effects of one or more phytopathogenic microorganisms (Mazhar & Hasnain, 2011). Exopolysaccharides (EPS) serve as a binding agent for soil particles. EPS producing cyanobacteria have previously been given a particular role in the stabilisation of soil particles as well as the regulation of penetration of water and retention in desert soils. Cyanobacteria strains are employed for salt-affected soil treatment because they are more tolerant and grow well in saline soil, can withstand up to 8 dS m^{-1}, and can be quickly mass cultured in the presence of light due to their higher photosynthesis efficiency. According to some studies, cyanobacteria can convert sodium clay into calcium clay. It provides oxygen to the roots of rice crops in saline-waterlogged conditions. It provides oxygen to the roots of rice crops in saline-waterlogged conditions. It has been reported that cyanobacteria could be an effective tool for salt-affected soil regeneration by producing a thick layer on the surface of saline soil and conserving and improving soil organic matter. Jan *et al.* (2017) reported that cyanobacteria (blue green algae) increased rice crop production in saline and high alkaline environments. Many free-living blue-green algae (cyanobacteria) fix atmospheric nitrogen through a process known as biological nitrogen fixation, and because they are photosynthetic, they do not compete for carbon and energy with crop plants and heterotrophic soil microflora. Despite the fact that nitrogen covers 79% of the atmosphere, plants cannot use it due to the presence of triple bonds in nitrogen molecules, which give them a stable nature. Cyanobacteria play an important role in converting this unutilized form of nitrogen to a usable form, which leads to an increase in plant growth and yield production (Garlapati *et al.*, 2019). Efficient atmospheric nitrogen-fixing cyanobacterial strains (*Nostoc linkia, Anabaena variabilis, Aulosira fertilisima, Calothrix* sp., *Tolypothrix* sp., *and Scytonema* sp.) were identified and used in rice production from various agroecological regions. The nitrogenous, as well as nitrate reductase activities of the algae, applied to the surface of plants, or the amino acids and peptides produced in the cyanobacterial filtrate, and/or other molecules that enhanced plant growth, could be attributed to the growth promotion in response to application of *Nostoc muscorum* (N-fixing cyanobacteria) (Grzesik *et al.,* 2017).

9.2.1.1.3 Azolla as Nitrogen Fixer

Azolla is a free-floating water fern with agronomic significance due to its ability to fix nitrogen. It develops a nitrogen-fixing symbiotic relationship with the cyanobacterium, *Anabaena azollae*, which lives in the foliar cavity of fern and most important is the ability of Azolla to naturally fix N and grow rapidly (Bhuvaneshwari & Singh, 2015). Nitrogen can be fixed by prokaryotes is a major challenge in this initiative, as crop growth is proportional to the amount of nitrogen available in soil. Consider the centuries-old practice of rice farmers in Central Asia of using the aquatic fern genus Azolla to uplift nitrogen supply to rice crops as a solution to this problem. The nitrogen-fixing bacteria *Anabaena azollae* has a symbiotic relationship with *Azolla* spp. It has been indicated that *Azolla* spp. produce 70 to 110 kg nitrogen/ha. Various researches have shown that applying *Azolla* spp. directly to agricultural soil can significantly decrease or completely eradicate the need for additional fertilization.

9.2.1.1.4 Frankia as Nitrogen Fixer

Frankia is a genus of soil actinomycetes in the Frankiaceae family that fix nitrogen both symbiotically and aerobically, whereas most rhizobias do not. The filamentous gram-positive *Frankia* sp. and the unicellular gram-negative paraphyletic rhizobia are quite distant phylogenetically, implying that these two major groups of nitrogen-fixing symbionts have obtained nitrogen-fixing mechanisms from different evolutionary origins. Frankia spp. are N_2-fixing actinobacteria with several distinguishing characteristics. Frankia fixes N_2 not only in free-living conditions but also in symbiosis with over 200 non-legume plant species from eight families known as actinorhizal plants. According to Karthikeyan (2016), *Casuarina equisetifolia* tree converts fixed ammonium ions into nitrogen oxides and amino acids as part of its symbiotic relationship with Frankia to produce proteins and other molecules (e.g. alkaloids). This tree offers carbon to the symbiotic bacteria in exchange for the fixed N. The symbiotic relationship of *Frankia* with *C. equisetifolia* fixes N at an estimated rate of 362 kg ha^{-1} $year^{-1.}$

9.2.1.1.5 Rhizobium as Nitrogen Fixer

Perhaps 80% of biologically fixed N in crop fields arises from a symbiotic relationship between leguminous plants and one of the Rhizobia species. Legumes can form symbiotic relationships with soil bacteria known as Rhizobia. These bacteria, when combined with legumes, can fix atmospheric nitrogen and are thus introduced into agricultural structures to enhance soil fertility, and plant growth, and reduce the use of chemical fertilizers. However, depending on the rhizobium species and its interaction with the environment, the anticipated benefits of nitrogen-fixing bacteria may be positive or negative. Isolation and screening of rhizobia stress-tolerant strains may improve plant growth and nitrogen fixation potential under stressful conditions. The most important aspect is the selection of effective Rhizobium strains (Mabrouk *et al.,* 2018). Symbiotic relationships between legumes and rhizobia, soil bacteria that stimulate the formation of nodules on plant roots, provide the most efficient contribution to biologically fixed nitrogen. Inside nodules, rhizobia differentiate into bacteroids that transfer atmospheric N_2 into ammonia for secretion to the plant host in exchange for dicarboxylates, primarily succinate and malate. Under field conditions, *Rhizobium* sp. inoculation results in a significant increase in growth and yield, and the number of nodules per root system is considerably higher in plants inoculated with *Rhizobium* sp. than in plants without Rhizobium sp. Rhizobia, in addition to their N_2-fixing activity with legumes, can enhance crop P nutrition by mobilizing inorganic and organic P. Many rhizobia isolates from various cross-inoculation groups isolated from soils in Iran can mobilize P from organic and inorganic sources. The mixture of Rhizobium and Phosphate Solubilizing Bacteria (PSB) had a synergistic effect on mungbean symbiotic parameters and grain yield.

9.2.1.2 Phosphate Solubilizing Microbes

Phosphorous is one of the most plenteous metallic elements in the crust of earth, and it can be observed in both organic and inorganic states in soils. It is used or taken up by plants in its inorganic form, orthophosphate (H_2PO_4- and HPO_4^{2-}). The main characteristics associated with phosphorous nutrition are photosynthesis, energy transfer, signal transduction, nitrogen fixation in legumes, crop quality, and resistance to plant diseases (Ingle & Padole, 2017). Phosphate Solubilizing Microorganisms, particularly Phosphate Solubilizing Bacteria, enhance the solubilization of a mixture of insoluble phosphorous through the invasion of phosphatase, natural acids, and phytase complexes, which is mostly present in a diverse range of soil microorganisms. *Sclerotium, Penicillium,* and *Aspergillus* are three species that change rapidly. *Micrococcus, Mycobacterium, Pseudomonas,* and *Flavobacterium, Bacillus, Pantoea, Rhizobia,* and *Burkholderia* are four bacteria that are constantly changing. Recently, the nematofungus *Arthrobotrys oligospora* was evaluated *in vitro* and *in vivo* for its ability to solubilize Kodjari rock phosphate (Burkina Faso, KRP), Togolese rock phosphate (Senegal TRP), and Tilemsi rock phosphate (Mali TIRP). The fungus solubilized all three types of rock phosphates and proved its capacity to solubilize additional phosphate from rock phosphates *in vivo* as well. The majority of the effect of using microorganisms as biofertilizers has been directed towards understanding biological nitrogen fixation (Khan *et al.,* 2009).

9.2.1.3 Sulphur Solubilizing Microbes

Sulphur is an essential macronutrient and a vital element of organic compounds found in plants, animals, microorganisms, and humans. Deterioration of sulphur compounds has thus become an unavoidable preliminary task in nature (Chaudhary & Goyal, 2019). The oxidation of lowered sulphur in soil is commonly regarded as a microbial process. Sulphur can be oxidized by a diverse range of microorganisms. Microbial sulphur oxidation is highly favorable to soil fertility, leading to the development of SO_4^{2}-, which can be used by plants, while the acidity produced by the oxidation process is used to solubilize other plant nutrients, particularly phosphorous, and improve alkaline soils. *Thiobacillus, Beggiatoa, Thiothrix, Thiomicrospira, Desulphuromonas,* and *Achromatium* are examples of oxidizing bacteria (Chaudhary *et al.,* 2019). Sulphur oxidizing fungus is a remarkably small research area, but there have been reports of it having applications in a variety of fields. The fungus either uses sulphur compounds as a substrate for energy production or dissolves sulphur compounds by secreting acids, making them a better option for sulphur oxidation. Oxidation of sulphur compounds has been reported over a wide range of fungal genera like *Absidia, Alternaria, Acremonium, Aspergillus, Amanita, Aureobasidium, Epicoccum, Cephalosporium, Fusarium, Hymenoscyphus, Geosmithia, Hypholoma, Mortierela, Monilia, Mucor, Myceliophthora, Paxillus, Paecilomyces, Penicillium, Pisolithus, Phanerochaete, Rhodotorulla, Suillus, Rhizopogon, Trichoderma*, and *Zygorhynchus,* etc (Chaudhary & Goyal, 2019).

9.2.1.4 Zinc Solubilizing Microbes

Zinc is a trace element that plants need to grow in relatively small concentrations (5 mg/kg to 100 mg/kg) in their tissues for germination. Malnutrition in soil is common, and it is induced by improper fertilizer application, intensive farming, and poor soil health. Zinc deficiency is expected to rise from 42% to 63% by 2025 if the underlying issues are not resolved. Zinc is required for the production of growth hormones. In plants, zinc deficiency causes inhibited shoot growth, affected membrane integrity, smaller leaves, chlorosis, and chronic exposure to light, heat, and fungal infections, as well as influencing grain yield, root growth, pollen development, and water intake and transport (Kamran *et al.,* 2017). According to a study, inoculating rice plants with a suitable selection of Zn-solubilizing bacterial strains (*Burkholderia* sp. and *Acinetobacter* sp.) advances growth patterns and rice yield and was shown to be more productive in obtaining Zn from the soil than non-inoculated plants (Vaid *et al.,* 2014).

9.2.1.5 Siderophore Producing Microbes

Many bacteria secrete ferric iron-specific ligands, which aid in iron sequestration and transport under iron-limiting conditions. Siderophores are low molecular weight iron-chelating secondary metabolites synthesized by numerous groups of microorganisms that help in iron scavenging. Endophytic bacteria produce siderophores, which help plants growth by providing iron. Microorganisms have created a particular mechanism for iron uptake by discharging siderophores to scavenge iron. Many bacterial genera including *Pseudomonas, Azotobacter, Bacillus, Rhizobium* and *Enterobacter* are known to synthesize siderophores (Maheshwari *et al.,* 2019). It also aids in the protection of phytopathogens caused by iron deficiency between the plant and pathogen. PGPR is associated with the roots and on the surface of root tissue, as well as within the soil that is induced by the root system. Non-ribosomal peptide bonds form the siderophores. It is a multidentate, organic, oxygen-donor ligand that is classified as hydroxamate, catecholate, and carboxylate, and is used to aid in the solubilization and chelation of iron transport to cells (Subramanium & Sundaram, 2020). These siderophore-producing microbes are environmentally friendly and long-lasting agents that may maintain plant stresses in degraded land. Because of the secretion of iron-chelating compounds, such as siderophore, the positive relationship between rhizosphere and microbes plays an important role in iron-stressed degraded land. In iron-stressed degraded soil, plant roots secrete siderophore to retain the iron level for their metabolic and physiological activities, but this is not achieved at the perfect level. SPM, on the other hand, produces a large number of iron-chelating compounds that can reduce plant stress in iron-stressed soil. Because salinity in the soil causes Fe deficiency, the two can have a negative impact on plant shoot and root growth, photosynthesis, transpiration rates, chlorophyll concentration, and stomatal conductance. Focusing on potential siderophore-producing PGPR that are also salt-tolerant could be useful as a technology through non-transgenic approaches for bringing salinity-affected lands under agriculture (Sultana *et al.,* 2021).

9.2.1.6 Plant Growth-Promoting Rhizobacteria (PGPR)

Kloepper invented the term PGPR in the 1970s, and it has since been known as (NPR) nodule-promoting rhizobacteria and plant health-promoting rhizobacteria (PHPR) found in the rhizosphere (Singh, 2018). Plant growth-promoting rhizobacteria (PGPR), which are microbial communities having the capability of colonizing plant roots, influence plant growth in a variety of indirect and direct forms in order to promote growth and/or prevent it from disease or insect attack. The number of PGPR identified in recent years has risen, owing primarily to the growing importance of the rhizosphere as a biosphere in the functioning of the global ecosystem. Plant growth has been revealed to be enhanced by bacteria such as *Pseudomonas, Azospirillum, Azotobacter, Klebsiella, Enterobacter, Alcaligenes, Arthrobacter, Burkholderia, Bacillus,* and *Serratia*. Several PGPR inoculants are currently on the market that emerge to promote growth via at least one mechanism: suppression of plant disease (referred to as bioprotectants), improved nutrient utilization (referred to as biofertilizers), or phytohormone production (biostimulants).

PGPR also has the capacity to synthesize the siderophore, HCN, phytohormones such as auxins (IAA), gibberellins, and cytokinins, and solubilize phosphate, thereby improving plant growth and acting as a biological control agent against tomato and potato bacterial wilt disease (Mekonnen & Kibret, 2021). Among others, rhizobia are the first group of bacteria attributed to the ability of PGPR to release IAA, which can aid in plant growth and pathogenesis. Traditionally, the mechanisms by which PGPRs promote plant growth have been classified as direct and indirect mechanisms that appear inside and outside the plant, respectively. They enhance plant nutrition by providing phytonutrients such as fixed nitrogen or solubilized minerals from the soil (such as P, K, Zn, Fe, and other essential mineral nutrients) and/or boosting plant growth and development by regulating phytohormone levels (such as auxins, cytokinins, gibberellins, abscisic acid, and ethylene) (Basu *et al.*, 2021). Indirect mechanisms are bacterial traits that prevent one or more plant pathogenic organisms, both fungi and bacteria, from functioning. ACC deaminase, antibiotics, cell wall

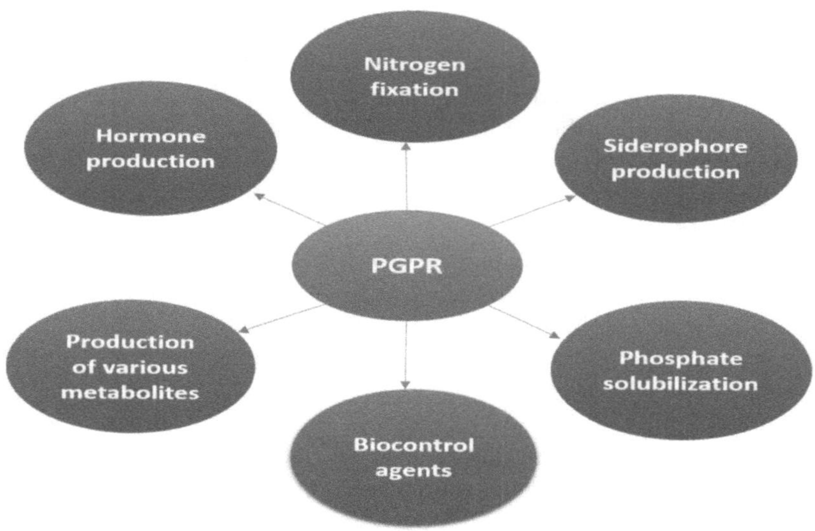

FIGURE 9.2 Functions of plant growth promoting rhizobacteria.

hydrolytic enzymes, competition, hydrogen cyanide, induced systemic resistance, quorum quenching, and siderophores are examples of indirect mechanisms. Biocontrol of some bacterial phytopathogens may be acquired through the selective use of bacteriophages, in addition to the traditional methods used by PGPB to control phytopathogens. Bacteriophages have been used to control phytopathogens such as *Ralstonia solanacearum* and *Erwinia amylovora* (Olanrewaju *et al.,* 2017). Figure 9.2 indicates the functions of plant growth-promoting rhizobacteria.

9.2.1.7 Arbuscular Mychorhiza (AM)

The name AMF comes from the fact that these fungi form distinctive finely branched hyphal structures called arbuscules within cortical cells of plant roots. The arbuscules are important for fungal multiplication because they can act as nutrient organs (or sites of nutrient exchange among symbionts) (Madhushan *et al.,* 2021). A mycorrhiza is a symbiotic relationship between a fungus and roots of a vasicular plant. The fungus infests the roots of host plant in this association, either intracellularly as in arbuscular mycorrhizal fungi or extracellularly as in ecto-mycorrhizal fungi. They play an important role in soil life and chemistry. Mycorrhizae are commonly classified as ectomycorrhiza (fungal hypha do not penetrate individual cells within the root) and endomycorrhiza (fungal hypha penetrate the cell wall and invaginate the cell membrane). Endomycorrhizae are classified into arbuscular, ericoid, arbutoid, monotropoid, and orchid mycorhizae. Arbuscular mycorrhizal (AM) fungi are common in soil ecosystems and form favourable symbioses with angiosperm and other plant roots (Sadhana, 2014). Arbuscular mycorrhizal fungi (AMF) are root autotrophic biotrophs that transfer mutual benefits with approximately 80% of plants. They are generally considered biofertilizers because they provide water, nutrients, and pathogen protection to the host in interaction for photosynthetic products. As a result, AMF are primary biotic soil components that, when lacking or depleted, can lead to less efficient ecological processes. In terms of sustainable agriculture, the process of re-establishing the natural level of AMF richness can be a feasible alternative to conventional fertilisation practises (Berruti *et al.*, 2016). Reciprocal relationships between mycorrhizae fungi and host plants are mutually beneficial (symbiosis mutualism). As a result, inoculation of mycorrhizae fungi can be regarded bio-fertilization for food crops, plantations, forestry, and greening crops. The symbiosis of arbuscular mycorrhizae fungi in the nursery can increase P uptake. However, in order to benefit from high symbiosis, it is necessary to understand the optimum symbiotic circumstances. The symbiosis of arbuscular mycorrhiza with plants is highly influenced by nutrient levels, inoculum dose, and fertilizers (Praharasti *et al.,* 2012). Exudates of extracellular polysaccharide and glomalin support in the formation of soil network hyphae. Glomalin, a polysaccharide, is the most essential factor to soil formation because it supports the growth of organic matter and the attachment of hyphae to the soil. Soil accumulation improves soil health and quality by increasing porosity, water retention capacity and gaseous exchange, protecting organic carbon, and promoting the growth of beneficial microfauna. Several studies show that AMF decrease the negative impact of heavy metal contamination in soil resulting from human activities and agrochemical product use. Calcium, aluminium, cadmium, selenium, and arsenic acid can all be assimilated by fungi. Several studies have shown that AMF tends to increase plant disease resistance against soilborne plant pathogens (Madhushan *et al.,* 2021). India has a diverse AM Fungi distribution. In 1974, Bakshi was the first to describe 14 spore types: *Glomus macrocarpum Tul* and *Tul var., G.geosporum, G.mosseae, Glomus* sp., *Sclerocystis coremioides, Sclerocystis* sp., *Gigaspora, Calospora, Acaulospora* sp., *Endogone gigantea, E. microcarpum,* Endogone 1, Endogone 2, and Endogone 3. In 1976, Gerdemann and Bakshi described two new species, *Glomus multicauli* and *Sclerocystis sinuosa* (Sadhana, 2014).

9.2.1.8 Phytohormones

Phytohormones are synthetic compounds that play an important role in natural growth and are found in low concentrations. These phytohormones shape the plant by influencing seed growth, flowering time, flower sex, leaf senescence, and fruit ripening. They also have an impact on gene expression and transcription, as well as cellular division and growth. Phytohormones also govern cellular processes, pattern formation, vegetative and reproductive growth, and stress responses in target cells. Thus, hormones regulate and determine all major activities such as leaf formation, flower formation, fruit development and ripening. The biosynthesis of plant growth regulators such as auxins, gibberellins, cytokinins, and ABA has been linked to the stimulation of plant growth and nutrient acquisition by beneficial rhizobacteria. Plant hormone levels in plant tissue are influenced by microbial regulators, which have been found to have effects similar to exogenous phytohormone applications.

9.2.1.8.1 Effect of Indole Acetic Acid

Auxins are essential phytohormones, and the auxin indole-3-acetic acid (IAA) has been found to enhance cell division, elongation, and differentiation (Egamberdieva *et al.,* 2017). Rhizobacterial stimulation of root growth is mostly thought to be via their ability to synthesize indole acetic acid (IAA – the most frequent, naturally occurring plant hormone of the auxin class). Many rhizosphere bacteria and cyanobacteria can generate auxins, and adding tryptophan to bacterial culture media offers a simple method to determine auxin production. Many rhizospheric and epiphytic bacteria have been found to be able to synthesize IAA, including *Azospirillum* spp., *Agrobacterium* spp., *Azotobacter* spp., *Alcaligenes* spp., *Entrobacter* spp., *Erwinia* spp., *Acetobacter* spp., *Rhizobium* spp., *Bradyrhizobium* spp. IAA synthesis is common in bacteria from the genera *Pseudomonas*, *Bacillus*, and *Xanthomonas* spp., as well as Methylobacteria (*Methylobacterium, Methylovorus, Aminobacter, and Paracoccus). Achromobacter, Flavobacterium, Arthrobacter, Klebisella, Rhodococcus, Mycobacterium, Sphingomonas, Stenotrophomonas,* and *Micrococoocus* have all been found to synthesize auxins. Many Streptomycetes and Archaea representatives for example, the *thermophylic* sp. synthesis IAA. Some cyanobacterial species synthesize IAA viz. *Nostoc, Anabaena, Calothrix, Hapalosiphon, Oscillatoria, Lyngbya, Phormidium* (Mohan & Kumar, 2019).

9.2.1.8.2 Effect of Gibberellic Acid

Gibberellin (GA) is an important hormone that has been found in soil microbes as well. Gibberellins are di-terpenoid variants with a carbon skeleton of C_{20} or C_{19}.

Hundreds of gibberellin structures have been observed to date. However, three to four strains of GAs generated by *Bacillus licheniformis* and *B. pumilus* have been identified. Gibberellins enhance seed germination, stem lengthening, and flowering in plants, as well as boosting photosynthetic rate. Inoculation of *A. lipoferum* with maize results in the production of gibberellins and the mitigation of drought stress. Several studies have found that applying growth regulators to crops, such as gibberellic acid, can help to mitigate the negative effects of salt stress. GA_3 inhibited NaCl-induced growth inhibition of shoot rice seedlings, and kinetin mitigated the effect of salinity on growth and plant growth regulator production in *Vigna sinensis* and *Zea mays*. Gibberellins are produced by fungi and a few bacteria in addition to higher plants. Gibberellins are currently produced by species of the fungal genera *Fusarium, Gibberella, Sphaceloma, Neurospora,* and *Phaeosphaeria* (Desai, 2017).

9.2.1.8.3 Effect of Abscisic Acid

Abscisic acid (ABA) is a sesquiterpenoid phytohormone that plays a role in plant developmental processes such as seed dormancy, cell division, and flower initiation and elongation. It is also formed as a result of biotic stressors such as pathogen infection and abiotic stressors such as drought or salt stress. ABA is currently used in agriculture to improve the colour of red table grapes and to create hybrid seeds. Furthermore, it has been demonstrated that ABA applications can improve citrus salt tolerance and drought tolerance in wheat cold/drought tolerance and chickpeas (Arnesen *et al.,* 2022). ABA performs another function in nutrient signalling in seedlings by mediating the regulatory effects of nitrate on root branching. Nitrate effects are dependent on the concentration as well, with low nitrate (1 mM) stimulating lateral root growth elongation and high nitrate (>10 mM) inhibiting elongation of lateral roots that have recently emerged throughout the root system. Increasing sugar has the opposite effect. Stress responses by plants prefer to be mitigated by the production of phytohormones. Among different endogenous phytohormones, the plant increases the response to salinity stress, and abscisic acid (ABA) biosynthesis occurs. The signalling pathway of ABA in plant development is important because it induces stomatal closure, which reduces water loss.

9.2.1.8.4 Effect of Cytokinin

Cytokinins (CKs) impact plant growth, development, and physiology in a variety of ways, including seed germination, apical dominance, flower and fruit growth, leaf senescence, and plant-pathogen interactions. CKs are isoprenoid-substituted adenines. Isopentenyltransferases (IPTs) are the first enzymes involved in catalysing the conversion of isoprenoid to other CKs such as cis-zeatin (cZ), trans-zeatin (tZ), and dihydrozeatin (DZ). The most abundant form of tZ is found in plants. CKs are further metabolized and inactivated via sugar conjugation or degradation by CK oxidases (CKXs) (Akhtar *et al.,* 2020). Plants do not produce CK alone; it can also be synthesized by plant-associated microorganisms such as bacteria and fungi. Fungal phytopathogens may also activate plant CK signaling/sensing via effector molecule secretion. Fungal CKs secreted by (hemi) biotrophic pathogens can act as virulence factors; however, most necrotrophic fungal plant pathogens do not secrete CKs during plant infection. Although a lifestyle-dependent role for CK signalling/perception in fungal plant pathogens has been proposed, little is known about CK perception, sensing, and signalling in fungal organisms (Anand *et al.,* 2022).

9.2.2 Role of Biofertilizers in Crop Production

9.2.2.1 An Ecofriendly Way to Replace Chemical Fertilizers

Biofertilizers maintain the soil environment rich in micro- and macronutrients by nitrogen fixation, phosphate, and potassium solubilization or mineralization, the discharge of plant growth regulating substances, the production of antibiotics, and the biodegradation of organic matter in the soil. Biofertilizers reduce ecological disruption. The use of biofertilizers replaces the use of chemical fertilizers, which are harmful to the environment. In support of sustainable agriculture, biofertilizers are gaining popularity as a feasible alternative to hazardous chemical fertilizers (Nosheen *et al.,* 2021).

9.2.2.2 Enhancement of Plant Tolerance to Environmental Stress

Abiotic stress has a negative impact on crop growth and productivity, particularly for crops grown in the field. This situation is deteriorating due to climate change and poor agricultural practices such as the overuse of chemical fertilizers and pesticides above suggested crop and soil threshold levels. Several previous reports highlight the difficulty of drought stress and its implications for crop breeding and yield. As a result, investigating the microbial potentials in soils and their communications can play an important role in reducing the rising threat of drought conditions. El Semary *et al.* (2020) used cyanobacterial microbial inoculums and discovered that using these microbes as biofertilizers at the strongest salinity level increases the negative effect of salinity alleviation to a greater extent. Phour & Sindhu (2020) utilized *Pseudomonas argentinensis* and *P. azotoformans* to reduce saline stress in Indian mustard (*Brassica juncea* L.).

9.2.2.3 Enhancement of Soil Fertility

According to many investigators, the use of bio-fertilizers improves soil fertility, crop yield, and final yield. Furthermore, their application in soil improves the biodiversity of soil while reducing the use of chemical fertilizers. Biofertilizers induce nutrients that promote the development of biochemical reactions in soils, which help to maintain plant health (Sneha et al.,2018).

9.3 MICROBES AS BIOCONTROL AGENT

Biological control of plant diseases refers to the inhibition of plant pathogen populations by living organisms. Beneficial microorganism isolates that are extremely effective against pathogens and can be increased on artificial media can be chosen from among the beneficial microorganisms. The use of such carefully chosen and mass-produced antagonists in high densities once or several times during a planting period is referred to as 'augmentative biological control'. Beneficial microorganisms as biological controls have gained prominence, and several biological control agents (BCAs) have been identified, including *Bacillus, Pantoea, Streptomyces, Trichoderma, Clonostachys, Pseudomonas, Burkholderia,* and certain yeasts. Biopesticides are currently being created and commercialized to combat leaf diseases, root diseases, and fruit storage diseases. Biological control is one such critical technique that is presently in the spotlight. Various biocontrol methods have been tested over time, and some have shown significant success and promise (Pandit *et al.,* 2022). There are three major categories of biopesticides. The first category is known as microbial pesticides, and it includes products derived from microorganisms such as bacteria, fungi, viruses, protozoa, and algae. Fungi, for example, can control weeds and eliminate specific insects. Pesticidal compounds in plants that result from plant genetic manipulation are included in the second group. The transgenic expression of the *Bt* gene, for example, allows plants to produce a bacterial protein that kills pests on their own (Samada & Tambunan, 2020). Several antagonistic organisms have recently been discovered for almost every life cycle stage of *S. sclerotiorum*. Fungi, bacteria, and cyanobacteria can all be antagonistic organisms.

9.3.1 Cyanobacteria as Biocontrol Agents

Cyanobacteria are also being used as a biocontrol agent in agriculture, especially for fungal plant diseases management. Fungal diseases in crops are one of the major concerns to agricultural production. Almost all plant pathogenic fungi spend part of their lives on their host plants and partly in soil or on plant debris. Plant pathogens are represented in all the major groups of fungi and symptoms caused by fungal diseases. Plant diseases need to be controlled to maintain the quality and abundance of food, feed and fiber produced by growers around the world. Different approaches are used to prevent, mitigate or control plant diseases using cyanobacteria (Anjaiah *et al.* 2003). In the crop production sequence, cyanobacterial (blue-green algae) inoculation has been shown to improve growth, nitrogen fixation, and yields. Cyanobacterial inoculation is identified to improve growth, nitrogen fixation, and yields in the rice-wheat cropping sequence; however, there are few reports on their contribution in disease reduction and fungal disease protection (Yanti & Hamid, 2021). The antimicrobial activity is determined by the algal species and the solvents used. Modern biotechnology developments are designed to improve the production of desired products in cyanobacteria for a variety of industrial applications. Biocidal metabolites produced by cyanobacteria include antitumor activity and enzyme inhibitors. Many studies show that cyanobacterial extracts have antifungal and actinomycetes pathogen activity in agar medium assays. Since the early 2000s, the orders of Nostocales and Oscillatoriales have received the most attention. *Microcystis aeruginosa, Anabaena* spp., *Fischerella* spp., *Nostoc* spp., *Scytonema* spp., *Lyngbya lutea, Oscillatoria* spp., *Phormidium tenue, Trichodesmium hildebrantii, Synechococcus elongates,* and *Synechocystis* sp. Mycelial growth of *Aspergillus* spp. was inhibited in an agar disc diffusion assay. The methanol extract of these cyanobacteria has consistently inhibited fungal growth (Righini *et al.*, 2022). *Aspergillus oryzae, Candida albicans, Penicillium notatum, Saccharomyces cerevisiae,* and *Trichophyton mentagrophytes* were all suppressed by crude ethanolic extracts from *Anabaena laxa*. The mode of action of broad-spectrum biocidal activity of cyanobacteria is similar to that of microbes. The potential and action of various cyanobacterial secondary bioactive compounds vary depending on species and stress. Cyanobacterial extracts can also aid bacterial, fungal, and nematodal bioagents fight back. *Nostoc* sp., a cryptophycin producer, is a natural substance of pesticides against insects, fungi, and bacteria. Roberti *et al.* (2015) also proved that the extract of the cyanobacterium *Anabaena* BEA300B inhibited pathogen sporulation.

9.3.2 Bacteria as Biocontrol Agents

Bacterial biocontrol agents (BCA) offer protection from pathogen infections through a variety of mechanisms. They may employ one or more pathways to prevent or reduce plant disease, interacting either directly or indirectly with the pathogen. BCA can directly interact with pathogens by secreting antimicrobial compounds, interfering with pathogen virulence and able to compete for nutrients and space. Many BCA synthesize and discharge metabolites that have antimicrobial activity by reducing pathogen growth or metabolic activity, such as lipopeptides, bacteriocins, antibiotics, biosurfactants, cell-wall degrading enzymes, or microbial volatile compounds (Bonaterra i Carreras *et al.*, 2022). Despite their distinct ecological niches, free-living rhizobacteria and endophytic bacteria share some pathways for promoting plant growth and controlling phytopathogens. The most well-known biocontrol mechanisms mediated by PGPB are competition for an ecological niche or a substrate, the production of inhibitory allelochemicals, and the formation of systemic resistance (ISR) in host plants to a wide range of pathogens or abiotic stresses (Compant *et al.*, 2005). Endophytes are appealing as biocontrol agents for a variety of reasons, including: i) it provides a natural means of control and can reduce reliance on hazardous chemicals. ii) being potentially self-sustaining and capable of spreading on their own after early establishment; and iii) minimizing the input of non-renewable resources while also enabling for long-term disease suppression in an environmentally friendly manner. Bacterial endophytes can colonise an ecological niche similar to that of plant pathogens, particularly vascular wilt pathogens, making them promising biocontrol agents against wilt diseases (Eljounaidi *et al.*, 2016). Pathogenic bacteria use quorum sensing to determine population size and time of entry into the apoplast or plant cell. N-acyl-homoserine lactones (AHLs) are frequently used as signalling molecules by Gram-negative bacteria. AHLs are synthesized by AHL synthase (luxI) and detected by a transcriptional regulator (luxR). Surprisingly, AHL molecules can also cause a reaction in the host plant (Legein *et al.*, 2020). PGPRs act as biopesticides based on the ability or behaviour of crops and biocontrol agents. PGPRs are bacterial species including *Alcaligenes, Azospirillum, Arthrobacter, Acinetobacter, Bradyrhizobium, Bacillus, Burkholderia, Enterobacter, Erwinia, Flavobacterium, Pseudomonas, Rhizobium, Azorhizobium, Bradyrhizobium, Allorhizobium, Sinorhizobium, Frankia, Mesorhizob, Azoarcus, Exiguobacterium, Methylobacterium, Paenibacillus,* and *Pantoe* (Lahlali *et al.*, 2022).

9.3.3 Fungus as Biocontrol Agents

Because fungi have a relatively high reproductive rate (both sexually and asexually), a short generation time, and are target-specific, the capacity for the application of fungal biological control agents against plant pathogens has significantly increased. Furthermore, in the absence of the host, they can maintain their existence by switching from parasitism to saprotrophism, allowing them to remain sustainable. Many fungal species have methods that allow them to effectively defend plants from plant-pathogenic fungi-caused diseases (Thambugala *et al.,* 2020). Fungal antagonists to use against pathogens that attack leaves and stems include: Lecanicillium, which is primarily an insect pathogenic fungus, but some strains have action against other fungi; *Ampelomyces quisqualis,* which is used against mildews; and *Nectria inventa* and *Gonatobotrys simplex*, which are Alternaria parasites. The fungus *Phlebiopsis gigantea* is employed to control *Heterobasidion annosum*, a fungal pathogen that generates rots in freshly cut stumps of pine trees and can disperse to intact trees via root-to-root contact. Phlebiopsis spores are painted on tree stumps or mixed into chainsaw lubricant. The fungus encompasses the same tissues as Heterobasidion and outcompetes it, but does not cause damage to the trees (Nega, 2014). *S. frugiperda* is an important crops pest that is hazardous and hard to control all over the world. Because fungal diseases affect more than 60% of insect deaths in nature, there have been numerous reports of fungal strains with efficient control potency for *S. frugiperda.* A survey in India discovered a large number of *S. frugiperda* infected with Metarhizium riley. Gomez *et al.* (2018) discovered that one of 97 *Beauveria bassiana* isolates from Mexico is 97% lethal to *S. frugiperda* second instar larvae. Trichoderma species have also been tested as BCA and used to control a variety of plant diseases as an alternative to synthetic pesticides. Biocontrol systems of Trichoderma are driven by the stimulation of multiple mechanisms, either indirectly through competition for space and nutrients, promoting plant growth and defensive mechanisms, and antibiosis, or directly through mycoparasitism. They can be found in both rhizospheric and non-rhizospheric soils, and they have endophytic relationships with many plants (Morais *et al.,* 2022).

9.4 CONCLUSION AND FUTURE PERSPECTIVES

This chapter explored biofertilizers as a green technology to improve crop production and soil health. Biofertilizers derived from microbes can enhance nutrient availability and revive soil fertility, promoting sustainable agriculture. While laboratory experiments are crucial for selecting potent biofertilizer candidates, field application requires further research. Educating farmers about the benefits of biofertilizers is essential for wider adoption. Developing persistent inoculants and integrating biological pest control methods are future directions for creating a more cost-effective and eco-friendly agricultural system.

ACKNOWLEDGMENTS

The authors are very much grateful for the research grant provided by the DST-SERB-ECRA scheme major research project sanctioned vide letter no. DST/SERB/

ECR/000446, under the Govt. of India. The chapter is also an outcome of the financial support from DST-INSPIRE (IF 190795) and RUSA-Phase 2.0 scheme. We acknowledge the funds from DST-PURSE, Govt. of India.

REFERENCES

Akhtar, S. S., Mekureyaw, M. F., Pandey, C., & Roitsch, T. (2020). Role of cytokinins for interactions of plants with microbial pathogens and pest insects. *Frontiers in Plant Science*, *10*, 1777.

Anand, G., Gupta, R., Marash, I., Leibman-Markus, M., & Bar, M. (2022). Cytokinin production and sensing in fungi. *Microbiological Research*, 262, 127103.

Ananya, A. K., & Ahmad, I. Z. (2014). Cyanobacteria 'the blue green algae' and its novel applications: A brief review. *International Journal of Innovation and applied studies*, *7*(1), 251.

Anjaiah, V., Cornelis, P., & Koedam, N. (2003). Effect of genotype and root colonization in biological control of fusarium wilts in pigeonpea and chickpea by Pseudomonas aeruginosa PNA1. *Canadian Journal of Microbiology*, *49*(2), 85–91.

Arnesen, J. A., Jacobsen, I. H., Dyekjaer, J. D., Rago, D., Kristensen, M., Klitgaard, A. K., & Borodina, I. (2022). Production of abscisic acid in the oleaginous yeast Yarrowia lipolytica. *FEMS Yeast Research*, *22*(1), foac015.

Basu, A., Prasad, P., Das, S. N., Kalam, S., Sayyed, R. Z., Reddy, M. S., & El Enshasy, H. (2021). Plant growth promoting rhizobacteria (PGPR) as green bioinoculants: recent developments, constraints, and prospects. *Sustainability*, *13*(3), 1140.

Berruti, A., Lumini, E., Balestrini, R., & Bianciotto, V. (2016). Arbuscular mycorrhizal fungi as natural biofertilizers: Lets benefit from past successes. *Frontiers in Microbiology*, *6*, 1559.

Bhat, T. A., Ahmad, L., Ganai, M. A., & Khan, O. A. (2015). Nitrogen fixing biofertilizers; Mechanism and growth promotion: a review. *Journal of Pure and Applied Microbiology*, *9*(2), 1675-1690.

Bhuvaneshwari, K., & Singh, P. K. (2015). Response of nitrogen-fixing water fern Azolla biofertilization to rice crop. *3 Biotech*, *5*, 523-529.

Bonaterra i Carreras, A., Badosa Romano, E., Daranas Boadella, N., Frances Ortega, J., Rosello Prados, G., & Montesinos Segui, E. (2022). Bacteria as biological control agents of plant diseases. *Microorganisms*, 10(9), 1759.

Chaudhary, S., & Goyal, S. (2019). Sulphur oxidizing fungus: A review. *Journal of Pharmacognosy and Phytochemistry*, *8*(6), 40-43.

Chaudhary, S., Tanvi, R. D., & Goyal, S. (2019). Different applications of sulphur oxidizing bacteria: A review. *International Journal of Current Microbiology and Applied sciences*, *8*(11), 770-778.

Chittora, D., Meena, M., Barupal, T., Swapnil, P., & Sharma, K. (2020). Cyanobacteria as a source of biofertilizers for sustainable agriculture. *Biochemistry and Biophysics Reports*, *22*, 100737.

Compant, S., Duffy, B., Nowak, J., Clement, C., & Barka, E. A. (2005). Use of plant growth-promoting bacteria for biocontrol of plant diseases: Principles, mechanisms of action, and future prospects. *Applied and Environmental Microbiology*, *71*(9), 4951-4959.

Debasis, M., Snezana, A., Panneerselvam, P., Manisha, S. A., Tanja, V., Ganeshamurthy, A. N., & Divya, J. (2019). Plant growth promoting microorganisms (PGPMs) helping in sustainable agriculture: Current perspective. *International Journal of Agricultural Sciences and Veterinary Medicine, Bd*, *7*(2), 50-74.

Desai, S. A. (2017). Isolation and characterization of gibberellic acid (GA3) producing rhizobacteria from sugarcane roots. *Bioscience Discovery*, *8*(3), 488-494.

Egamberdieva, D., Wirth, S. J., Alqarawi, A. A., Abd_Allah, E. F., & Hashem, A. (2017). Phytohormones and beneficial microbes: Essential components for plants to balance stress and fitness. *Frontiers in Microbiology*, *8*, 2104.

El Semary, N. A. H., Alouane, M. H. H., Nasr, O., Aldayel, M. F., Alhaweti, F. H., & Ahmed, F., 2020. Salinity stress mitigation using encapsulated biofertilizers for sustainable agriculture. *Sustainability*, 12(21), 9218.

Elavarasi, P., Yuvaraj, M., & Gayathri, P. (2020). Application of bacteria as a prominent source of Biofertilizers. *Biostimulants in Plant Science*, 89825, 1-8.

Eljounaidi, K., Lee, S. K., & Bae, H. (2016). Bacterial endophytes as potential biocontrol agents of vascular wilt diseases - Review and future prospects. *Biological Control*, *103*, 62-68.

Fukami, J., Cerezini, P., & Hungria, M. (2018). Azospirillum: Benefits that go far beyond biological nitrogen fixation. *AMB Express*, *8*(1), 73.

Garcia-Fraile, P., Menendez, E., & Rivas, R. (2015). Role of bacterial biofertilizers in agriculture and forestry. *Aims Bioengineering*, *2*(3), 183-205.

Garlapati, D., Chandrasekaran, M., Devanesan, A., Mathimani, T., & Pugazhendhi, A. (2019). Role of cyanobacteria in agricultural and industrial sectors: An outlook on economically important byproducts. *Applied Microbiology and Biotechnology*, *103*, 4709-4721.

Gomez, I., Ocelotl, J., Sanchez, J., Lima, C., Martins, E., Rosales-Juarez, A., & Soberon, M. (2018). Enhancement of Bacillus thuringiensis Cry1Ab and Cry1Fa toxicity to Spodoptera frugiperda by domain III mutations indicates there are two limiting steps in toxicity as defined by receptor binding and protein stability. *Applied and Environmental Microbiology*, *84*(20), e01393-18.

Grzesik, M., Romanowska-Duda, Z., & Kalaji, H. M. (2017). Effectiveness of cyanobacteria and green algae in enhancing the photosynthetic performance and growth of willow (Salix viminalis L.) plants under limited synthetic fertilizers application. *Photosynthetica*, *55*, 510-521.

Hamed, S. M., El-Gaml, N. M., & Eissa, S. T. (2022). Integrated biofertilization using yeast with cyanobacteria on growth and productivity of wheat. *Beni-Suef University Journal of Basic and Applied Sciences*, *11*(1), 112.

Ingle, K. P., & Padole, D. A. (2017). Phosphate solubilizing microbes: An overview. *International Journal of Current Microbiology and Applied Sciences*, *6*(1), 844-852.

Jan, Z., Ali, S., Sultan, T., Wasiullah, W., & Ahmad, W. 2017. The role of cyanobacteria in availability of major plant nutrients and soil organic matter to rice crop under saline soil condition. *Sarhad Journal of Agriculture*, *33*(4), 566-572.

Kamran, S., Shahid, I., Baig, D. N., Rizwan, M., Malik, K. A., & Mehnaz, S. (2017). Contribution of zinc solubilizing bacteria in growth promotion and zinc content of wheat. *Frontiers in Microbiology*, *8*, 2593.

Karthikeyan, A. (2016). Frankia strains for improving growth, biomass and nitrogen fixation in Casuarina equisetifolia seedlings. *Journal of Tropical Forest Science*, 28(3), 235-242.

Khan, M. S., Zaidi, A., & Wani, P. A. (2009). Role of phosphate solubilizing microorganisms in sustainable agriculture - A review. *Sustainable Agriculture*, 27 , 29-43.

Kumar, N. (2016). Effect of algal bio-fertilizer on the Vigna radiata: A critical review. *International Journal of Engineering Research and Applications*, *6*(2), 85-94.

Lahlali, R., Ezrari, S., Radouane, N., Kenfaoui, J., Esmaeel, Q., El Hamss, H., & Barka, E. A. (2022). Biological control of plant pathogens: A global perspective. *Microorganisms*, *10*(3), 596.

Legein, M., Smets, W., Vandenheuvel, D., Eilers, T., Muyshondt, B., Prinsen, E., & Lebeer, S. (2020). Modes of action of microbial biocontrol in the phyllosphere. *Frontiers in Microbiology*, 11, 1619.

Mabrouk, Y., Hemissi, I., Salem, I. B., Mejri, S., Saidi, M., & Belhadj, O. (2018). Potential of rhizobia in improving nitrogen fixation and yields of legumes. *Symbiosis*, *107*(73495), 1-16.

Madhushan, K. W. A., Herath, B. M. M. D., Karunarathna, S. C., & Yapa, P. N. (2021). Role of arbuscular mycorrhizal fungi in sustainable crop production and forestry in Sri Lanka - A review. *Studies in Fungi*, *6*(1), 437-449.

Maheshwari, R., Bhutani, N., & Suneja, P. (2019). Screening and characterization of siderophore producing endophytic bacteria from Cicer arietinum and Pisum sativum plants. *Journal of Applied Biology and Biotechnology*, *7*(5), 7-14.

Mazhar, S., & Hasnain, S. (2011). Screening of native plant growth promoting cyanobacteria and their impact on Triticum aestivum var. Uqab 2000 growth. *African Journal of Agricultural Research*, *6*(17), 3988-3993.

Mekonnen, H., & Kibret, M. (2021). The roles of plant growth promoting rhizobacteria in sustainable vegetable production in Ethiopia. *Chemical and Biological Technologies in Agriculture*, *8*(1), 1-11.

Mohan, A., & Kumar, B. (2019). Plant growth promoting activities of cyanobacteria growing in rhizosphere of agriculturally fertile soil. *IOSR Journal of Biotechnology and Biochemistry*, *5*(4), 28-36.

Morais, E. M., Silva, A. A. R., Sousa, F. W. A. D., Azevedo, I. M. B. D., Silva, H. F., Santos, A. M. G., ... & Araujo, F. D. D. S. (2022). Endophytic Trichoderma strains isolated from forest species of the Cerrado-Caatinga ecotone are potential biocontrol agents against crop pathogenic fungi. *PLoS One*, *17*(4), e0265824.

Naqqash, T., Malik, K. A., Imran, A., Hameed, S., Shahid, M., Hanif, M. K., & Van Elsas, J. D. (2022). Inoculation with Azospirillum spp. acts as the liming source for improving growth and nitrogen use efficiency of potato. *Frontiers in Plant Science*, *13*. 929114

Nega, A. (2014). Review on concepts in biological control of plant pathogens. *Journal of Biology, Agriculture and Healthcare*, *4*(27), 33-54.

Nosheen, S., Ajmal, I., & Song, Y. (2021). Microbes as biofertilizers, a potential approach for sustainable crop production. *Sustainability*, *13*(4), 1868.

Olanrewaju, O. S., Glick, B. R., & Babalola, O. O. (2017). Mechanisms of action of plant growth promoting bacteria. *World Journal of Microbiology and Biotechnology*, *33*, 1-16.

Onyia, C., Okoh, M. A., & Okoh, I. (2020). Production of plant growth-promoting bacteria biofertilizer from organic waste material and evaluation of its performance on the growth of corn (Zea mays). *American Journal of Plant Sciences*, 11(02), 189-200.

Pandit, M. A., Kumar, J., Gulati, S., Bhandari, N., Mehta, P., Katyal, R., & Kaur, J. (2022). Major biological control strategies for plant pathogens. *Pathogens*, *11*(2), 273.

Phour, M., & Sindhu, S. S. (2020). Amelioration of salinity stress and growth stimulation of mustard (Brassica juncea L.) by salt-tolerant Pseudomonas species. *Applied Soil Ecology*, *149*, 103518.

Praharasti, A. S., Kusumaningtyas, A., & Helbert, S. U. (2012). Effects of bio-fertilizer and vesicular-arbuscular mycorrhiza (VAM) application on growth and productivity of sweet-corn crop (Zea mays Saccharata). *Balancing Efforts on Environment Usage in Economy and Ecology*, 153, 1-8.

Righini, H., Francioso, O., Martel Quintana, A., & Roberti, R. (2022). Cyanobacteria: a natural source for controlling agricultural plant diseases caused by fungi and oomycetes and improving plant growth. *Horticulturae*, *8*(1), 58.

Roberti, R., Galletti, S., Burzi, P. L., Righini, H., Cetrullo, S., & Perez, C. J. B. C. (2015). Induction of defence responses in zucchini (Cucurbita pepo) by Anabaena sp. water extract. *Biological Control*, *82*, 61-68.

Sadhana, B. (2014). Arbuscular Mycorrhizal Fungi (AMF) as a biofertilizer - A reviewInternational Journal of Current Microbiology and Applied sciences, *3*(4), 384-400.

Samada, L. H., & Tambunan, U. S. F. (2020). Biopesticides as promising alternatives to chemical pesticides: A review of their current and future status. OnLine Journal of Biological Sciences, *20*, 66-76.

Seenivasagan, R., & Babalola, O. O. (2021). Utilization of microbial consortia as biofertilizers and biopesticides for the production of feasible agricultural product. *Biology*, *10*(11), 1111.

Singh, I. (2018). Plant Growth Promoting Rhizobacteria (PGPR) and their various mechanisms for plant growth enhancement in stressful conditions: A review. *European Journal of Biological Research*, *8*(4), 191-213.

Singh, R., Parihar, P., Singh, M., Bajguz, A., Kumar, J., Singh, S., & Prasad, S. M. (2017). Uncovering potential applications of cyanobacteria and algal metabolites in biology, agriculture and medicine: Current status and future prospects. *Frontiers in Microbiology*, *8*(2017), 515.

Sneha, S., Anitha, B., Sahair, R. A., Raghu, N., Gopenath, T. S., Chandrashekrappa, G. K., & Basalingappa, M. K. (2018). Biofertilizer for crop production and soil fertility, *Academia Journal of Agricultural Research*, 6(8): 299–306

Subramanium, N., & Sundaram, L. (2020). Siderophore producing Pseudomonas spp. isolated from rhizospheric soil and enhancing iron content in Arachis hypogaea L. plant. *Journal of Agricultural Technologies*, *16*, 429-442.

Sultana, S., Alam, S., & Karim, M. M. (2021). Screening of siderophore-producing salt-tolerant rhizobacteria suitable for supporting plant growth in saline soils with iron limitation. *Journal of Agriculture and Food Research*, *4*, 100150.

Thambugala, K. M., Daranagama, D. A., Phillips, A. J., Kannangara, S. D., & Promputtha, I. (2020). Fungi vs. fungi in biocontrol: An overview of fungal antagonists applied against fungal plant pathogens. *Frontiers in Cellular and Infection Microbiology*, *10*, 604923.

Vaid, S. K., Kumar, B., Sharma, A., Shukla, A., & Srivastava, P. (2014). Effect of Zn solubilizing bacteria on growth promotion and Zn nutrition of rice. *Journal of Soil Science Plant Nutrition*, *14*, 889-910.

Yanti, Y., & Hamid, H. (2021, March). Development of the PGPR and cyanobacteria consortium for growth promotion and control Ralstonia syzigii subsp. indonesiensis of tomato. In *IOP Conference Series: Earth and Environmental Science* (Vol. 709, No. 1, p. 012085). IOP Publishing.

10 Scientific Validation and Techniques for Understanding Natural and Organic Farming Perspective

Shivanshu Garg, Pooja Nain, Pankaj Kumar, and Himanshu Punetha

10.1 INTRODUCTION

Since the inception of cultivation practices, agriculture has been a moral commitment to sustain future generations, evolving from traditional methods to modern techniques driven by scientific understanding. Traditional agriculture, rooted in basic practices, gave way to uniform procedures of sowing, ploughing, irrigation, and harvesting, relying on intuitive phenotypic observations. However, the advent of scientific progress led to the emergence of modern agriculture, characterized by hi-tech machinery and chemical inputs aimed at boosting productivity, albeit at the cost of soil health. Organic and natural farming emerged as sustainable alternatives, prioritizing soil rejuvenation and ecological balance over commercial gains. While conventional agriculture feeds the growing global population, its reliance on chemical inputs poses a threat to soil fertility and environmental stability. Therefore, the validation of farming methods through scientific inquiry is imperative for the promotion of sustainable practices (Willer et al. 2021).

10.2 FOUNDATIONS OF CONCEPT

The concepts behind every technique developed so far in agriculture pertain to its value to experimental analysis and reproducibility of the experiment at different locations. The growth of a crop is abided by the rules of nature about different locations. Although the chemistry behind the metabolism of plants is universally alike, the economic benefits are meant to be considered at the same time along with enough productivity not to range just from hand to mouth. Currently, various methodologies are available for the generation of crop produce in the horticulture and agriculture sector whose producing potential has to be matched by natural and organic farming with

DOI: 10.1201/9781003343707-10

respect to conventional farming. Here comes the role of integrated farming, which is in itself a version of conventional farming governed by low-scale use of chemicals and fertilizers. However, this approach is not as good as natural and organic farming whereby inputs are biological agents themselves only. Nonetheless, the path from cultivation to productivity contains a number of steps. Each step has its own value. To add to that value conventional farming, natural and organic farming are the motors driving the vehicle. Conventional farming utilizes more fuel and gives better productivity. But the cost of that fuel has to be met by nature by imparting its biological values. As we all know, as long as nature is healthy, humans are alive and healthy. To maintain human health, we have to keep nature and its biological components alive. Scientific methodologies and techniques have to be implemented in such a way that they fulfill the needs of farmers with ease, and farmers on a large scale can utilize these complex methodologies drawn into simple concepts of the utilization of the technique laid by scientific foundation. Several biological components of nature as driving agents of natural and organic farming exist and their understanding at the level of scientific mechanisms is essential for farmers to get economic benefits by managing crop and land.

10.3 GLOBAL ACCEPTANCE

Mankind has realized the importance of organic agriculture and is currently practicing it in 188 countries where agricultural land covering 72.3 million hectares (mha) is organically managed under the collective hands of more than 3.31 million farmers. The realization of conserving biodiversity hotspots from harmful chemicals has made Australians practice it at 35.70 mha. The concerns have also been shown by Argentina and Spain with 3.80 mha and 2.85 mha land demarcated for use of organically produced products, respectively. With increasing awareness of health among people all around the world, the mass population is keen to consume natural crop produce, free of any pesticides and thus has developed interest in adoption of scientific methods and techniques that may enhance natural farming at domestic and commercial level. The Research Institute of Organic Agriculture (German: Forschungsinstitut für biologischenLandbau, FiBL established 1973), the pioneering institute at an international level for the promotion of organic agriculture, did a collective survey of organic farmland available worldwide and showed the increase in organic farmland from 0.13 mha to 1.1 mha from 1973 to 2019. Not only have people produced crops organically, but they have also aimed for balance among other agricultural aspects like beekeeping (as pollination agents), aquaculture (to meet protein demands), and protecting forest areas (to maintain biodiversity) (FiBL 2019). The communion of greenhouse gases with food generation is a core representation of agriculture where, on one handit meets the survival demands and on the other hand its implementation comes with adverse consequences for climate (UNF 2021).

With increasing concentration of ozone layer-depleting gases in atmosphere, the entrance of harmful UV-B radiations into the earth's atmosphere has caused human health issues such as tumor formation. Moreover, harmful UV-B radiation distorts the genetic architecture of crops by chromosomal mutation, thus impacting human and crop health at the same time. The United Nations Framework Convention on Climate Change marked all the abovementioned gases as potential candidates for the greenhouse effect and also stated that N_2O and CH_4 are produced majorly as by-products of agricultural services (SEEG 2022).

10.4 BIOTECHNOLOGICAL CONSIDERATIONS

An antagonistic kind of relationship has deprived concomitant behavior between organic agriculture and biotechnological tools (Husaini and Sohail 2018). The critics of biotechnology have always opposed organic agriculture and vice versa. A false belief is rooted deep down in smallholder farmers about organically generated crops and biotechnology use (Purnhagen et al. 2021). Years of heated debate and countless meetings have resulted in the non-allowance of transgenic crops according to the needs of farming suitability. Even with the advent of rising biotechnology tools in developing the crops, so far only a marginal decrease in pesticide consumption has been noticed. This has led farmers to think that there is little to no use for biotechnological practices (IFOAM 2016). Contrary to this, a major biotechnological intervention, CRISPR-Cas 9 technology, abbreviated as, "Clustered regularly interspaced short palindromic repeats-Caspase 9," allows modifications at ends of knock-in/knock-out coders of genome (Doudna and Charpentier 2014).

10.4.1 Modern Breeding: Approaches to Organic Farming

The transfer of a piece of DNA (a gene) in order to modify the genome of a concerned organism, from one to another, is termed genetic modification (Schaart et al. 2015). Two major universal ways exist that bring about the change in the genome of an organism:transgenesis and mutagenesis. Transgenesis involves the implementation of DNA segments from one species to the other species falling under the same genera and the recipient organism expresses the trait character of the source (gene) organism. Mutagenesis does not involve the exogenous gene segment transfer. Rather, the existing genome is mutated at coding sites in order to give an altogether distinct phenotypic effect (Abiri et al. 2016). Newer biotechnological techniques may be applied to the breeding of organically produced crops without the involvement of transgenesis through CRISPR-Cas9; cisgenesis (in which the DNA segment coding for protein synthesis with native promoter elements, introns, and stop codons without the need of T-border elements); and intragenesis (for which a sexually compatible group is needed in order to form new DNA segments, coding for new set of instructions) (Farmeurope et al. 2017).

To begin with, several examples have been laid out by scientists in which they show how gene manipulation can be done within the plant without exogenous DNA coding segments needing to be taken into consideration. Moreover, experimental model organisms for plants like *Arabidopsis thaliana* (Thale Cress Grass) and *Nicotiana tobaccum* (tobacco) have been tested before the implementation of CRISPR-Cas9 in crop plants. This concept of biotechnology eradicates the question of involvement of the foreign genes in edible crops as animal-sourced genes were thought to destroy

TABLE 10.1
Crop traits modified through CRISPR-Cas9

S. No.	CRISPR-Cas9 Implemented	Crop-Trait	References
1.	Gene insertion-Gene exchange	Climate-resilient Maize	(Shi et al. 2017)
2.	Gene knockout	a. Browning-resistant Mushroom	(Cai et al. 2022)
3.		b. Enhanced omega-3 fatty acid in germanseasame	(Lyzenga et al. 2019)
4.	Gene disruption	Climate resilient Soyabean	(Waltz 2016)
5.	Targeted fragment deletion	Browning resistant Apple	(Nishitani et al. 2016)
6.	Targeted mutagenesis	GABA (Gamma amino butyric acid) rich Tomato	(Gramazio et al. 2020)

holiness in crops a vegetarian diet (Jansing et al. 2019). Crops sourced out in the market in this manner are listed in Table 10.1.

The production of human insulin in bacteria in 1977 (Itakura et al. 1977) is still a controversial talk of town among critical figures of biotechnology, even though it has saved lives to an impeccable scale. To shed light on the importance of the topic of food security, the following years of 1980 saw the rise in the development of transgenic *Nicotiana tobaccum* (Tobacco), *Zea mays* (Maize), and *Triticum aestivum* (Wheat) in the United States of America with approval of Flavr Savr™tomato, the first conceptualized genetically modified food in 1994 (Kramer and Redenbaugh 1994). The Flavr Savr™tomato has been in use for 29 years and even without any reported issues related to safety and health concern many people still falsely think that it may cause a hazardous effect on health and may even cause death from the use of such transgenic crops. These kinds of false accusations on biotechnological inputs puts food security in danger (Husaini and Sohail 2018).

10.5 FARMING THE CULTIVATION: ORGANIC AND NATURAL

Organic agriculture as cited by terms of the International Federation of Organic Agricultural Movements [IFOAM] (2005) is "A production system that sustains the health of soils, ecosystems, and people." There is strict prohibition of compounds synthesized through chemical means like fertilizers, pesticides, synthetic growth hormones, and transgenic utilities (Treadwell et al. 2015), which are not harmful in every respect. However, in conventional farming, the usage of compounds like lime sulfur (responsible for a decrease in soil pH), boric acid (linked to a decrease in soil pH and harmful to the beneficial microbe of soil), sodium per-carbonate (responsible for blocking ion channels in plants), rotenone (links with Parkinson's development and blocks electron transport chain), and bromo-methane (responsible for death of pollinating insects) are still in use at global scale to enhance productivity (Seaman and Sideman 2013; Ruddock 2016).

10.5.1 Global Promoters of Organic Farming

The International Federation of Organic Agriculture Movements (IFOAM), established in 1972, defined principles of organic farming as C.H.E.F. (Care, Health, Ecology, and Fairness) with respect to the mode of farming done by humans. Their vision and mission is to"foster the broad adoption of truly sustainable agriculture, value chains, and consumption in line with the principles of organic agriculture." The Research Institute of Organic Agriculture, also known as FiBL, founded in 1973, is facilitating the development of sustainable agriculture in continents all over the world ranging from Asia to Africa, Latin America to Eastern Europe working intensively at the grass-root level with local bodies. Its coordinated long-term trials include research on organic farming systems, improving the efficacy of economic outputs. FiBL made its worldwide commitment towards understanding scientific values pertaining to agricultural practices globally with a clearly defined vision in 11 missions commencing from 2018 and going up to 2025 (Table 10.2) (FiBL 2019).

Likewise, to deliver the goal of self-sufficiency through securing education and health for rural communities, the International Institute of Rural Reconstruction (IIRR), a non-profit organization, founded in 1960, offers programs concerning safety of the environment and ensuring food security for the world's population. Their mission is to "Empower rural people globally and gain them the socio-economic equities by their own creativeness and leadership quality generation in every aspect of life." In order to sustain life, IIRR has come out boldly on at global scale (FiBL 2019).

10.5.2 Implementing the Techniques: Natural and Organic Farming

These are the skills of learning the techniques and implementing them at field levels in order to gain sustainability to proceed with natural and organic farming. Local factors, including the availability of raw materials and resources, play a critical role in determining the success of these farming practices. It is the background knowledge that if a farmer has dealt with in-depth, the implementation of tasks would be easy to achieve. The deprivation of nutrients in the soil is the most important concern a farmer has to look into before going for natural and organic farming practices. The presence of a beneficial microbial community is the factor a soil analyst starts with when defining good field conditions from at farmer's perspective.

10.5.2.1 The Pioneering Steps

Farm analysis comprises a number of factors to consider. The distribution of crops suited to agro-ecological regions is considered as a priority. Meanwhile, getting soil analysis done and gathering information about its erosion level, presence/absence of contaminants like heavy metal ions, the abundance/lack of organic matter content and micronutrients along with soil texture and water-holding capacity are key points

TABLE 10.2
FiBL missions along with objectives

S. No.	Mission	Objective
1.	Plant breeding for agro-ecological and organic cropping systems	Breeding crops according to geographical locations and to test the resilience of different cropping systems
2.	Use of digital technology in modern organic farming	To cover lag in technological inputs responsible for crop production for the sake of local farmers
3.	Organic crop protection for the entire farming sector	Provision of insurance claims
4.	Optimum nitrogen management for crops, livestock and the environment	Soil health analysis and irrigation channels generation in water deprived areas
5.	Animal welfare, productivity and sustainability – solving the conflict	Sustaining practices of aquaculture and cattle rearing
6.	Climate change and organic farming	Effective in lowering green house gas emissions
7.	Improving the sustainability of agriculture and the value chain	Supplying commodities in market for economic benefits of farmers concerned
8.	Making value chains transparent and securing the long-term trust of consumers	Multiple distributors and local field knowledge
9.	Organic advisory services of the future – the digital advisor and the personal discussion	Mass gathering programs concerned with awareness about organic farming
10.	Organic farming's contribution to sustainable development in the countries of the global South	Development of organic farming in countries like India, China, Indonesia, Brazil, Nigeria and Mexico
11.	FiBL campus – Agro-ecology and organic farming need interdisciplinary and trans-disciplinary education and training	The extension of information for management of farming practices at ends of younger generation to save the future of ecology

of analysis. From an irrigation point of view, the average annual rainfall and levels of varying temperatures in response to frost and humidity are important considerations. The availability of mulching and manuring material with access to local labor and markets are helpful in building up a keen place for natural and organic farming practices. For example, "Fanya Juu"(a Kishwahili language word meaning "throw it upwards") terraces are enacted upon an encounter with high soil erosion levels. It includes digging up the trenches alongside contours simultaneously forming up the binds by soil dug from the trench. Further *Pennisentum purpureum* (Napier Grass) rooting stabilizes these bunds along with forest trees like *Pinus roxburgii* (chirpine) and *Eucalyptus deltoides* (eucalyptus). Within a year, in semi-arid areas where Fanya

Juu is practiced, formation of bench terraces becomes fruitful along with high efficiency of soil water conservation. The addition of green manures further supports the yield and growth parameters of plants utilized in natural and organic farming practices. Similarly, when strong wind and high solar heat influence are high, green manures serve their purpose of maintaining the cool and moist soil micro-climate. This further adds to their advantage of use in natural and organic farming practices. The sure short points to be kept in mind by a farmer about managing the farmland are the biodiverse nature of the farm, the fertile nature of soil, minimal pest attacks, and a destined market for the successful sale of produce (IFOAM 2003).

Any of the crops planted in the field are susceptible to damage until procured carefully. However, in that category there are also two kinds. One is leguminous crops and cereals, having the natural capability of robustness, costing little to produce, and utilizing moderate nutrient demands. Peas, beans, sorghum, millets, and maize fall under this category. Second is short-term crops like vegetables with a more susceptible nature of being attacked by pests and susceptibility to abiotic stresses. So whenever a farmer starts practicing natural and organic farming, the first step should be taken at a larger scale for the first kind and at a smaller scale for the second kind. Moreover, green manure is best suited for the long-term fertile nature of the soil and is equally good for both of the above-categorized crops. However, to bring resilience to the field in terms of preventing soil erosion, water erosion, and nutrient loss, mulching is a crucial part of farming practices in conventional, natural, and organic agriculture. It not only keeps the soil moist but also improves the infiltration potential of rain and irrigation water due to which there is no formation of crust and pores for water entry remain open. Suppression of weed growth, oppression of extra heat, and provision of nutrient availability are advantageous aspects of mulching. Other than these advantages, the crucial one is the retainment of organic matter in soil due to addition of mulch. This speeds up the process of conversion of organic matter into humus, which ultimately leads to enhanced crop growth irrespective of crop-roration practices (Seaman and Sideman 2013; IFOAM 2003).

10.5.2.2 Scientific Temperament Implementations

Conferred with the highly prestigious Ramon Magsaysay Award, Desikottam Award, and Earth Council Award, Masanobu Fukuoka, a Japanese man, brought about a straw revolution in which he went to grow rice based on mulching practices. He is also known as a pioneer of natural farming, which he described in his terms as "to do nothing farming." This basis differentiated natural and organic farming aspects because in organic farming, tillage is required but according to Fukuoka, the description of natural farming is "no tillage." This means *Trifolium repens* (white clover) and the winter crop of *Secale cereale* (rye) are sown in between before harvesting the rice crop and re-sowing of the rice crop. As rice is harvested forth, the straw is used as a loose layer of mulch. Thereafter, both crops are grown until their harvest, and no extra tillage is required for next sowing season of rice. Hence, the commencement of natural farming paves the less effort and time-saving way for cultivation along with advantages over pest attacks (IFOAM 2003).

Providing optimum water is more profoundly done by planting pits side by of crops known by names of *tassa*in Niger and*zaï*in Burkina Faso. These are hand-dug pits with 20 cm depth and 20 cm diameter filled with water. The roots of the crop find water and grow in that direction, a classical example of geotropism is thus shown by plants. The only problem that might be faced is when soil is dry, a lot of water can be seeped and thus lost. However, moisture is naturally retained in the soil. Over time, these planting pits retain fertility of soil due to accumulation of organic matter and nutrients. Other types of earth bunds that are employed in these farming practices are semi-circular, and are known as half-moon micro-catchments. The main goal of generating them is to rehabilitate degraded dry land. Their embodiment framework is destined for desert margins, and in Sahel (a region in Africa) they are known by the names of "Demilunes." As their structural framework allows them to catch water down a slope, the water gathers there and at that point crops like millets, cowpea, and sorghum are sown. Moreover, these crops have deep roots ranging from 0.8 meter to 1.2 meter in soil. As a result, once the first cropping season is over and the crop is harvested thereafter, the soil structure goes on improving (FiBL 2011).

The conclusive evidence has shown the importance of intercropping and crop rotation with respect to soil fertility and an overall reduction in pest attacks. The soils when regulated with nutrients deposition naturally, and the aspects of natural and organic farming are met. To do this, introducing winter cover crops before late planted crops are sown in favor of nitrogen and organic matter content. Likewise, seedbed preparation becomes easy if winter-killed cover crops like oat-pea are grown before early season crops. Due care should be given to repetition of a single type of crop sowing in cumulative seasons. When a broad range of insect pest attacks becomes prevalent, different crop combinations should be employed in simultaneous sowing seasons. Another consideration to be taken into account about vegetables is to grow spinach, lettuce and pea before harvesting tomatoes. The science behind this is that tomatoes utilize excessive nutrients and can be grown on less nutrient-efficient soil. Similarly, cultivating lettuce and legumes before potatoes enriches soil with extra nitrogen, which is essential for potato growth. In all, all the crops should be grown before potatoes as they utilize much of the available nutrients in soil. All the above points are meant to be considered for natural and organic farming practices (FAO 2011).

10.5.2.3 Pivotal Role of Nature

Only ground natural rock-based minerals (Table 10.3) (IFOAM 2003) are utilized in organic farming practices. However, their solubility does not fall under the control limit of humans. So, their distribution is questionable, and moreover, only antidote doses of minerals one for another are available to reclaim or overcome the previous distribution.

To ward off pests, there should be measures taken into account in the selection of adapted and resistant varieties, preparation of clean seed and planting material, utilizing suitable cropping systems (intercropping and mixed cropping), balancing nutrient management system and organic matter, applying suitable soil cultivation

TABLE 10.3
Natural rock-based minerals in organic farming

S. No.	Fertilizer	Source	Characteristics & Applicability
1.	Rock Phosphate	Pulverized rocks on river beds have phosphorus	Faster adsorption towards other minerals and slower adsorption towards organic matter content in soil
2.	Lime	Limestone, Algal biomass	Essential for maintaining low pH and source of trace elements
3.	Stone Powder	Rocks (Pulverized only)	Fine-scale attributes give more adsorbing capability for heavy elements
4.	Plant Ash	Dried/burned organic matter	Rich sources of Potassium (K) and Calcium (Ca) easily available to plant

methods, using good water management with subsequent selection of optimum planting time and spacing and finally, and using proper sanitation measures like discarding infested/infected plant parts from soil and plant itself. Pre-preparation for organic farming practices includes hot water treatment of seeds to evade any possible chance of diseases like spots (black and red), black rot, and black leg, especially in the case of cruciferous vegetables. In this way, seed-borne pathogens like *Xanthomonas spp., Colletotrichum spp., Alternaria spp., Septoria spp., Pseudomonas spp., and Phoma spp.* are eradicated. The average time for treatment with hot water is 18 minutes with varying temperatures from 47–55°C for broccoli, cauliflower, mustard, pepper, tomato, eggplant, khol rabi, turnip, lettuce, celery, radish, and kale. While for banana suckers and potato tubers, it is 55°C for 10 minutes (Shiva et al. 2004).

10.5.3 Technicalities at the End of Natural Farming

Farming practices are under the direct influence of soil properties. In Indian soils, there is enough quantity of zinc (Zn) and boron (B) like elements for a plant to sustain biochemical activities derived from these. In contrast to this, severe nitrogen (N) deficiency in more than 58% of Indian soils with 35% having moderate deficiency reflects the necessity of N demand to be met for better results in productivity. So if natural farming and organic farming are said to be less productive, it is not the sole fault of these farming practices. These are the scenarios of micro- and macroenvironments in and around the soil, which are the actual deciders of productivity and plant health. In addition to N deficiency, more than 50% of arable land in India has poor phosphorus (P) content (Chaudhary et al. 2015).

10.5.3.1 Componential Significance

Natural farming has its roots deep down in the underlying concepts of formulations known by the name of beejamritha, jeevamritha, angiastra, brahmastra, and neemastra. Of the fivecomponents mentioned above, the first two ensure seed and plant health, respectively. The next three are responsible for imparting disease-resistance properties to plants in natural farming. Three simple functions of Jeevamritha are to ward off pests, make beneficial microbial interaction fluent, and to increase the bioavailability of nutrients to plants already present in the soil.

Standardized procedure for Jeevamritha preparation (per acre of land): (Kumar et al. 2020)

- To 200 litres of water, add 10 Kg (Kilogram) of cow dung and 10 liters of cow urine
- To it add jiggery and pulse flour. The addition of soil is done on purpose to facilitate solid matter
- Incubate for 48 hours (hrs.) at room temperature
- Jeevamritha is ready to be applied for

Predatory actions of insect pests and microbial communities towards seed sown in soil are enacted upon by counteraction of beejamritha as it contains the natural insecticide, fungicide, and bactericide in the form of cow urine and cow dung.

Standardized procedure for Beejamritha preparation (per acre of land): (Kumar et al. 2020)

- Fix 2 Kg of cow dung to a cloth and let it be dipped in cow urine for subsequent 12 hrs. at room temperature
- The next step is to make dry dense concentrated balls by squeezing them with cloth and subsequent addition of 25% of lime to the total mass
- Storage at room temperature for further use
- Apply to plant seeds by proper soaking and further drying before sowing

From time immemorial, *Azadirachta indica* (neem plant) has been a plant of choice against various ailments for humans. Nonetheless, it has far better potential in treating pest-related problems of plants. The use of neem leaves in neemastra makes its name and function evident.

Standardized procedure for Neemastra preparation (per acre of land): (Kumar et al. 2020)

- Make powder of neem leaves and add sufficient water to make a paste weighing 5 Kg
- To it add 3 Kg Cow dung and 15 lt cow urine
- Handful of soil is added to make a solid slurry
- Preparation is ready to be applied to plant parts infected with disease and displaying symptoms

Brahmastra carries out the broad range eradication of pests. It utilizes five different kinds of plant leaves (walnut, chili, mango, custard apple, and neem) instead of a single one (only neem leaves as in neemastra). The rest of the formulation is the same as of neemastra. Likewise, agniastra employs the use of garlic, chilli, tobacco, and neem leaves. The variation in utilization of raw components makes agniastra effective against hard exoskeleton possessing insect pests (majorly beetles like stem

TABLE 10.4
Mulching practices employed in natural farming

S. No.	Mulching Type	Details of Function
1.	Soil Carrying Mulch	Generated by not disturbing already present topsoil, whereby it serves function of moisture optimization and aeration in soil
2.	Straw Carrying Mulch	Generated by dried biomass of residues derived from rice, maize and wheat mainly. The dry matter serves as the basis of the multiplication of soil microbes
3.	Live Material Mulch	Generation of proper channels of nitrogen fixation, phosphorus, and sulfur cycles by monocotyledonous (rice and wheat) and dicotyledonous (pulses) crops when planted in multi-planting systems

and root borers), and on the other hand, brahmastra wards off technically all kinds of insect pests. Any farming practice in agriculture carries mulching practices. It is the kind and nature of mulch used in natural farming (Table 10.4) (Kumar et al. 2020)that differentiates it from aspects of conventional farming.

10.6 MICRO-ENVIRONMENT OF FARMING

A zone inhabiting biochemical secretions from plant roots present within soil is termed as a rhizosphere. It is the place where biological interactions are organized to transform a living system out of it. These interactions might be influential in both positive and negative terms (Kumar et al. 2015).Several prominent examples exist in nature thatdescribe the behavior of plant secretions in front of microbial communities and biochemical pathways for these chemicals illustrate the amazing set of interactions. Like internodal growth has been a key aspect for cereals productivity from time immemorial. This sort of task is accomplished by *Azotobacter chroococcum,* which assists cereals in biostimulation by means of producing gibberellins. PGPRs (plant growth-promoting rhizobacteria) work in symbiotic associations with the plants, both specifically and non-specifically. The principal work of action includes an increase in levels of cytokinin, antioxidants, over-expression of ACC deaminase gene, increase in the production of volatiles, and IAA (indole acetic acid) metabolites. The volatiles thus imparted lead to increase in Na^+ translocation along with improved measures of tolerance of plant towards drought and salt stress. Also, PGPRs bring about changes in root morphology heading towards improved transportation of metabolites. This contributes overall to the elevation in nitrate and phosphate molecules, thus improvising fertility re-storage in the plant concerned (Zhang et al. 2019).

All important edible food crops produced by mankind such as maize *(Zea mays),* wheat *(Triticum aestivum L.),* and rice *(Oryza sativa)* are biofertilized by microbial community, of which *Azospirillum brasilense* is a notable member as it aids in phosphate solubilization, helping plants to build up their cell wall and generate nucleotides, sugars, and other important biochemical services about which phosphate is known for (Lucy et al. 2004). Other than biostimulation and biofertilization, the PGPRs also play a key role in bioprotection and bioremediation. For example, *Bacillus subtilis* produces lipopeptides and broad substrate-specific catalases. Moreover, the same species is also responsible for degrading xenobiotics and allelochemicals in a soil's micro-environment. Targeting ammonia synthesis for biofertilization has been a key aspect of *Bacillus subtilis* during maize cultivation (Ouhaibi-Ben Abdeljalil et al. 2016; Ait-Kaki et al. 2014; Tahir et al. 2017). Certain ornamental plant species are known to be biostimulated by *Staphylococcus saprophyticus* through the manufacturing of IAA through auxin gene induction (Manzoor et al. 2019).

10.6.1 Microbiological Considerations

The beneficial microbes create nutrient channeling between plant and soil thereby increasing the soil's fertility and crop productivity. These microbes are biologically diverse performing different functions according to their diverse nature. They are called biofertilizers (Table 10.5). Implementing the use of biofertilizers in place of synthetic chemical fertilizers at a larger scale will eventually open up new gates for industrialized businesses. Imparting health to climate and plants, biofertilizers are thus ecosystem-saving agents (Mishra and Dash 2014).

10.6.1.1 Biofertilizers: The Regulators in Action

The manipulation of the human mind towards the significant aspect of biofertilizers involves steps like isolation and characterization of microbes intended to serve as biofertilizers in the field. The harmful pathogenic microbes also inhabit the same soil, so this step becomes crucially important from the farming perspective. The metabolites produced by the microbial community are checked onto the plants in controlled experiments and their use is further demarcated for betterment of plant health. Basically, plants utilize biofertilizers to maintain the biochemical dynamism of life (Salar et al. 2017). XIn addition to the supply of nutrients to plants from soil, biofertilizers are also engaged in the suppression of soil-borne pathogenic diseases in crops. The habitual place of residing the plant of biofertilizers is the root nodule part, wherein they extract electrons from molecules of di-nitrogen to get it modified into ammonia, which by complex metabolic pathways serve as the basis for biochemical synthesis of nitrogen-containing biomolecules like proteins and associative factors of vitamins (Nyoki and Ndakidemi 2018; Belhadi et al. 2018).

The potential application of algal biofertilizer includes reducing synthetic nitrogen fertilizer in land already present in order to reclaim the original natural soil (Subramanian et al. 1994). The enhancements in seed germination rate (Shukla and Gupta 1967) and soil qualitative factor enhancement (Delattre et al. 2016), serving as the source of phytohormones like auxins, cytokinins, jasmonic acid, ethylene, brassinosteroids, and abscisic acid (Romanenko et al. 2015) are accomplished by application of algal biofertilizers in field. The prominent examples

TABLE 10.5
Biofertilizer types and prominent examples

S. No.	Name of Biofertilizer with Characteristic Properties	Prominent Examples	References
1.	Free-living Nitrogen Fixers	*Clostridium spp., Klebsiella spp., Desulfovibrio spp., Anabaena spp., Rhodospirillum spp.*	(Singh et al. 2014; Unkovich et al. 1997)
2.	Symbiotic Nitrogen Fixers	*Azospirillumbrasilense, Azospirilliumamazonense, Acetobacter diazotrophicus*	(Singh et al. 2014)
3.	Associative Nitrogen Fixers	*Frankia, spp., Trichodesmium spp., Mesorhizobium spp., Allorhizobium spp., Sinorhizobium spp.*	(Singh et al. 2014; Rodriguez and Fraga 1999)
4.	Phosphate Solubilizing	*Bacillus megaterium, Pseudomonas striata, Rhizoctonia solani, B. polymyxa*	(Rodriguez and Fraga 1999; Chen et al. 2006)
5.	Potassium Solubilizing	*Paenibacillusglucanolyticus*	(Singh et al. 2014)
6.	Phosphate Mobilizing	*Pezizellaericae, Amanita spp. Laccariaspp, Boletus Spp,*	(Singh et al. 2014; Unkovich et al. 1997)
7.	Silicate Solubilizing	*Thiobacillus thioxidans,* Bacillus globisporus	(Sheng et al. 2008)
8.	Plant Growth-Promoting Rhizobacteria	*Pseudomonas fluorescens, Agrobacterium rhizogenes, Erwinia amylovora*	(Singh et al. 2014)
9.	Azolla and Blue Green Algae Biofertilizer	*Anabaena azollae*	(Soma et al. 2018)

of algal fertilizers includes sources of single-cell proteins like *Spirulina maxima,* which also serve to promote seed germination rate in *Arabidopsis thaliana* (Thale Cress grass) and *Triticum aestivum* (wheat) in particular (Michalak et al. 2016).The survival-favoring biochemical interaction also exists for algal and PGPR biofertilizer species, wherein*Scenedesmus obliquus* (algal counterpart) favors the growth rate of *Rhizobium japonicum* (PGPR strain) (Fingerhut et al. 1984). Diversity in the nature of species has been attributed to ten types of major biofertilizers,which can further be divided into specific nutrient channeling capabilities (left) and species-specific type (right) (Figure 10.1).

Some algal biofertilizers have been found to induce growth and increase yield in specific crops. For example, *Dunaliella salina* was found to improve growth and yield of wheat (El Arroussi and Elbaouchi 2016). Similarly, *Chlorella* spp. and *Spirulina* spp. for potato & pea (Ronga et al. 2019) and *Spirulina platensis* for pepper and beet (Dias et al. 2016). Aside from the promotion of leaf rate, seed germination rate, and root growth rate, a scenario has been observed in nature where a particular

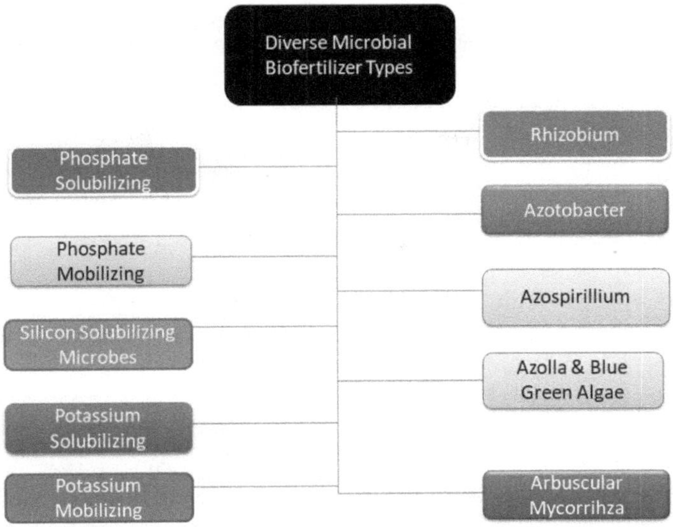

FIGURE 10.1 Diverse microbial biofertilizer types.

type of biofertilizer, *Chlorococcum humicolum,* inhibits both the necrotrophic fungus (*Botrytis cineraria*) in strawberry and also the powdery mildew (*Erysiphe betae*) in turnips and tomato (Kulik 1995).

10.6.1.2 PGPRs and Heavy Metal Elements

Mercury (Hg), arsenic (arsenic), cadmium (Cd), and lead (Pb) are the sort of elements that hamper the flow of nutrients between soil and plants and as well as among the metabolic enroutes through which these nutrients distribute through ion channels of the plant to keep it alive and viable in nature. The reason for prolonged persistence is thought to be attributed by their high atomic weight. Other notable effects of these heavy metallic elements are disturbing the soil's pH and its overall texture (Hamid et al. 2021).The natural presence of microbial populations in those soils where heavy metals have been accumulated is owed to the contribution in the form of symbiotic associations, where both counterparts get the benefit. In this symbiotic association the heterotrophic microbes utilize the carbon and other heavy element content of organic matter provided by the plant and plants utilize harmless biomolecules produced by microbes in the former process, which leads to better health for plants (Zafar-ul-Hye et al. 2013). Prominent examples exist like those of the Brassicaceae family having 3709 species in all, with notable members being the *Brassica napus* and *Brassica juncea* that lower the lead, nickel, and zinc content in soil with symbiotic companions like *Bacillus megaterium* (Reichman 2014) and *Bacillus subtilis* SJ101 (Zaidi et al. 2006), respectively. To add to their advantage, both plants utilize different approaches for the betterment of overall soil health. While the *B. napus* with its symbiotic PGPR *Bacillus megaterium* enhances the dry matter yield of plant and thus improves soil health, the *B. juncea* symbiotic partner *Bacillus subtilis* SJ 101 actually accumulates the nickel and prevents its passage to the plant metabolic system as well as to the soil. Another approach that is utilized by plants and PGPRs is to alleviate the toxic element concentration like that of chromium. It is evident from the data available that *Pisum sativum* working with *Rhizobium sp.* and *Microbacterium sp.*do this bioremediation part (Mishra et al. 2016). The PGPRs have been found to be associated with the manipulation of genetics of plants that helps plants to overcome stress from heavy metals. Notable examples are *Astragalus sinicus* with its PGPR symbiont*Mesorhizobium huakuii* sub sp. *rengei* B3. This microorganism actually enhances the expression of phytochelatin synthase (*PCS*) gene in plants, resulting in an overall increase in the ability of plant cells to bind to cadmium. Then this plant can be uprooted and the other commercial cultivars are thus protected from the harmful devastating effects of Cadmium (Sriprang et al. 2003).

10.7 CONCLUSION AND FUTURE PERSPECTIVES

Natural and organic farming holds life longevity benefits for humans. The implementation of these farming practices requires knowledge about scientifically designed principles. Scientific explanations are well known but complete biochemical mechanism of action is not known. Farming of these practices is economic and simple at the same time. No large labor requirements are needed to be met whether it is farm machinery or mass labor. The pre-treatment of seed and generation of specific nutrient potential can be easily managed at household levels. A single piece of structural framework is not likely to get wasted in these aspects of farming practices. People who are totally unaware of natural farming practices should be imparted knowledge about the usage of easy, reliable and scholastic scientific formulations like jeevamritha and beejamritha. These natural formulations protect nature in the long run and ensure the safety of a natural environment for future generations to come. Another factor that hinders the implementation of natural farming at large scale is the crop productivity, which is much less, and without the input of biotechnology the economic benefits cannot be gained. This raises the concern of food insecurity for the world's ever-increasing population.

Validation of higher case scenarios concerning pest management, soil health management, and water management have been well optimized with the scientific layout of techniques for natural and organic farming. The role played by international organizations for uplifting natural and organic farming has resulted in better ecological frameworks to feed humanity with pure food. The methodological concepts utilized in natural & organic farming are free from health hazards as are commonly encountered in conventional farming. Moreover, health benefits derived from cow dung and cow urine to humans are advantagesin these environments. Whether it is cleansing up of soil by PGPRs involved in sequestering heavy elements or it is pathogen removal from distant plant parts, the role of biofertilizer is a classical example of the biochemical dynamic flow of metabolites between living forms. A deep understanding of interaction biochemistry might help in revolutionizing the field of climate change, crop productivity levels, and crop losses incurred due to pathogen attacks. Due to increased pesticide use, pollutants are increasing widely. With the implementation of natural and organic farming practices these pollutants can be minimized but at the same time crop productivity should not be reduced. The complete understanding of scientific methods involved and their utilization has made clear the beneficial role of natural farming. The ill effects of climate change can be minimized by preservation of natural resources like land, clean air, and water, and all three are enhanced in their utilization by the means of natural farming. This makes natural and organic farming essential in achieving sustainability goals of agriculture and climate simultaneously. To address the challenges of soil erosion and soil fertility, the contribution of these farming practices to our current understanding of scientific techniques needs to be much clearer and more straight-forward.

REFERENCES

Abiri, R., Valdiani, A., Maziah, M., Shaharuddin, N. A., Sahebi, M., Yusof, Z. N., Atabaki, N., and Talei, D. (2016). A Critical Review of the Concept of Transgenic Plants: Insights into Pharmaceutical Biotechnology and Molecular Farming. *Current Issues in Molecular Biology* 18: 21–42.

Ait-Kaki, A., Kacem-Chaouche, N., Ongena,M., Kara-Ali,M., Dehimat, L. and Kahlat, K.2014. In vitro and in vivo characterization of plant growth promoting Bacillus strains isolated from extreme environments of eastern Algeria. *Applied Biochemistry and Biotechnology.* 172: 1735–1746.

Belhadi, D., Lajudie, P., Ramdani, N., Roux, C., Boulila, F., Tisseyre, P., Boulila, A., Benguedouar, A., Kaci, Y. and Laguerre, G. 2018. Vicia faba L. in the Bejaia region of Algeria is nodulated by Rhizobium leguminosarumsv. viciae, Rhizobium laguerreae and two new genospecies. *System & Applied Microbiology*. 41:122–130.

Cai, Z-X., Chen, M-Y., Lu, Y-P., Guo, Z-J., Zeng, Z-H. and Liao, J-H. 2022. Metabolomics and transcriptomics unravel the mechanism of browning resistance in agaricusbisporus. *Plos One*.17: 25–65.

Chaudhary, S.K., Biswas, P.P. and KapilH. 2015. Soiland nutrient management policies. In: *State of Indian Agriculture–Soil* (Eds.) Pathak,H., Sanyal, S.K. and Takkar, P.N. New Delhi:National Academy of Agricultural Sciences, pp. 332–342.

Chen, Y.P., Rekha, P.D., Arun, A.B., Shen, F.T., Lai, W.A. and Young, C.C. 2006. Phosphate solubilizing bacteria from subtropical soil and their tricalcium phosphate solubilizing abilities. *Applied Soil Ecology*. 34:33–41.

Delattre, C., Pierre, G., Laroche, C. and Michaud, P. 2016. Production, extraction and characterization ofmicroalgal and cyanobacterial exopolysaccharides. *Biotechnology Advances*. 34:1159–1179.

Dias, G.A., Rocha, R.H.C., Araújo, J.L., De Lima, J.F. and Guedes, W.A. 2016. Growth, yield, and post harvest quality in eggplant produced under different foliar fertilizer (Spirulina platensis) treatments. *Semina-CienciasAgrarias*. 37:38–93.

Doudna, J.A. and Charpentier, E. 2014. The new frontier of genome engineering with CRISPR-Cas9. *Science*. 346: 6213.

ElArroussi, H. and Elbaouchi, A. 2016. Halophilic microalgae Dunaliella salina extracts improve seed germination and seedling growth of Triticum aestivum L. under salt stress. *Acta Horticulturae.* 1: 13–26.

FAO. 2011. Climate change and food systems resilience in Sub-Saharan Africa. Edited by Lim LiChing, Sue Edwards and Nadia El-Hage Scialabba, pp. 361–377.

Farmeurope, Madre, Y. andAgostino, V. D. 2017. New Plant-Breeding Techniques: What Are We Talking About?https://edepot.wur.nl/509404. (Accessed: March 10, 2023).

FiBL. 2011. African Organic Agriculture Training Manual – Pest, Disease and Weeds. Version 1.0 June 2011. Edited by Gilles Weidmann and Lukas Kilcher. Research Institute of Organic Agriculture FiBL, Frick https://forestphytophthoras.org/sites/default/files/educational_materials/African%20Organic%20Agriculture%20Training%20Manual%2009-14.pdf. (Accessed: March 10, 2023).

FiBL.2019.Area data on organic agriculture in Europe 2000–2017. The Statistics.FiBL.org website maintained by the Research Institute of Organic Agriculture (FiBL), Frick, Switzerland. Available at www.fibl.org/en/shop-en/2020-organic-world-2019m (Accessed: March10, 2023).

Fingerhut, U., Webb, L.E. andSoeder, C.J. 1984. Increased yields of Rhizobium japonicum by an extract of the green alga, Scenedesmus obliquus (276-3a). *Applied Microbiology & Biotechnology.*19:358–360.

Gramazio, P., Takayama, M. and Ezura, H. 2020. Challenges and prospects of new plant breeding techniques for GABA improvement in crops: Tomato as an example. *Frontiers in Plant Science*. 11: 1–16.

Hamid, B., Zaman, M., Farooq, S., Fatima, S., Sayyed, R.Z. and Baba, Z.A. 2021. Bacterial plant biostimulants: A sustainable way towards improving growth, productivity, and health of crops. *Sustainability.* 13: 28–56.

Husaini, A.M. and Sohail, M. 2018. Time to redefine organic agriculture: Can't GM crops Be certified as organics?*Frontiers in Plant Science.* 9: 4–23.

IFOAM. 2003. *Training Manual for Organic Agriculture in the Tropics*. Edited by Frank Eyhorn, Marlene Heeb, Gilles Weidmann, pp.190–209, www.fao.org/fileadmin/templates/nr/sustainability_pathways/docs/Compilation_techniques_organic_agriculture_rev.pdf. (Accessed: March 10, 2023).

IFOAM. 2016. Organics international position on genetic engineering and genetically modified organisms. Brussels: Draft for Public Consultation. Available at: www.ifoam.bio/why-organic/shaping-agriculture/four-principles-organic. (Accessed March 10, 2023).

Itakura, K., Hirose, T., Crea, R., Riggs, A. D., Heyneker, H. L. and Bolivar, F.1977. Expression in Escherichia coli of a chemically synthesized gene for the hormone somatostatin.*Science.* 198: 1056–1063.

Jansing, J., Schiermeyer, A., Schillberg, S., Fischer, R. and Bortesi, L. 2019. Genome editing in agriculture: Technical and practical considerations. *International Journal of Molecular Sciences*. 20: 12–31.

Kramer, M.G. and Redenbaugh, K. 1994. Commercialization of a tomato with an antisense polygalacturonase gene: The FLAVR SAVR tomato story. *Euphytica*. 79: 293–297.

Kulik, M.M. 1995. The potential for using cyanobacteria (blue-green algae) and algae in the biological control of plant pathogenic bacteria and fungi. *European Journal of Plant Pathgology*. 101: 585–599.

Kumar, A., Vandana, R. S., Singh, M. and Pandey, K. D. 2015. Plant growth promoting rhizobacteria (PGPR). A promising approach for disease management. In *Microbes and Environmental Management.* New Delhi: Studium Press, Singh, S., and Singh, D.P.pp. 195–209.

Kumar, R., KumarS, YashavanthB.S., Meena, P.C., Ramesh, P., Indoria, A.K., Kundu, S., Manjunath, M. 2020. Adoption of Natural Farming and its Effect on Crop Yield and Farmers' Livelihood in India. *ICAR-National Academy of Agricultural Research Management, Hyderabad, India.* Niti Aayog, Government of India.

Lucy, M., Reed, E. and Glick, B. R. 2004. Applications of free-living plant growth-promoting rhizobacteria. *Antonie van Leeuwenhoek*. 86: 1–25.

Lyzenga, W.J., Harrington, M., Bekkaoui, D., Wigness, M., Hegedus, D.D. and Rozwadowski, K.L. 2019. CRISPR/Cas9 editing of three CRUCIFERIN C homoeologues alters the seed protein profile in camelina sativa. *BMC Plant Biology*. 19: 1–16.

Manzoor, M., Gul, I., Ahmed, I., Zeeshan, M., Hashmi, I. and Amin, B.A.Z. 2019. Metal tolerant bacteria enhanced phytoextraction of lead by two accumulator ornamental species. *Chemosphere*. 227: 561–569.

Michalak, I., Chojnacka, K., Dmytryk, A., Wilk, R., Gramza, M. and Roj, E. 2016. Evaluation of supercritical extracts of algae as biostimulants of plant growth in field trials. *Frontiers in Plant Science*. 7: 15–91.

Mishra, J., Prakash, J., and Arora, N. K. 2016. Role of beneficial soil microbes in sustainable agriculture and environmental management. *Climate Change andEnvironmental Sustainability*. 4: 137–149.

Mishra, P. and Dash, D. 2014. Rejuvenation of biofertilizer's for sustainable agriculture and economic development. *Consilience: The Journal of Sustainable Development*. 11: 41–61.

Nishitani, C., Hirai, N., Komori, S., Wada, M., Okada, K. and Osakabe, K. 2016. Efficient genome editing in apple using a CRISPR/Cas9 system. *Scientific Reports*. 6: 1–8.

Nyoki, D. and Ndakidemi, P. A. 2018. Rhizobium inoculation reduces P and K fertilization requirement in corn-soybean intercropping. *Rhizosphere*. 5: 51–56.

Ouhaibi-Ben Abdeljalil, N., Vallance, J., Gerbore, J., Rey, P. and Daami- Remadi, M. 2016. Biosuppression of Sclerotinia stem rot of tomato and biostimulation of plant growth using tomato-associated rhizobacteria. *Journal of Plant Pathology & Microbiology*. 7: 2–18.

Purnhagen, K.P., Clemens, S., Eriksson, D., Fresco, L.O., Tosun, J. and Qaim, M. 2021. Europe's farm to fork strategy and its commitment to biotechnology and organic farming: Conflicting or complementary Goals?*Trends in Plant Science*. 26: 600–606.

Reichman, S.M. 2014. Probing the plant growth-promoting and heavy metal tolerance characteristics of Bradyrhizobium japonicum CB1809. *European Journal of Soil Biology*. 63: 7–13.

Rodriguez, H. and Fraga, R. 1999. Phosphate solubilizing bacteria and their role in plant growth promotion. *Biotechnology Advances*. 17: 319–339.

Romanenko, K., Kosakovskaya, I.V. and Romanenko, P.A. 2015. Phytohormones of microalgae: Biological role and involvement in the regulation of physiological processes. Pt I. Auxins, abscisic acid, ethylene. *International Journal on Algae*. 17: 275–289.

Ronga, D., Biazzi, E., Parati, K., Carminati, D., Carminati, E. and Tava, A. 2019. Microalgal biostimulants and biofertilisers in crop productions. *Agronomy*.9: 192–212.

Ruddock, V. (2016). Permitted Chemicals List for Organic Farming. Available at: http://organic.lovetoknow.com/Permitted_Chemicals_List_for_Organic_Farming. (Accessed: March 10, 2023).

Salar, R.K., Purewal, S.S. and Sandhu, K.S. 2017. Bioactive profile, free-radical scavenging potential, DNA damage protection activity, and mycochemicals in Aspergillus awamori (MTCC 548) extracts: A novel report on filamentous fungi. *3Biotech*. 7: 164–180.

Schaart, J., Riemens, M., Van de Wiel, C., Lotz, B., Smulders, R.andWageningen University and Research. 2015. Opportunities of New Plant Breeding Techniques. https://edepot.wur.nl/357723. (Accessed: March 10, 2023).

Seaman, A. and Sideman, E. 2013. *Resource Guide for Organic Insect and Disease Management*, 2nd Edn. Ithaca, NY: Cornell University, pp.202–321.

SEEG. 2022. Total gasses emission. Available at: https://plataforma.seeg.eco.br/total_emission (Accessed: March 10, 2023).

Sheng, X.F., Zhao, F., He, L.Y., Qiu, G. and Chen, L. 2008. Isolation and characterization of silicate mineral solubilizing Bacillus globisporus Q12 from the surfaces of weathered feldspar. *Canadian Journal of Microbiology*. 54: 1064–1068.

Shi, J., Gao, H., Wang, H., Lafitte, H. R., Archibald, R. L. and Yang, M. 2017. ARGOS8 variants generated by CRISPR-cas9 improve maize grain yield under field drought stress conditions. *Plant Biotechnology Journal*. 15: 207–216.

ShivaV., Pande,P. andSingh,J. 2004. *Principles of Organic Farming: Renewing the Earth's Harves*. New Delhi:Navdanya,pp.1–199.

Shukla, A.C. and Gupta, A.B. 1967. Influence of algal growth-promoting substances on growth, yield and protein contents of rice plants. *Nature*.213:744.

Singh, S., Singh, B.K., Yadav, S.M. and Gupta, A.K. 2014. Potential of biofertilizers in crop production in Indian agriculture. *American Journal of Applied Nutrition & Fertilization Technology*. 4: 33–40.

Soma, K., Burg, S.W.K., Hoefnagel, E.W.J., Stuiver, M. and Heide, M. 2018. Social innov-ation –A future pathway for blue growth. *Marine Policy*. 87: 363–370.

Sriprang, R., Hayashi, M., Ono, H., Takagi, M., Hirata, K. and Murooka, Y. 2003. Enhanced accumulation of Cd^{2+} by a Mesorhizobium sp. transformed with a gene from Arabidopsis thaliana coding for phytochelatin synthase. *Applied & Environmental Microbiology*. 69: 1791–1796.

Tahir, H. A., Gu, Q., Wu, H., Raza, W., Hanif, A. and Wu, L. 2017. Plant growth promotion by volatile organic compounds produced by Bacillus subtilis SYST2. *Frontiers in Microbiology*. 8: 171–182.

Treadwell, D., Riddle, J., Barbercheck, M., Cavanaugh-Grant, D. and Zaborski, E. 2015. *What Is Organic Farming? Cooperative Extension System*. Amherst, MA: University of Massachusetts. Available at: https://ag.umass.edu/risk-management/organic-farmers. (Accessed: March 10, 2023).

UNF. 2021. Climate issues to watch in 2022: A year for more action and bigger ambition. Available at: https://unfoundation.org/author/un-foundation-climate-and-environment-experts/ (Accessed: March 10, 2023).

Unkovich, M.J., Pate, J.S. and Sanford, P. 1997. Nitrogen fixation by annual legumes in Australian Mediterranean agriculture. *Australian Journal of Agricultural Research*. 48: 267–293.

Waltz, E. 2016. CRISPR-edited crops free to enter market, skip regulation. *Nature Biotechnology*. 34: 582–615.

Willer, H., Trávníček, J., Meier, C. and Schlatter, B. 2021. The world of organic agriculture statistics and emerging trends. Available at: www.cabdirect.org/cabdirect/abstract/2016 3135703. (Accessed: March 10, 2023).

Zafar-ul-Hye, M., Ahmad, M. and Shahzad, S. M. 2013. Short Communication Synergistic effect of rhizobia and plant growth promoting rhizobacteria on the growth and nodulation of lentil seedlings under axenic conditions. *Soil & Environment*. 32: 79–86.

Zaidi, S., Usmani, S., Singh, B. R. and Musarrat, J. 2006. Significance of Bacillus subtilis strain SJ-101 as a bioinoculant for concurrent plant growth promotion and nickel accumulation in Brassica juncea. *Chemosphere*. 64: 991–997.

Zhang, X., Baars, O. and Morel, F. M. 2019. Genetic, structural, and functional diversity of low and high-affinity siderophores in strains of nitrogen fixing Azotobacter chroococcum. *Metallomics*. 11: 201–212.

Index

Note: Entries in **bold** refer to tables, entries in *italics* refer to figures.

FSC
www.fsc.org
MIX
Papier | Fördert gute Waldnutzung
FSC® C083411

Kolibri 360 GmbH
Ferdinand-Jühlke-Straße 7
99095 Erfurt, Deutschland
produktsicherheit@kolibri360.de